T0134459

Advances in Intelligent Systems and Computing

Volume 399

Series editor

Janusz Kacprzyk, Polish Academy of Sciences, Warsaw, Poland
e-mail: kacprzyk@ibspan.waw.pl

About this Series

The series "Advances in Intelligent Systems and Computing" contains publications on theory, applications, and design methods of Intelligent Systems and Intelligent Computing. Virtually all disciplines such as engineering, natural sciences, computer and information science, ICT, economics, business, e-commerce, environment, healthcare, life science are covered. The list of topics spans all the areas of modern intelligent systems and computing.

The publications within "Advances in Intelligent Systems and Computing" are primarily textbooks and proceedings of important conferences, symposia and congresses. They cover significant recent developments in the field, both of a foundational and applicable character. An important characteristic feature of the series is the short publication time and world-wide distribution. This permits a rapid and broad dissemination of research results.

Advisory Board

More information about this series at http://www.springer.com/series/11156

Suzana Loshkovska · Saso Koceski
Editors

ICT Innovations 2015

Emerging Technologies for Better Living

Editors
Suzana Loshkovska
Faculty of Computer Science
Ss. Cyril and Methodious University
Skopje
Macedonia

Saso Koceski
Faculty of Computer Science
University Goce Delcev
Stip
Macedonia

ISSN 2194-5357 ISSN 2194-5365 (electronic)
Advances in Intelligent Systems and Computing
ISBN 978-3-319-25731-0 ISBN 978-3-319-25733-4 (eBook)
DOI 10.1007/978-3-319-25733-4

Library of Congress Control Number: 2015952757

Springer Cham Heidelberg New York Dordrecht London

Printed on acid-free paper

Springer International Publishing AG Switzerland is part of Springer Science+Business Media (www.springer.com)

Preface

The ICT Innovations conference is a framework where academics, professionals, and practitioners interact and share their latest results and interests related to basic and applied research in ICT. The conference is organized by the Association for Information and Communication Technologies (ICT-ACT), which supports the development of information and communication technologies in Macedonia, the Balkan region and beyond.

The 7th ICT Innovations 2015 conference gathered 316 authors from 36 countries reporting their scientific work and novel solutions. Only 26 papers were selected for this edition by the international program committee consisting of 215 members from 55 countries, chosen for their scientific excellence in their specific fields.

ICT Innovations 2015 was held in Ohrid, Macedonia during October 1–4, 2015. The special conference topic was “Emerging Technologies for Better Living” and it was mainly focused on scientific topics and technologies that have transformed the working and the living environments making them safer, more convenient, and more connected. The conference gave an overview of the emerging technologies, systems, applications, and standardization activities for better living, and identified their opportunities and challenges. The conference also focused on variety of ICT fields: Enhanced living environments, Data Mining & Information Retrieval, Bioinformatics & Biomedical Engineering, Connected Health Technologies, Digital Signal & Image Processing, Artificial Intelligence, Internet & Web Applications, Distributed & Parallel Processing, Internet of things, Robotics & Automation, Pattern Recognition, Assistive technologies, E-health, Context-aware Systems, Pervasive Technologies, Ambient Intelligence, Ubiquitous Computing, Embedded Systems, Innovative Media and Tools.

We would like to express sincere gratitude to the authors for submitting their contributions to this conference and to the reviewers for sharing their experience in the selection process. Special thanks to Katarina Trojacanec and Ivan Kitanovski for their technical support in the preparation of the conference proceedings.

September 2015

Suzana Loshkovska
Saso Koceski

Organization

ICT Innovations 2015 was organized by the Macedonian Society of Information and Communication Technologies (ICT-ACT).

Conference and Program Chairs

Suzana Loshkovska	University Ss. Cyril and Methodius, Macedonia
Saso Kocevski	University Goce Delcev-Stip, Macedonia

Program Committee

A.Velastin Sergio	Universidad de Santiago de Chile, Chile
Achkoski Jugoslav	Military Academy “General Mihailo Apostolski”, Macedonia
Ackovska Nevena	University Ss.Cyril and Methodius, Macedonia
Ahsan Syed	Technische Universität Graz, Austria
Aiello Marco	University of Groningen, Netherlands
Akhtar Zahid	University of Udine, Italy
Albert Dietrich	University of Graz, Austria
Aliu Azir	Southeastern European University of Macedonia, Macedonia
Alor Hernandez Giner	Instituto Tecnologico de Orizaba, Mexico
Alti Adel	University of Setif, Algeria
Alvarez Sabucedo Luis	Universidade de Vigo. Depto. of Telematics, Spain

Dobre Ciprian	University Politehnica of Bucharest, Romania
Drlik Martin	Constantine the Philosopher University in Nitra, Slovakia
Drusany Staric Kristina	University medical centre Ljubljana, Slovenia
Dzeroski Saso	Jožef Stefan Institute, Slovenia
Ellul Joshua	University of Malta, Malta
Fati Suliman Mohamed	Universiti Sains Malaysia, Malaysia
Fels Deborah	Ryerson University, Canada
Fetaji Majlinda	Southeastern European University of Macedonia, Macedonia
Filiposka Sonja	University Ss.Cyril and Methodius, Macedonia
Frasheri Neki	Polytechnic University of Tirana, Albania
Fujinami Kaori	Tokyo University of Agriculture and Technology, Japan
Gajin Slavko	University of Belgrade, Serbia
Gama Joao	University Porto, Portugal
Ganchev Ivan	University of Limerick, Ireland
Ganchev Todor	Technical University Varna, Bulgaria
Garcia Nuno	Universidade da Beira Interior, Portugal
Gavrilov Andrey	Laboratory Hybrid Intelligent Systems, Rusia
Gawanmeh Amjad	Khalifa University, United Arab Emirates
Gialelis John	University of Patras, Greece
Gievska Sonja	The George Washington University, USA
Gjorgjevikj Dejan	University Ss.Cyril and Methodius, Macedonia
Gligoroski Danilo	Norwegian University of Science and Technology, Norway
Goleva Rossitza	Technical University of Sofia, Bulgaria
Gomes Abel	Univeristy of Beira Interior, Portugal
Gramatikov Saso	University Ss.Cyril and Methodius, Macedonia
Gravvanis George	Democritus University of Thrace, Greece
Grguric Andrej	Ericsson Nikola Tesla - Research and Innovations Unit, Croatia
Grosu Daniel	Wayne State University, USA
Guralnick David	International E-Learning Association
Gushev Marjan	University Ss.Cyril and Methodius, Macedonia
Haddad Yoram	Jerusalem College of Technology, Israel
Hadzieva Elena	University of Information Science and Technology (UIST) "St. Paul the Apostle", Macedonia
Hao Tianyong	Columbia University, USA
Hoic-Bozic Natasa	University of Rijeka, Croatia
Holmes Violeta	University of Huddersfield, UK
Hsieh Fu-Shiung	Chaoyang University of Technology, Taiwan
Huang Yin-Fu	National Yunlin University of Science and Technology, Taiwan

Huraj Ladislav	University of SS. Cyril and Methodius, Slovakia
Huynh Hieu Trung	Industrial University of Ho Chi Minh City, Vietnam
Iantovics Barna Laszlo	Petru Maior University of Tg. Mures, Romania
Ictz Vacius	Kaunas University of Technology, Lithuania
Ilarri Sergio	University of Zaragoza, Spain
Isomursu Minna	VTT Technical Research Centre, Finland
Ivanovic Mirjana	University of Novi Sad, Serbia
Jiang Yichuan	Southeast University, China
Kalajdziski Slobodan	University Ss.Cyril and Methodius, Macedonia
Kalinov Alexey	Cadence Design Systems, Russia
Kaloyanova Kalinka	University of Sofia - FMI, Bulgaria
Karaivanova Aneta	Bulgarian Academy of Sciences, Bulgaria
Kawamura Takahiro	The University of Electro-Communications, Japan
Knepper Richard	Indiana University, USA
Kocarev Ljupcho	University Ss.Cyril and Methodius, Macedonia
Koceska Natasa	University Goce Delcev, Macedonia
Kocev Dragi	Jožef Stefan Institute, Slovenia
Kokol Peter	University of Maribor, Slovenia
Kon-Popovska Margita	University Ss.Cyril and Methodius, Macedonia
Kraljevski Ivan	VoiceINTERconnect GmbH, Germany
Kulakov Andrea	University Ss.Cyril and Methodius, Macedonia
Kulkarni Siddhivinayak	University of Ballarat, Australia
Kumar Das Ashok	International Institute of Information Technology, India
Kumar Singh Brajesh	Faculty of Engineering and Technology, RBS College,India
undu Anirban	Kuang-Chi Institute of Advanced Technology, Singapore
Kuribayashi Minoru	Kobe University, Japan
Kurilovas Eugenijus	Vilnius University, Lithuania
Kurti Arianit	Linnaeus University, Sweden
Kwiatkowski Jan	Wroclaw University of Technology, Poland
Lamas David	Tallinn University, Estonia
Lastovetsky Alexey	University College Dublin, Ireland
Le Khac Nhien An	University College Dublin, Ireland
Li Rita Yi Man	Hong Kong Shue Yan University, Hong Kong
Lim Hwee-San	Universiti Sains Malaysia, Malaysia
Lindh Thomas	KTH, Sweden
Ljubi Igor	Croatian Institute for Health Insurance, Croatia
Machado Da Silva José	FEUP, Portugal
Madevska Bogdanova Ana	University Ss.Cyril and Methodius, Macedonia
Madjarov Gjorgji	University Ss.Cyril and Methodius, Macedonia

Malcovati Piero	University of Pavia, Italy
Marengo Augostino	Università degli Studi di Bari Aldo Moro, Italy
Markovski Smile	University Ss.Cyril and Methodius, Macedonia
Martinovska Cveta	University Goce Delcev, Macedonia
Mastrogiovanni Fulvio	University of Genoa, Italy
Michalak Marcin	Silesian University of Technology, Poland
Mihajlov Dragan	University Ss.Cyril and Methodius, Macedonia
Mileva Aleksandra	University Goce Delcev, Macedonia
Mileva Boshkoska Biljana	Faculty of information studies in Novo Mesto, Slovenia
Mishev Anastas	University Ss.Cyril and Methodius, Macedonia
Mishkovski Igor	University Ss.Cyril and Methodius, Macedonia
Mitreski Kosta	University Ss.Cyril and Methodius, Macedonia
Mocanu Irina	PUB, Romania
Moen Anne	University of Oslo, Norway
Mrabet Radouane	Mohammed V - Souissi University, Morocco
Nicolau Viorel	Dunarea de Jos University of Galati, Romania
Nicolin Alexandru	Horia Hulubei National Institute of Physics and Nuclear Engineering, Romania
Noguera Manuel	Universidad de Granada, Spain
Norcio Anthony	UMBC: An Honors University In Maryland, USA
Nosovic Novica	University of Sarajevo, Bosnia and Herzegovina
Ognjanović Ivana	Univerzitet Donja Gorica, Montenegro
Panov Pance	Jožef Stefan Institute, Slovenia
Pantano Eleonora	University of Calabria, Italy
Paprzycki Marcin	IBS PAN and WSM, Poland
Parycek Peter	Danube-University Krems, Austria
Pastorino Matteo	Life Supporting Technologies - UPM, Spain
Patel Shushma	London South Bank University, UK
Pedersen Christian Fischer	Aarhus University, Denmark
Perälä-Heape Maritta	Centre for Health and Technology (CHT), Finland
Petcu Dana	West University of Timisoara, Romania
Pinheiro Antonio	Universidade da Beira Interior, Portugal
Pinkwart Niels	Humboldt Universität zu Berlin, Germany
Pleva Matus	Technical University of Košice, Slovakia
Podobnik Vedran	University of Zagreb, Croatia
Pop Florin	University Politehnica of Bucharest, Romania
Popeska Zaneta	University Ss.Cyril and Methodius, Macedonia
Porta Marco	University of Pavia, Italy
Potolea Rodica	Technical Univeristy of Cluj-Napoca, Romania
Rege Manjeet	Rochester Institute of Technology, USA
Regina Castelo Branco Kalinka	Institute of Mathematics and Computer Sciences, Brasil

Reiner Miriam	Technion – Israel Institute of Technology, Israel
Ristevski Blagoj	University St Clement of Ohrid, Macedonia
Ristov Sasko	University Ss.Cyril and Methodius, Macedonia
Roose Philippe	LIUPPA, France
Saini Jatinderkumar	Narmada College of Computer Application, India
Sas Corina	University of Lancaster, UK
Savovska Snezana	University St Clement of Ohrid, Macedonia
Schreiner Wolfgang	Research Institute for Symbolic Computation (RISC), Austria
Schwiebert Loren	Wayne State University, USA
Scotney Bryan	University of Ulster, UK
Šendelj Ramo	Univerzitet Donja Gorica, Montenegro
Siládi Vladimír	Matej Bel University, Slovakia
Silva Josep	Universitat Politècnica de València, Spain
Silva Manuel	Instituto Superior de Engenharia do Porto, Portugal
Smolders Roel	VITO, Belgium
Sonntag Michael	Johannes Kepler University Linz, Austria
Spinsante Susanna	Università Politecnica delle Marche, Italy
Stojanovic Igor	University Goce Delcev, Macedonia
Stoyanov Stanimir	University Paisii Hilendarski, Bulgaria
Stulman Ariel	The Jerusalem College of Technology, Israel
Subramaniam Chandrasekaran	Kumaraguru College of Technology, Coimbatore, India
Sun Chang-Ai	University of Science and Technology Beijing, China
Thiare Ousmane	Gaston Berger University, Senegal
Trajanov Dimitar	University Ss.Cyril and Methodius, Macedonia
Trajkovic Ljiljana	Simon Fraser University, Canada
Trajkovik Vladimir	University Ss.Cyril and Methodius, Macedonia
Trajkovski Igor	University Ss.Cyril and Methodius, Macedonia
Trcek Denis	University of Ljubljana, Slovenia
Tseng Yuh-Min	National Changhua University of Education, Taiwan
Tudruj Marek	Polish-Japanese Institute of Information Technology, Institute of Computer Science, Polish Academy of Sciences, Poland
Valderrama Carlos	UMons University of Mons, Electronics and Microelectronics Dpt., Belgium
Velinov Goran	University Ss.Cyril and Methodius, Macedonia
Vlahu-Georgievska Elena	University St Clement of Ohrid, Macedonia
Vrdoljak Boris	University of Zagreb, Croatia
Wac Katarzyna	University of Geneva, Switzerland
Wibowo Santoso	Central Queensland University, Australia

Wozniak Michal	Wroclaw University of Technology, Poland
Xu Lai	Bournemouth University, UK
Xu Shuxiang	University of Tasmania, Australia
Yue Wuyi	Konan University, Japan
Zavoral Filip	Charles University Prague, Czech Republic
Zdravev Zoran	University Goce Delcev Macedonia
Zdravkova Katerina	University Ss.Cyril and Methodius, Macedonia
Zeng Xiangyan	Fort Valley State University, USA

Organizing Committee

Vladimir Trajkovikj	University Ss. Cyril and Methodius, Macedonia
Ivan Chorbev	University Ss. Cyril and Methodius, Macedonia
Elena Vlahu-Gorgievska	University St Clement of Ohrid, Macedonia
Snezhana Savoska	University St Clement of Ohrid, Macedonia
Blagoj Delipetrev	University Goce Delcev-Stip, Macedonia
Elena Hadzieva	University of Information Science and Technology St. Paul the Apostle, Macedonia

Technical Committee

Sonja Filiposka	University Ss. Cyril and Methodius, Macedonia
Katarina Trojacanec	University Ss. Cyril and Methodius, Macedonia
Tomche Delev	University Ss. Cyril and Methodius, Macedonia
Ivan Kitanovski	University Ss. Cyril and Methodius, Macedonia

Contents

Invited Keynote Paper

Proceeding Papers

Research and Innovation in ICT

With Examples in the Field of eHealth and Wellbeing

Andrej Kos, Urban Sedlar, Matevz Pustišek

University of Ljubljana, Faculty of electrical Engineering
{andrej.kos, urban.sedlar, matevz.pustisek}@fe.uni-lj.si

Abstract. The ICT environment has changed drastically in last 5 years. The paper describes the challenges of enabling critical mass of knowledge for multidisciplinary research and innovation in the current field of ICT. In introduction we describe our team, the importance of ICT and particular broadband internet access. Internet of things is also described, bringing great diversity of new domains the ICT has to encompass and support. The main part of the article addresses the ICT environment changes and different approaches towards research and innovation adaptation detailed with examples. As the research and innovation environment changes, we as a team have to adapt. The challenges and solutions are presented, based on concrete examples from the field of eHealth and wellbeing.

Keywords: Research, Innovation, ICT, Approaches, Ecosystem of talents

1 Introduction

The importance of research and innovation in ICT to support growth of all industry sectors is mentioned in many strategic documents. In the article we look behind the stage and present the complexity of the ecosystem of talents. We pay special attention to issues and solutions of conducting research and innovation in the broad field of ICT enabled Internet of things, taking into account type of research, type of funding, team members, critical mass knowledge and the actual university-industry environment.

1.1 Our team

The Laboratory for telecommunications (LTFE) and Laboratory for Multimedia (LMMFE) team's mission is: (i) excellence in research and industry cooperation, (ii) educating young talents and professionals, (iii) developing cool ideas and new knowledge and (iv) innovation for life and business.

There are 50 members in the LTFE and LMMFE team, not counting graduate and undergraduate students. About 50% of our budget is based on direct industry projects,

S. Loshkovska and S. Koceski (eds.), *ICT Innovations 2015*,
Advances in Intelligent Systems and Computing 399,
DOI: 10.1007/978-3-319-25733-4_1

15% from university for academic pedagogical activities and 35% from national and mostly European projects.

Our main teaching, research and development directions are communication networks and services - with focus on network operator grade systems, multimedia - including development of applications for smart and digital television and user experience evaluations, and various web and mobile applications in the domain of eLearning, eHealth and wellbeing.

In addition, we strongly aim to extend our research innovation beyond the obvious scope of an academic research group. We primarily exercise research innovation at university level (teaching, projects, Makerlab [1], Demola [2]) and entrepreneurship level (LUI [3], Hekovnik [4]), but also build the innovation ecosystem around secondary level schools (Openlab [5]) and around primary level schools (ZOTKS [6]). We act as research, development and training institution for local industry (ICT Academy [7]), and as ICT innovation center in Slovenia, having strong national and regional linkages. In this term we are unique in the region and often cited as best practice example [8,9].

1.2 Broadband networks

ICT has become the key element for growth in almost every industry sector. Therefore broadband communications technologies and services, including Internet, today play a similar role as roads and railways in the previous century and broadband has strong positive effects on the increase in GDP and job creation. A 10% increase in broadband penetration results in a 1-1.5% increase in the GDP annually and 1.5% labor productivity gains. There are many socio-economic benefits from high speed networks. To be able to fully exploit the potential of high speed broadband i.e. the increased innovation, improved welfare, customer benefits, new jobs, reduced environmental impact, better demography trends, increased productivity and increased GDP, new services, education of users and demand stimulation is need [10,11].

The Digital Agenda for Europe (DAE) [12] aims to help Europe's citizens and businesses to get the most out of digital technologies. It is the first of seven flagships initiatives under Europe 2020, the EU's strategy to deliver smart sustainable and inclusive growth. Launched in May 2010, the DAE contains 101 actions, grouped around seven priority areas. The goals connected to broadband and the use of digital technologies were set as stated below:

- the entire EU to be covered by broadband internet access by 2013,
- the entire EU to be covered by broadband internet access above 30 Mbps by 2020,
- 50 % of the EU to subscribe to broadband internet access above100 Mbps by 2020.

1.3 Internet of Things

Internet of Things (IoT) is an ecosystem where the virtual world integrates seamlessly with the real world of things. The IoT enables applications with high social and business impact, and once the enabling technologies are stable, it is expected new ones will emerge as well. Businesswise, the most important aspect is creation of value-added services that are enabled by interconnection of things, machines, and smart objects and can be integrated with current and future business development process. IoT as part of broader ICT has become part of practically all domains i.e. buildings, energy, entertainment, healthcare, wellbeing, lifestyle, industrial, transportation, retail, security and public safety and other.

2 ICT environment change

In last five years the ICT innovation and research area has changed drastically. Ten years ago, the ICT research was focused on communication networks on one side, and information technology support on the other side. During this time classical communication and information technologies and services, i.e. networks, protocol, connectivity, have become a commodity. Broadband networks, wired and wireless, as well as databases with open data accessible via open APIs are expected to be available all the time and everywhere, similar as 230V plugs in the wall.
There is of course still plenty of basic research and innovation within core ICT domains, however it is present in fewer areas, i.e. mobile, quality of experience, software defined networking/radio, security, etc. The majority of research has moved to applicative, interdisciplinary areas, where ICT plays just a part of the solution, as shown in Figure 1.

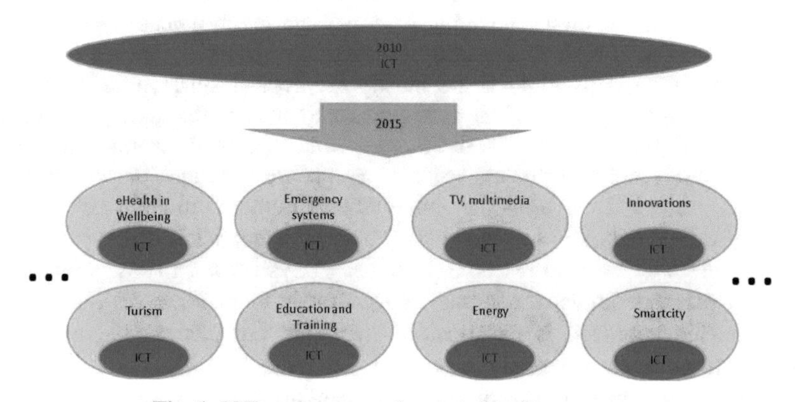

Fig. 1. ICT environment change in last five years

Conducting research and innovation, where ICT has the role of supporting service, means that researches have to acquire domain specific knowledge from a/many different domain/s.

The solution is to focus on just one or a few domains and build long-term critical mass of research and innovation expertize and financing. There are time, money and competence needed to adapt and there are many possible focuses to choose from. The chances to be successful in proposals for national or European project funding are on average 10% and falling. Adding to this less research and innovation funds, many research groups have decided to apply for projects in many/more fields/domains. Such amount of proposals causes tremendous stress for researches and evaluators. The outcome is also that the research might be determined by projects that a group wins, not the focus the group has set.

Taking these facts into account, the internal organization was adapted order to follow the above mentioned trend and to achieve results mentioned above.

3 Approaches to research

There is basic and applicative research in the field of ICT. With ICT being a support service to almost all other industries, the research is becoming more and more applicative.

The question that arises is, how to finance the research. The first option is to "apply for national and European project", hope for best results, and perform and hopefully continue the research already being "in the club" and having big projects. This type of research is excellent for long-term planning and stability of a research group. However, the efforts to get and run the project and tackle all the formal details, makes the group conducting the actual research less sharp and innovative, possibly going into doing "research projects" for salary.

Another option is "let's just do research". This approach brings in lots of motivation and focus, high innovation potential and much quicker results as just the research is in focus and all the other burdens do not exist (except for long term stability and financing). Students are mostly involved in this type of research. This approach is sustainable only if the researches/students are fluctuating and there exist funds for mentoring support. The limiting factor can be need for new equipment (not already existing) and nevertheless at least small source of funds that normally comes from other projects. Innovators and entrepreneurs involved in and motivated for this kind of research are highly searched for at start-ups, companies, and even universities.

Next option is "let's do research, if industry pays for it". Lately, with decreasing budgetary research funds, a common suggestion is to increase the industry funded research. This approach is excellent, if there are companies able to absorb the research, if both parties really want and need the research, and if there is an established culture of research for industry. There are cases where companies see this kind of cooperation just as "low cost development" (however it really isn't) or "with a contract I pay for development and get also the research and IPR". This issue is also strategically important for the universities as a whole.

Another option is "PhD positions paid from budget", which is typically long term on one hand and formally very restrictive in terms of candidates and mentors on the other hand. PhD positions are typically won on tenders. So the available position cannot be planed more than a few months in advance., therefore we often cannot promptly respond to emerging research challenges, because timing is mostly dictated by formal procedures of official funding entities. Within this timing it is also hard to find a winning combination of a candidate, mentor and topic. We see that there is less and less candidates willing to go through the process of PhD, however, they would indeed perform the quality research.

4 Examples of ICT supported research and innovation in the field of eHealth and wellbeing

4.1 Stress measurements and sensing

Stress is a physiological, psychological and behavioral response to every change people must adapt to. ICT enables new options to monitor and prevent stress. For a thorough and more reliable identification of chronic stress the measurements need to be conducted continuously, throughout a longer time period. This can be done with the use of different electronic measurement devices that periodically, based on physiological indicators, can evaluate the current level of stress. Such indicators (markers) include: electro-dermal activity (EDA), various pulse samples, blood pressure and respiration activity. On today's market, we can already find small sensor devices that allow these types of measurements.

We built a presence based context-aware chronic stress recognition system. It obtains contextual data from various mobile sensors and other external sources in order to calculate the impact of ongoing stress. By identifying and visualizing ongoing stress situations of an individual user, he/she is able to modify his/her behavior in order to successfully avoid them. Clinical evaluation of the proposed methodology has been made in parallel by using electro-dermal activity sensor.[13]

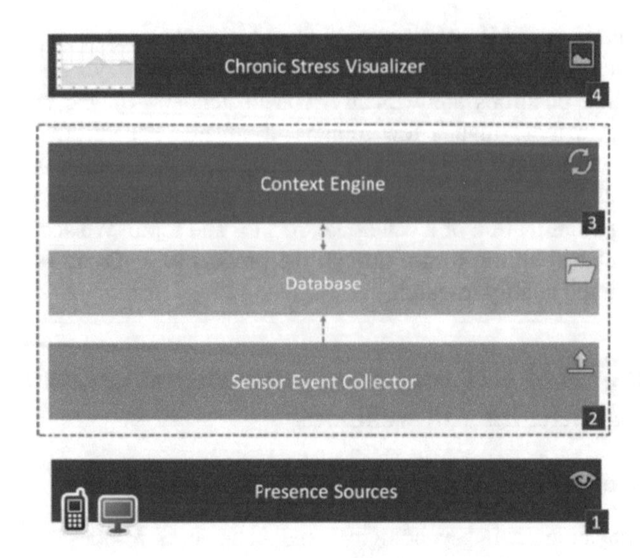

Fig. 2. Context-aware chronic stress recognition system

The sensor classification and presence based context-aware chronic stress recognition system were developed within the scope of young researcher program financed by Slovenian Research Agency. We had problems finding appropriate candidate for this position even when the research grant had been already assured. In the end we acquired and excellent foreign student. Because of very strict execution timeframe, which does not efficiently considers some facts that are out of the control of the researcher and his mentors (e.g. duration of peer review process for journal publications required for Ph.D. candidates [14]), student was left with very little time to get accustomed to new living and working environment.

4.2 DeSA

DeSA [15] is a healthy lifestyle mobile application for bio-sensing and healthy lifestyle management for diabetes patients and for other users that want to monitor their lifestyle. It allows users to track multiple health and fitness parameters, with special focus on diabetes management. It was tested in a pilot trial conducted in Norway hospital with diabetes patients, including the security features of the application i.e. secure sending of data to the physician and sending to an electronic health record (EHR).

The application can be installed on any iOS mobile device (iPhone, iPod, iPad). Glucose measurements can be performed by using the 2in1 smart glucometer, which utilizes the headphones connector to establish connectivity with the smartphone. The small portable glucometer makes mobility easier for diabetes patients. Since the glucometer does not need an additional power source, the users do not need to worry about batteries. The measured data are automatically stored into the local database and can be immediately presented in multiple charts or in the text format logbook.

Similar logging is also implemented for weight, insulin, stress, and food intake. The food logging feature is also useful for monitoring and storing different parameters e.g. carbohydrates, fat, sugar, protein, calories and water. A simple pie chart represents the daily food log and helps users find the desired balance between the intakes of different nutrients. DeSA activity tracking uses the in-built iPhone sensors to monitor steps, while Fitbit cloud synchronization. The application helps users reach their daily activity goals, which are set according to predefined settings or user ages.

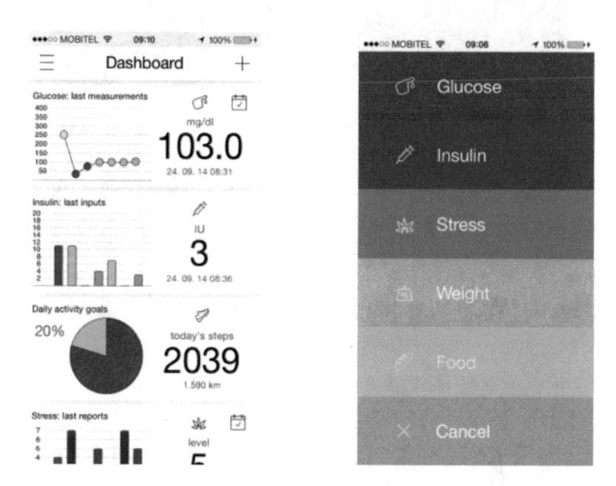

Fig. 3. DeSA application

The application was developed within the scope of EU 7. FW project Future Internet Social and Technological Alignment Research - FI-STAR, grant FP7-604691. Research conducted within a large EU funded project enabled deep and long-term (three years) involvement of several research fellows in the selected research domain. But the scope of our work was strongly directed by project objectives and sometimes unforeseen developments in the enabling technologies and principles, and thus to some extent limiting flexibility in following new research opportunities.

4.3 Welly

Welly is a well-being application, developed with the purpose of motivating users (especially younger population) to get involved in more physical activity. Application depends on the gamification effects for source of motivation to adapt to new and healthier lifestyle in real world. In order to use the application, the user needs to connect a Fitbit sensor, which is tracking user's activity by counting the number of steps the user has taken. The Fitbit sensor periodically sends the data about user's activity to the Fitbit cloud. As soon as the new data appears in the Fitbit cloud, a notification is sent to the importer, which generates an event on the platform. The received data then propagates through the platform and is stored in the database. The final visualization of user's activity progress is implemented as a mobile (iOS) application, which features a cartoon-like graphical avatar named Welly. Each step the user makes is reflected on Welly's progress, and by gathering steps the user can help the cute stranded alien travel around the world in search of his home. The more active the user is, the faster Welly progresses through levels. Users can compare their progress with other players and post their results to Facebook and Twitter, further amplifying the gamification effect. A screenshot of the Welly application and the Fitbit sensor is presented Figure 4 [16].

Fig. 4. Welly application screenshot and Fitbit sensor

The Welly application was developed within the scope of UL FE innovation environment (Makerlab). It was a response to an emerged innovation challenge that rapidly resulted in a small prototype solution. However, without a broader research or business backup it remained in this prototype phase.

4.4 Froc

Froc is a high chair for toddlers and kids. It is designed as an adjustable high chair for playful kids and their frisky parents. It's characterized by smart features, superior stability and natural materials. It was developed with parents in mind, who want to ensure a completely safe, yet relaxed and carefree childhood for their kids. It is based on rethinking the concept of wooden high chair for children. Majority of wooden highchairs are either Tripp Trapp or its clones, based on 40 years old design. Froc chair wants to be unique, practical and attractive, more suitable for modern style of living.

The Froc 2.0 chair version, currently being developed and tested, is equipped with the Internet of things electronic and communications system for measuring the weight of the child. It allows to link with a smartphone via Bluetooth 4.0 protocol. The applications open a great variety of additional features, such as child growth monitoring, food intake monitoring, healthy sitting positions, etc.

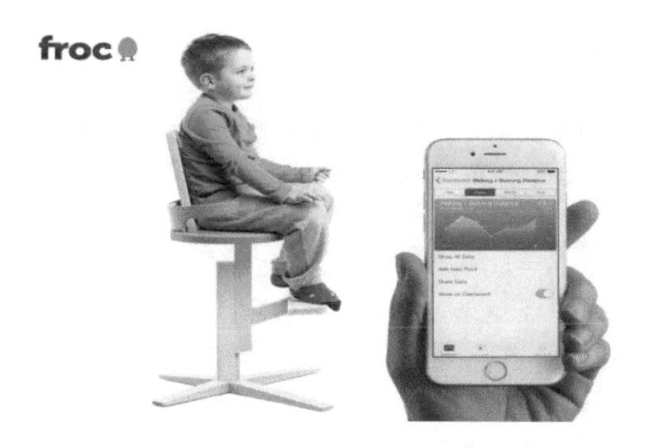

Fig. 5. Smart Froc – The world's first smart chair for kids

Froc is a product of Slovenian company Rimarket [17]. Internet of things electronic and communications subsystem is designed, integrated and developed by UL FE (Makerlab). Kickstarter campaign is being prepared for fund raising.

5 Conclusion

We presented the challenges and some of our solutions when conductions the research and innovation in ICT with examples in the field of eHealth and wellbeing. Having a relatively large research and innovation group in a small national environment and bigger EU environment poses many challenges in terms of research type, funding and IPR, personal growth and focus.

We came to the conclusion that the solution to sustain and grow the LTFE and LMMFE team and critical mass of knowledge is the combination of all mentioned approaches. That enables us to (i) be focused enough to be competitive on the national and EU research market, (ii) be flexible enough to be able to pivot in the fast changing ICT research and innovation area with (iii) the ability to support the industry.

Going more focused is the mission of start-ups, going more general would make the team noncompetitive on the research market. The biggest challenge we face is the personal growth of each team member and the integration of all mentioned (sometimes conflicting) research and innovation approaches.

The key element of success is to establish and sustain an ecosystem of talents, with young motivated researches and innovators (students) entering and the start-ups exiting and later cooperating. We presented cases of four very distinct approaches towards research and innovation. But in our experience, the innovativeness of the results is more dependent on the talents and motivation, rather than on form of research organization and its' financing. So a right mix of industry, national and EU projects with project based learning, on top of every day hard work is of most importance.

Acknowledgement

The authors wish to acknowledge the support of the research program "Algorithms and Optimisation Procedures in Telecommunications", financed by the Slovenian Research Agency.

6 References

1. MakerLab Ljubljana (2015) MakerLab Ljubljana. https://www.facebook.com/MakerLabLjubljana. Accessed 20 Jun 2015
2. Demola (2015) Introducing Demola Slovenia. http://slovenia.demola.net/about. Accessed 20 Jun 2015
3. Ljubljana University Incubator, LUI (2015) Welcome to LUI. http://lui.si/welcome-to-lui/. Accessed 20 Jun 2015
4. Hekovnik (2015) Startup School. http://hekovnik.si/. Accessed 20 Jun 2015
5. Openlab (2015) About Openlab. http://www.openlab.si/en. Accessed 20 Jun 2015
6. ZOTKS (2015) Association for Technical Culture of Slovenia http://www.zotks.si/www/portal/sl/default.asp. Accessed 20 Jun 2015
7. ICT academy initiative (2015) About ICT Academy. http://www.ict-academy.eu/en/ict-academy/about-ict-academy. Accessed 20 Jun 2015
8. Kos A (2011) Telekomunikacije zelo vplivajo na druge panoge. *Finance*, ISSN 1318-1548, 69:24-25 Ljubljana
9. Kos A (2012) Prenos znanstvenih in raziskovalnih dosežkov IKT v industrijski razvoj. *Finance*, ISSN 1318-1548, 80:13 Ljubljana
10. Kos A, Isaković M, Peternel B (2013) Primeri dobrih praks gradnje odprtih širokopasovnih omrežij - projekt ENGAGE. V: Devetindvajseta delavnica o telekomunikacijah, Brdo pri Kranju, Slovenija, 27- 28. May 2013
11. Simič N (2013) Infrastruktura za izpolnitev Digitalne agende in kaj po tem - primer Slovenije : VITEL ISSN 1581-6737. Elektrotehniška zveza Slovenije, Ljubljana
12. Digital Agenda for Europe, DAE (2015) Digital Agenda for Europe. http://ec.europa.eu/digital-agenda/en. Accessed 20 Jun 2015
13. Peternel K, Pogačnik M, Tavčar R, Kos A (2012) A presence-based context-aware chronic stress recognition system. 12(11):15888-15906. Sensors doi: 10.3390/s121115888
14. Omerovic S, Tomazic S, Milutinovic M, Milutinovic V(2009) A Methodology for Written and Oral Presentation of Research Results, Journal of Professional Issues in Engineering Education and Practice, ASCE, USA, Vol. 135, Issue 3/4
15. Future Internet Social and Technological Alignment Research - FI-STAR (2015) DeSA destress application available on iTunes. https://www.fi-star.eu/news/view/article/desa-destress-application-available-on-itunes.html. Accessed 20 Jun 2015
16. Kos A, Sedlar U, Volk M, Peternel K, Guna J, Kovačić A, Burger G, Bešter J, Tomažič S, Pogačnik M (2015) Realtime eHealth visualisation and actuation platform, International Journal of Embedded Systems 7(2):104-114
17. froc (2015) Smart Froc, the World's First Smart High Chair!. http://froc.eu/. Accessed 20 Jun 2015

Semantic Policy Information Point – preliminary considerations

Michał Drozdowicz[1], Maria Ganzha[1,2], Marcin Paprzycki[1]

[1] Systems Research Institute Polish Academy of Science, Warsaw, Poland
{michał.drozdowicz, maria.ganzha, marcin.paprzyck}@ibspan.waw.pl
[2] Department of Mathematics and Information Sciences,
Technical University of Warsaw, Warsaw, Poland

Abstract. Internet of Things (IoT), as a new paradigm for information management requires a number of novel solutions. The aim of this note is to consider methods and approaches needed to facilitate autonomous resource access. The access policies have not only to become capable of dealing with device-to-device interactions, but have to be flexible enough to deal with enormous heterogeneity of entities that are to interact in the IoT. Specifically, we will briefly summarize the existing approaches to access management and outline our approach to provisioning of the needed functionality.

1 Introduction

With the rising prevalence of connected devices, including networks of sensors, there is a growing interest in providing solutions for capturing, storing and processing the vast amounts of collected data. In this context, topics such as interoperability within the Internet of Things (IoT) also gained a lot of attention. A different issue that remains open, and at the same time growing in significance, is that of privacy and security of the data; on all levels of this fast growing ecosystem.

As what concerns regulation of access to the data and operations (services) exposed by the elements of the IoT, there are many similarities with the typical Web resources and services. There is an "entity," possibly described with several assigned attributes or roles, that requests access to "collected data" and/or physical or virtual resource(s) (or specific "services" available within such resources). In response, based on some declarative or imperative rules, such request is granted (or denied). Observe an important difference between the "standard web-provided" resources and such resources materializing in the IoT context. In the first case, typically, human-computer interactions are the core use-case. In the latter, the focus is on device-to-device interactions. Note also that in the case of the IoT, there are multiple reasons why the simple approaches, such as attribute or role based access control methods, may not scale well-enough and use of other solutions may be required.

The main aspects that make the IoT unique, when comparing to typical resources and services accessible in the Web, are:

S. Loshkovska and S. Koceski (eds.), *ICT Innovations 2015*,
Advances in Intelligent Systems and Computing 399,
DOI: 10.1007/978-3-319-25733-4_2

- Huge number of resources / producers ([8,9])
- Fast growing number of consumers ([10,13,23])
- Enormous heterogeneity of data and service formats and descriptions ([6,7, 11,14])
- Unprecedented dynamics of (often short-lived) interactions between constantly changing parties ([17,18,22])
- Machine-machine interactions – especially on the "lower level" where, typically, one device "consuming" data produced by another device, while the role of the "human" is almost completely marginalized.

The aim of this note is to briefly summarize most common approaches of dealing with the aforementioned challenges and to introduce a semantically enriched access control policy system. To this effect, in the next sections we summarize: (i) policy-based access control, (ii) XACML language, (iii) semantic approaches to access control and (iv) semantic extensions to the XACML. We follow with an outline of the proposed approach to the (re)design of the *Policy Information Point*.

2 Policy based access control

Study of pertinent literature reveals the, in many cases, access control is embedded into the logic of the service or the resource provider, and intertwined with the business logic. However, in an environment consisting of a very large number of different services, such approach leads to an unmaintainable, inconsistent set of rules. In other words, it is rather difficult to be able to understand what policies do exist in the system (they are "hidden" within the devices) and how they relate to each other (e.g. [9,17,22]).

A better approach would be to move access control decisions outside of the services and devices, e.g. into a centralized authorization component or a set of such components. One way to design such a subsystem would be to use an "engine" that uses declarative policies, specifying the conditions under which a given request is accepted, or when access is denied. Obviously, such approach would have to face the problem of having a "centralized" solution (i.e. potential bottleneck of a large-scale distributed system). However, this problem is in many ways mainly a technical one. As soon as we separate the logic and authorization rules from the logic of the applications, we can see them as being a common part of all (sub)systems within the IoT. In this case, the analysis of a specific request can be "scaled down" to the devices (note that, in this case, all devices would subscribe to the common rules – or a part of these rules that they need to operate). This, in turn, makes the problem technically solvable (using existing techniques; e.g. from the fields of Grid / P2P computing [16].

3 XACML

Let us now consider ways of "representing" rules defining access policies. Currently, one of the most common policy specification languages is the eXtensible

Access Control Markup Language (XACML; [2]). It is a declarative language, and a standard for implementation of processing engines developed and maintained by the OASIS group. The standard uses the XML as its internal format, but many implementations handle information transfer in other formats, such as, for instance, the SAML ([1]).

At its core, XACML is an implementation of the Attribute Based Access Control (ABAC; [26]), basing its authorization decisions on values of attributes. In the XACML, these are grouped into four categories:

- *Subject* – an entity (possibly a person) requesting access,
- *Resource* – the entity, access to which is under control,
- *Action* – the action that the *Subject* requests to be performed on the *Resource*,
- *Environment* – other attributes that bring additional context.

The decisions are made based on policies consisting of rules. In the case when multiple policies are applicable to the same request, a policy combining algorithm, defined in the, so called, policy set that encompasses all existing / defined policies, is used to produce the final result.

The way that the XACML engines make decisions on incoming requests is based on a two-step attribute evaluation. First, the conditions defined in the *Target Element* of the policy or rule are checked, to limit the number of rules that have to be processed. Second, the *Condition* is evaluated and, based on the result of such evaluation, the rule or policy decision is made.

The reference architecture of an XACML processing system contains the following major components:

- *Policy Enforcement Point* (PEP) – responsible for the actual act of enabling or preventing access to the resource. It also coordinates the execution of, so called, *Obligations*, which are additional operations that should be performed when a decision has been made (e.g. logging the request for auditing purposes).
- *Policy Information Point* (PIP) – a source of attribute values.
- *Context Handler* – which converts requests and responses between native formats and the XACML canonical representation and coordinates, with the PIPs, gathering of the required attribute values.
- *Policy Decision Point* (PDP) – which evaluates policies and issues the final authorization decisions.
- *Policy Administration Point* (PAP) – which defines, stores and manages policies.

Diagram 1 depicts the sequence of messages in a typical access control decision reaching process that takes place in the considered architecture. Handling obligations has been omitted for the clarity of presentation.

While the XACML has many advantages, such as expressiveness, extensibility and reusability of policies, it also has some important drawbacks. More complex policies that involve relations between the attributes or calculation of values

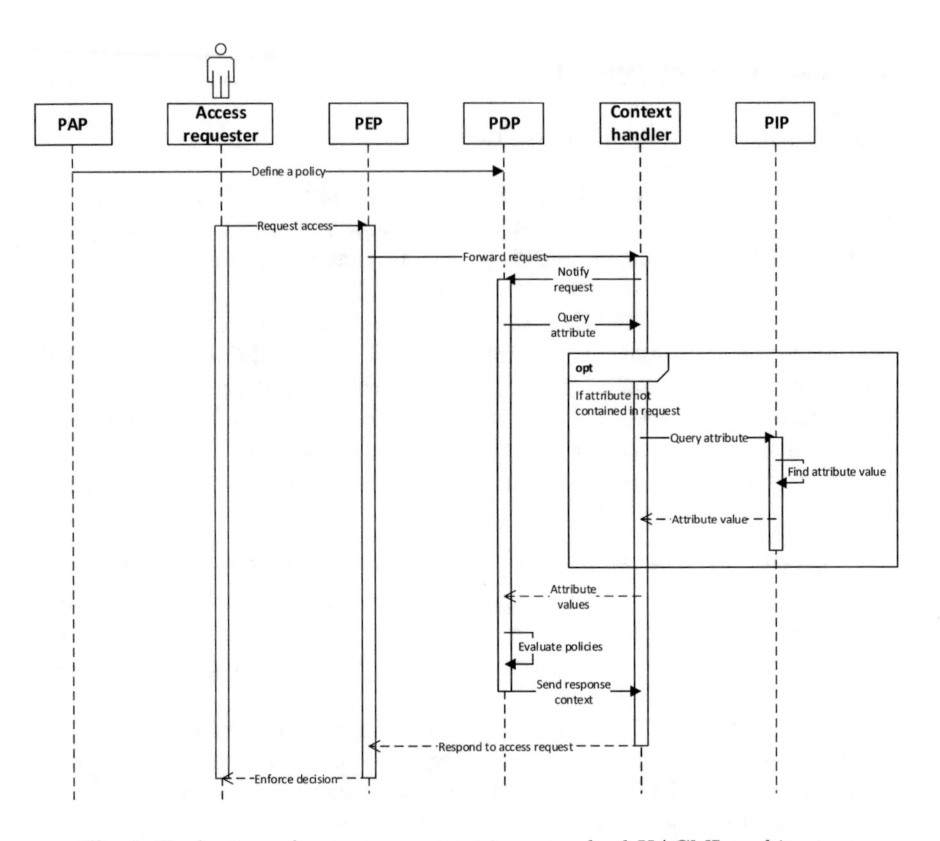

Fig. 1. Evaluation of an access request in a standard XACML architecture

based on other attributes, can be extremely convoluted and hard to maintain. In this respect, what XACML lacks most is the possibility to (i) *reason about the domain* containing the attributes, and (ii) *infer additional data* in an automatic way. Moreover, the XACML standard deals only with policy definition and enforcement, while it does not provide solutions for attribute management.

4 Semantic approaches to access control policies

The functionality that the XACML is missing can be naturally handled using techniques from the area of semantic data processing. As a matter of fact, there have been projects tackling the challenge of access control by defining policies using special ontologies and developing the decision engines based on semantic reasoners.

The *Rei Project* ([28]) was created to address security concerns in the Semantic Web, in particular in mobile and pervasive computing scenarios. Its engine utilized policies defined in OWL-Lite and provided ability to reason about permissions and obligations, policy conflict resolution, permission delegation and possibility to define variables that could be referenced in the policy rules. The project was later modified, as Rein, to use the N3 language as its base and provide extensible, reusable meta-policies for federation networks. Currently, the project is no longer under active development. Specifically, the Rei Project ended in 2005, while the latest version of the Rein specification is from 2006.

The *KAoS Framework* ([27]) is similar to the Rei in its language choices – it is using OWL-DL for policy specification. It differs in its primary application area, which is multi-agent and distributed systems. As a matter of fact, the policy authorization engine was just a part of a more general agent platform that included also directory services and simplified GUI tools for the policy creation and management. The KAoS framework can be considered abandoned since at least 2008.

In contrast to the previous two projects, *Ponder* ([20]), which actually predated them and laid grounds for many of their features, did not apply any specific ontological language for its policies. Instead, it utilizes a custom declarative language that is later processed by the Java based engine. It provided reasoning capabilities, however, not as powerful as OWL-based reasoners used in Rei and KAoS. Similarly to the other projects, this one also has been abandoned (by approximately 2002).

Overall, while quite interesting, these projects never gained enough traction and did not achieve significant adoption. In this respect the plain XACML was much more successful.

5 Semantic extensions to the XACML

As a consequence of its popularity, the XACML has also became the base for research projects dealing with various aspects of access control. Some were centered around the challenges of spatio-temporal constraints applied to policies (see, [31]), while others were dealing with deficiencies of the language in some more sophisticated elements of Role Based Access Control (see, [24,25]).

There were also more general extensions of the language, aimed at taking advantage of the semantic reasoning capabilities of the description logic languages. In [19], the authors extended the SAML and XACML languages with the possibility to include embedded RDF/XML documents.

In [30], a similar approach was introduced, however, it did not assume any inclusion of semantic information directly into the policies. The policies and ontologies were clearly separated – the role of the semantic reasoner was restricted to providing additional data to the standard (XACML-based) decision making process. Moreover, using this approach the policy author did not need be concerned with knowledge about ontological techniques as these affected only the "domain part" of the system. In the context of the IoT, these properties are extremely useful – the policies may be stated in the vocabulary of the system performing the authorization, while it is going to be up to the ontological layer of the system to perform any necessary mapping or merging of concepts from incoming requests.

6 Semantic Policy Information Point

Due to the aforementioned strengths of the approach proposed in [30], we have decided to follow the general (top-level) ideas outlined in that paper (while applying slightly different solutions on the "technical level"). The result is development of the *Semantic Policy Information Point* (SemanticPTI) for the XACML reference architecture that is capable of providing values to unknown attributes, by inferring them from the ontologies describing the domain of the system.

The general algorithm used by the SemanticPIP is described by the following steps:

1. When the *Context Handler* queries for an attribute value, the SemanticPIP translates the request context into a temporary ontology. For each attribute category (e.g. Subject, Resource, Action, Environment) described in the request, an OWL *Class* and an OWL *Individual* of the same type are created. Next, each attribute value from the request is converted into a *Data property axiom* of the appropriate type.
2. The temporary ontology is merged with the existing (in the system) ontologies specifying the details of the domain.
3. The Pellet ([12]) reasoner is used to reason about all ontology properties that are not explicitly specified in the request.
4. For each recognized (by the reasoner) property, a SPARQL query is issued on the ontology to retrieve the attribute value and type. Listing 1.1 shows a sample query used for selecting the attribute. Here, the `categoryId` parameter is replaced, during the runtime, with the unique identifier of the *Individual* created in step 1, while the `attributeId` is simply the fully qualified *id* of the requested attribute.

5. Finally, the response from the SPARQL is converted back into the format acceptable by the *Context Handler*.

```
PREFIX rdf: <http://www.w3.org/1999/02/22-rdf-syntax-ns#>
SELECT ?val WHERE
{
 <categoryId> <attributeId> ?val
}
```

Listing 1.1. SPARQL query

In comparison to the solution developed by Priebe et al. ([30]) our solution does not change the reference architecture of the XACML system. Instead it implements the contract of the PIP component. This is also reflected in the fact that the SemanticPIP only queries for, and returns the attribute values that are requested by the *Context Handler*, reducing the burden placed on the SPARQL engine.

The component is built as an attribute finder extension to the *Balana Framework* ([5, 15]) XACML engine developed by the WSO2 as a continuation of the popular Sun's XACML Implementation. The engine was chosen due to its maturity and widespread use, as part of the WSO2 Identity Server product package.

7 Conclusions and future work

Te aim of this note was to outline issues involved in moving the rule-based resource security and access policies from standard web services towards the Internet of Things ecosystem. The proposed Semantic Policy Information Point has been implemented and is being thoroughly tested. There are two scenarios where it is going to be applied. First, a non-IoT one, originating from the *Agents in Grid* project (see, [21, 29]). The IoT application will be focused on heterogenous policy application for a sensor network in the field of transport and logistics. The SemanticPIP development plans include also support for the additional XACML profiles, namely the Role Based Access Control Profile ([3]) and the Hierarchical Resource Profile ([4]).

References

1. Security assertion markup language (SAML) v2.0. https://www.oasis-open.org/standards, 2005.
2. extensible access control markup language (XACML) version 3.0. http://docs.oasis-open.org/xacml/3.0/xacml-3.0-core-spec-os-en.html, 2013.
3. XACML v3.0 core and hierarchical role based access control (RBAC) profile version 1.0. http://docs.oasis-open.org/xacml/3.0/rbac/v1.0/xacml-3.0-rbac-v1.0.html, 2014.
4. XACML v3.0 hierarchical resource profile version 1.0. http://docs.oasis-open.org/xacml/3.0/hierarchical/v1.0/xacml-3.0-hierarchical-v1.0.html, 2014.

5. Balana – Open source XACML 3.0 implementation. http://xacmlinfo.org/category/balana/, 2015.
6. Bringing Big Data to the Enterprise. http://www-01.ibm.com/software/data/bigdata/, 2015.
7. Collaborative open market to place objects at your service. http://www.compose-project.eu/, 2015.
8. CSA for Global RFID-related Activities and Standardisation (CASAGRAS2). http://www.iot-casagras.org/, 2015.
9. Internet-of-Things Architecture. http://www.iot-a.eu/public, 2015.
10. Iot@work. https://www.iot-at-work.eu/, 2015.
11. Open source cloud solution for the internet of things. http://www.openiot.eu/, 2015.
12. Pellet. https://github.com/Complexible/pellet, 2015.
13. uBiquitous, secUre inTernet-of-things with Location and contEx-awaReness (BUTLER: Smart life. http://www.iot-butler.eu/, 2015.
14. "web of objects" ITEA 2 Project. www.web-of-objects.com/, 2015.
15. WSO2 Balana Implementation. https://github.com/wso2/balana, 2015.
16. Nick Antonopoulos, Georgios Exarchakos, Maozhen Li, and Antonio Liotta. *Handbook of Research on P2P and Grid Systems for Service-Oriented Computing: Models, Methodologies and Applications.* IGI Global, 2010.
17. Erin Anzelmo, Alex Bassi, Dan Caprio, Sean Dodson, Rob van Kranenburg, and Matt Ratto. Internet of Things. discussion paper. Institute for Internet and Society, Berlin, October 2011.
18. L. Atzori, A. Iera, and G. Morabito. The internet of things: A survey. In *Computer Networks: The International Journal of Computer and Telecommunications Networking*, volume 54, page 2787âĂŞ2805. Elsevier North-Holland, Inc., 2010.
19. Ernesto Damiani, Sabrina De Capitani di Vimercati, Cristiano Fugazza, and Pierangela Samarati. Extending policy languages to the semantic web. In *Web Engineering*, pages 330–343. Springer, 2004.
20. N. Damianou, N. Dulay, E. C. Lupu, and M. Sloman. Ponder: A language for specifying security and management policies for distributed systems. Technical report, Imperial College, UK, Department of Computing, 2001.
21. M. Drozdowicz, M. Ganzha, K. Wasielewska, M. Paprzycki, and P. Szmeja. Agreement technologies. chapter Using Ontologies to Manage Resources in Grid Computing: Practical Aspects. Springer, 2013.
22. D. Evans. The internet of things: How the next evolution of the internet is changing everything. Technical report, Cisco Internet Business Solutions Group (IBSG), 2011.
23. FASyS. Absolutely safe and healthy factory. http://www.fasys.es/en/index.php, 2015.
24. Rodolfo Ferrini and Elisa Bertino. Supporting rbac with xacml+ owl. In *Proceedings of the 14th ACM symposium on Access control models and technologies*, pages 145–154. ACM, 2009.
25. Nurmamat Helil and Kaysar Rahman. Extending xacml profile for rbac with semantic concepts. In *Computer Application and System Modeling (ICCASM), 2010 International Conference on*, volume 10, pages V10–69. IEEE, 2010.
26. Vincent C Hu, David Ferraiolo, Rick Kuhn, Adam Schnitzer, Kenneth Sandlin, Robert Miller, and Karen Scarfone. Guide to attribute based access control (abac) definition and considerations. *NIST Special Publication*, 800:162, 2014.

27. M. Johnson, P. Chang, R. Jeffers, J. Bradsha, V-W. Soo, M. Breedy, L. Bunch, S. Kulkarni, J. Lott, N. Suri, and A. Uszok. KAoS semantic policy and domain services: An application of DAML to web-services-based grid architectures. In *Proceedings of the AAMAS 03: Workshop on Web Services and Agent-Based Engineering*, 2003.
28. L. Kagal, T. Finin, and A. Johshi. A policy language for pervasive computing environment. In *Proceedings of IEEE Fourth International Workshop on Policy (Policy 2003)*, pages 63–76. Los Alamitos, CA: IEEE Computer Society, 2003.
29. Kamil Lysik, Katarzyna Wasielewska, Marcin Paprzycki, Maria Ganzha, John Brennan, Violeta Holmes, and Ibad Kureshi. Combining aig agents with unicore grid for improvement of user support. In *The First International Symposium on Computing and Networking - Across Practical Development and Theoretical Research, Dogo SPA Resort, Matsuyama, Japan, December 4-6, 2013.*, pages 66–74, 2013.
30. T. Priebe, W. Dobmeier, and N. Kamprath. Supporting attribute-based access control with ontologies. In *Availability, Reliability and Security, 2006. ARES 2006. The First International Conference on*, page 8, 2006.
31. Que Nguyet Tran Thi and Tran Khanh Dang. X-strowl: A generalized extension of xacml for context-aware spatio-temporal rbac model with owl. In *Digital Information Management (ICDIM), 2012 Seventh International Conference on*, pages 253–258, Aug 2012.

A Roadmap to the Design of a Personal Digital Life Coach

Nuno M. Garcia

University of Beira Interior, Faculty of Engineering, Covilhã, Portugal
Instituto de Telecomunicações, Covilhã, Portugal
Universidade Lusófona de Humanidades e Tecnologias, Lisbon, Portugal
ngarcia@di.ubi.pt

Abstract: The timely and exact detection of abnormal events on the life of someone is one of the cornerstone of for successful Ambient Assisted Living strategies, either because these events need to be avoided, or because it is necessary that these events occur as a mean to implement a program for training a particular lifestyle. Technology has given us a plethora of sensors and actuators that can be used to deploy an infrastructure that would allow a successful detection of daily living activities, but often this would imply to use technology intensive, privacy invading solutions that despite its efficiency would render them non-adoptable by users. In this paper, we will explore the opportunities and challenges that AAL poses in the field of identification of daily activities, and how such a solution can be designed as to be an user-adoptable part of a monitoring and training of lifestyles.

Keywords: Personal Digital Life Coach, Mobile Application, Smartphone, Smartphone Sensors, Sensor Data Fusion, Sensor Data Imputation.

1 Introduction

The creation of a "Personal Digital Life Coach" is still a somehow distant goal. Yet, the creation of such piece of digital craftsmanship, would be of extreme help, not only for those of us who may be older and ill, but also for the older and healthy and, why not, for the young and ill or healthy, as this would allow each one of us to profit for the advices of a digital omnipresent friend, who would know our habits and life style, our ambitions, goals, and limitations, and would allow us to get expert advice on each step of our lives.

A Personal Digital Life Coach (or PDLC for short) will monitor our actions and activities, be able to recognize its user state of mind, and propose measures that not only will allow the user to achieve his/her stated goals, but also to act as an intermediate health and well being agent between the user and his/her immediate care givers.

The PDLC here proposed is therefore one of many possible scenarios. This paper addresses what would be the possible architectures, requisites, sensory capabilities, and interfaces with the user and with the community of users of a PDLC.

S. Loshkovska and S. Koceski (eds.), *ICT Innovations 2015*,
Advances in Intelligent Systems and Computing 399,
DOI: 10.1007/978-3-319-25733-4_3

Before describing the steps that need to be taken to devise such a system, it is convenient to elaborate on the advantages that such system may bring to the user. As it is widely known, many diseases are the result of unhealthy lifestyles. Paramount of unhealthy lifestyles are tobacco addiction (or other addictions) and sedentary life styles, such as the ones that do not integrate an adequate amount of exercise. Among these diseases one can find those related to obesity, including some types of diabetes, cardiovascular related diseases, including high blood pressure, and others such as lung and larynx cancer. The treatment of life style related diseases is responsible for a large share of the nation's health budgets, and this has become a concern for policy makers and governments worldwide.

This paper describes the one of the possible concepts for such a tool, and one of possible roadmaps that will allow its construction in a, hopefully, not so distant future. The definition of the stakeholders, primary and secondary, is also discussed.

The remainder of this paper is organized as follows: this paragraph concludes the introduction (Section 1); section 2 presents the identification of the stakeholders of a PDLC solution and the primary and secondary initial meta-requirements; section 3 discusses the possible architectures for a PDLC solution; section 4 presents a proposal of roadmap, and section 5 terminates this paper with relevant conclusions.

2 Stakeholders and Meta-Requirements for the PDLC

The creation of a "Personal Digital Life Coach" is still a somehow distant goal. Yet, the creation of such piece of digital craftsmanship, would be extremely useful, not only as a mean of assistance to those of us who are more fragile and need assistance and guidance, but also to those of us who being healthy, can still profit from some expert advice.

It must be stated that a PDLC is to be a shift in the paradigm of current Ambient Assisted Living (AAL) and Enhanced Living Environments (ELE) platforms, *i.e.*, instead of the solution being focused on the needs of a given user or a given user profile, the solution is focused on the needs of "the User". This acknowledges that a human user may live a life that is so complex and diverse, that is extremely complicated to fit its life pattern in the definition of a generic user profile. Traditional AAL approaches are often described as "the tool that does X to the users that suffer/need Y", and this is an approach that has been enough for the old manner of living the world, what can be described as "the wheelchair approach". Wheelchairs are undoubtedly very useful devices for people who have temporary or permanent disabilities that impair their mobility, and unfortunately, often wheelchairs are the only solution. Yet, by solving the mobility problem, the wheelchair can bring a completely new set of other problems that, being of lesser importance to the user, still may need to be addressed. Moreover, wheelchairs are solutions that, given the alternatives, are high adoption ratios among its potential users; nevertheless, this is very often the only solution.

Human societies are also complex, and have been fast changing in the last decades [1, 2] particularly in what is generally accepted as the cell of human societies, the family. The concept of family as therefore evolved, and as authors in [1] acknowledge, there is an increase in the plurality and diversity of the concept itself. The Internet and its contribution to the feeling of connectedness [3, 4] also contributes to the notion that a family is no longer the aggregate of people who, sharing sociological bonds, share a common life space.

If, on one hand, the family is the natural support for the individual, not only he/she who may be fragile, but also to him/her who needs advice or company, on the other hand, most western countries have also developed some form of social providence or welfare state, in the sense that medical care and life supporting means are usually not denied to anyone who may need them. The immediate cost of the welfare to the state's treasures is not to be demised, although some argue that the reason to establish and support a welfare state is not the value of its policies, but the values of its principles. Philosophical and political issues apart, nations and governments have come to realize that they too are interested parties in the wellbeing of each individual citizen.

As a smaller group of the states welfare and wellbeing policies, even if these have private instead of public actors, are the persons who perform the acts that allow the states to achieve the overall goal of keeping its citizens happy and healthy.

Considering a PDLC as an instrumental tool for the goals of keeping one's health and wellbeing, a goal that is shared by the person itself, his/her family and ultimately, the society he/she lives in, there are several stakeholders for a PDLC. As the centre and as primary stakeholder, the individual, now promoted to the category of PDLC user him/herself. As secondary stakeholders, we identify the user's family (friends, relatives, colleagues). As tertiary stakeholders, we have the caregivers, the health and wellbeing professionals, and, because the nature of this stakeholder is collective, we have the company, the organization, and the state the user belongs to. Although this is a very linear manner of describing potential stakeholders, it must also be noted that secondary and tertiary stakeholders share a fuzzy border, as for example, family often acts as caregivers, and caregivers often become friends.

Having defined who the stakeholders are, we can now define what will be the primary and secondary initial meta-requirements for a PDLC, while keeping in mind that a personal digital life coach is a tool that implies that there is a life to be lived, *i.e.*, the user is still able, to a minimum extent, to conduct and decide on his/her daily life choices.

As primary meta-requirement, a PDLC is a tool that is able to:

1. identify a diverse range of daily activities
2. identify patterns of behaviour and detect when abnormal or unexpected behaviour occurs
3. estimate the amount of energy its user is expending on each activity
4. assess (estimate? infer?) the mood of the user
5. input and record the user's personal goals, *e.g.*, to loose weight, to quit an addition, to work fewer hours
6. keep track of the effective behaviour and compare it with the expected behaviour, given not only previous behavioural patterns but also the user's recorded goals

7. monitor and record some of the user's life signals, *e.g.*, his/her electrocardiography (ECG) signal, or his/her blood pressure, or the user's amount of movement, *etc.*
8. alert the immediate and secondary caregivers when a life threatening event occurs
9. keep the collected data in the most strict confidential manner, only disclosing data that it is allowed to, to the relevant stakeholders
10. integrate error safe mechanisms that allows it to assume that the advices and data that is feeding to its user are flawed and are potentially dangerous.

As secondary meta-requirement, a PDLC is a tool that:

1. works in a non-intrusive, user adoptable manner
2. allows the user to interface with it using natural interfaces such as voice and gestures
3. interfaces with the user using voice and text (email, SMS, *etc.*)
4. allows the communication of relevant and authorized data with other PDLCs as a mean to reap the benefits from big data applications
5. communicates with or is fed by several agents that collect data in the users devices, such as, Internet home gateway router, social network feeds and activity report agents, car computer, *etc.*.
6. is able to work in full compliance of national and international laws, particularly those that regulate medical devices, and those that regulate the privacy, safety and security of personal information.

The definition of the stakeholders and of the meta-requirements for a PDLC allows the conclusion that there are a number of important milestones to conquer before such a solution may appear on the shelves of a digital application store.

3 Architecture for a PDLC Solution

The first step to enable the creation of a PDLC, is to devise a system that has the following characteristics:

1. is a computational platform with some processing power (although "some" is not a scientific expression, we still lack the techniques to estimate how much processing power a complex to-be developed application will need)
2. is small and powerful enough to perform data storage, data fusion, data imputation, and can be connected to the Internet
3. at least part of the system can be carried by it's user most of the time
4. integrates a set of sensors that can sense and identify most of the things the user is doing most of the time
5. can communicate with actuators and other hardware that the user interacts with
6. and, is safe and guarantees the user that his/her data are secure and cannot be disseminated without the user's consent.

Additionally, the system will be non-intrusive, and will provide its user with services in a manner that it will contribute to easy user adoption. Or in other words, the centre of the solution is the user, not the technology nor the application.

In the line of what Weiser has defined for ubiquitous computing [5], a PDLC will fail because it does not have enough data, or because the data it used was not correctly weighted, and these errors need to be acknowledged and foreseen in the planning of the tool.

Out-of-the-shelf smartphones can presently perform most of these tasks, with the added bonus that most of the times the device is carried by its user. But this is not yet a closed issue, as there are still too many variables to address, for example, there are different usage styles for smartphones, *i.e.* while it seems some users will use the smartphone on the pocket of their trousers or jackets, others will likely carry it in their purses or bags.

A PDLC will be a system that can detect not only the expectable events on a person's daily life, but also the abnormal events or the undesirable events, making it possible to, in view of a user's expected pattern of behavior, and his/her announced and defined goals, provide the user with expert advice that allows the user to correct, if not at the precise moment, the unadvisable behavior.

Considering the amount of data that a PDLC has to collect and process, it becomes evident that several degrees of efficiency and complexity for a PDLC must be defined. The simpler one will be a standalone application, probably installed in a smartphone or some other wearable device that performs a basic level of monitoring and allows for some basic life coaching decisions. In fact, there are already several systems who monitor the user while he/she exercises, and allow him/her to keep track of the energy expenditure, some including social networking components [6]. The more complex solution will probably be a multi-agent software, spanning several of the users hardware interaction points, and allowing the collection of the fused data in a central system where decisions are made and advices are conveyed to the user. By hardware interaction points we refer to the hardware that supports the user's activities, *e.g.*, a smartphone, of course, but also, the television set, the home router, or the users car.

The data for this collection may be stored and processed locally, in the case of the simpler architectures, or may be stored and processed in some form of cloud. Of relevance will be the fact that a complex PDLC will have the ability to share, in a secure, safe and confidential manner, the data from its users with other systems, allowing for the application of algorithms for big data processing.

A PDLC will also be able to perform data imputation as a manner to fill in the gaps in the collected data, either because the sensors were busy for some other application, or because there were errors in the transmission, or because of noisy or lossy environments in the collection or transmission phases. Also, a PDLC will also be able to perform data fusion, as a mean to minimize uncertainty regarding the identification of a given user's state of mind or a particular user's task. As mentioned before, the PDLC must assume that the data and advices that conveys to its user may cause him/her harm, and therefore must integrate not only the necessary fail safe mechanisms to avoid so, but also mechanisms that allow frequent feedback from its user, to allow the confirmation or infirmation of the results of its processes. These mechanisms, and also the mechanisms that will support decision-making, data imputation and data fusion may be founded in artificial intelligence algorithms and techniques.

4 A Roadmap

This paper describes, at a conceptual level, what would be the roadmap to design a Personal Digital Life Coach, a tool that would act like a "best friend" for its user, giving advices and tips on the choices of his/her daily living, with the ultimate objective to allow the training of healthier and happier life styles.

The roadmap for the development of such a device includes the following steps:

1. the development of software that can be installed in personal mobile devices, such as smartwatches or smartphones, and that is able to identify daily activities of a user
2. the development of software that can recognize patterns of daily activities and also identify meaningful deviations to these patterns
3. the development of agents that can reside in the hardware that the user interacts with, and that can send to a central processing application their relevant usage data
4. the development of software that can integrate all the data that a user generates into meaningful insights of his/her personal life style
5. the development of software that given the identification of the user's behaviour, can create acceptable behaviour alternatives that allow the user to achieve his/her previously defined lifestyle objectives.

This roadmap also includes, at some point:

1. the identification of some of the user's emotions, *e.g.* by means of sensors and usage data patterns
2. the identification and association of the user's data to its relevant caregivers, family and friends
3. the capability to "understand" clusters of data from these users, into a social behaviour scenario, allowing the realization of the social integration of the user, or its absence
4. the capability to integrate failsafe mechanisms and frequent user interaction mechanisms that allow the user to confirm or infirm the algorithm's output, and therefore, establish a feedback loop that allows for additional algorithm output correctness.

Some of these checkpoints are not yet at our reach, and others have a long research ahead. Nevertheless, in the field of hypothesis, a PDLC would be a very useful application.

5 Conclusions

The notion of having a "friend in the computer", being this computer some wearable computational platform is a risk, primarily because users are not keen in digitalizing their entire life, but also because we still have to assess what is the real level of benefits of prejudice that computational applications have brought to human development. As always, not all humans are alike, and if for some, a PDLC would be a significant help, to others it would just be meaningless.

This paper presented the concept of a Personal Digital Life Coach, an application or set of applications that will allow a user to receive advices and information that allow him/her to train his/her lifestyle into a more healthier and maybe happier one.

As discussed previously, it is my belief that we still have a long way to go to master all the technologies that will allow us to build a PDLC. Yet I have no doubt that this will eventually happen, some day.

Acknowledgments

The authors acknowledge the contribution of COST Action IC1303 AAPELE Algorithms, Architectures and Platforms for Enhanced Living Environments, and COST Action IC1307 IV&L Integrating Vision and Language.

Authors affiliated with the *Instituto de Telecomunicações* also acknowledge the funding for the research by means of the program FCT project UID/EEA/50008/2013. (*Este trabalho foi suportado pelo projecto FCT UID/EEA/50008/2013*).

References

1 Kuijsten, A. C.: Changing family patterns in Europe: A case of divergence?. European Journal of Population/Revue Européenne de Démographie, vol. 12, pp. 115-143 (1996)

2 Rindfleisch, A., Burroughs, J. E., Denton, F.: Family structure, materialism, and compulsive consumption. Journal of consumer research, pp. 312-325 (1997)

3 IJsselsteijn , W., van Baren , J., van Lanen. F.: Staying in touch: Social presence and connectedness through synchronous and asynchronous communication media. Human-Computer Interaction: Theory and Practice (Part II), vol. 2, pp. 924-928 (2003)

4 Kuwabara, K., Watanabe, T., Ohguro Ohguro, T., Itoh, Y., Maeda, Y.: Connectedness oriented communication: Fostering a sense of connectedness to augment social relationships. in Applications and the Internet, 2002.(SAINT 2002). Proceedings. 2002 Symposium on, pp. 186-193 (2002)

5 Weiser, M.: Some computer science issues in ubiquitous computing. Communications of the ACM, vol. 36, pp. 75-84 (1993)

6 Sousa, P., Sabugueiro, D., Felizardo, V., Couto, R., Pires, I., Garci, N.: mHealth Sensors and Applications for Personal Aid. in Mobile Health, ed: Springer, pp. 265-281 (2015)

Load Balancing of Distributed Servers in Distributed File Systems

Ravideep Singh[1], Pradeep Kumar Gupta[2], Punit Gupta [1], Reza Malekian[2], Bodhaswar T. Maharaj[2], Darius Andriukaitis[3], Algimantas Valinevicius[3], Dijana Capeska Bogatinoska[4], and Aleksandar Karadimce[4]

[1] Deptt. of CSE, Jaypee University of IT, Waknaghat, Solan, HP, 173 234, India, {ravideep36, punitg07}@gmail.com

[2] Department of Electrical, Electronic, and Computer Engineering, University of Pretoria, South Africa, 0002
{pkgupta, reza.malekian}@ieee.org, sunil.maharaj@up.ac.za

[3] Department of Electronics Engineering, Faculty of Electrical and Electronics Engineering, Kaunas University of Technology, Kaunas, Lithuania
{darius.andriukaitis, algimantas.valinevicius}@ktu.lt

[4] Faculty of Computer Science and Engineering, University of Information Science and Technology St. Paul the Apostle, Ohrid, Republic of Macedonia, 6000
{dijana.c.bogatinoska, aleksandar.karadimce}@uist.edu.mk

Abstract. With the rapid growth in technology, there is a huge proliferation of data in cyberspace for its efficient management and minimizing the proliferation issues. Distributed file system plays a crucial role in the management of cloud storage which is distributed among the various servers. Many times some of these servers get overloaded for handling the client requests and others re-main idle. Huge number of client requests on a particular storage server may in-crease the load of the servers and will lead to slow down of that server or dis-card the client requests if not attended timely. This scenario degrades the over-all systems performance and increases the response time. In this paper, we have proposed an approach that balances the load of storage servers and effectively utilizes the server capabilities and resources. From our experimental results and performance comparison of proposed algorithm with least loaded algorithm we can conclude that our approach balances the load, efficiently utilize the server capabilities and leverage the overall system performance.

Keywords: Distributed file system, Cloud storage, Performance, Overload, Load balancing

1 Introduction

With the rapid growth in the world of technology, a tremendous amount of data is being proliferated over the Internet. Cloud computing provides cloud storage to man-age this data on the Internet and act as a repository in which the data is maintained, and made available to the consumer located geographically. Key building technologies for cloud include the distributed file systems,

S. Loshkovska and S. Koceski (eds.), *ICT Innovations 2015*,
Advances in Intelligent Systems and Computing 399,
DOI: 10.1007/978-3-319-25733-4_4

virtualization, MapReduce programming paradigm and so forth [1]. The management of data in cloud storage requires a special type of file system known as distributed file system (DFS), which had functionality of conventional file systems as well as provide degrees of transparency to the user, and the system such as access transparency, location transparency, failure transparency, heterogeneity, and replication transparency [2]. DFS provides the virtual abstraction to all clients that all the data located closest to him. Generally, DFS consists of master-slave architecture in which master server maintains the global directory and all metadata information of all the slave servers. Whereas, slave represents a storage server that stores the data connected to master server and other storage servers as well. This storage server handles the thousands of client requests concurrently, in DFS. The load distribution of requests on these storage servers is uneven and lead to overall performance degradation. Resources are not exploited adequately, because some server gets too many requests and some remain idle. In a distributed file system, load can be either in terms of requests handled by a server or storage capacity of that server or both. In this paper, we have proposed an approach for load balancing of client requests to be handled by a server. We have proposed a strategy to balance the load of requests for overloaded servers in a distributed file system. During load balancing parameters like CPU utilization, storage utilization, buffer space, and network bandwidth plays a key role. Load balancing using these parameters might be difficult so an intelligent way is required to handle each server efficiently. The remainder of this paper is organized into various sections where section 2 focuses on the related work of various load balancing algorithms and approaches. Section 3 describes about the problem statement using which we have designed the algorithms and performed experimental work. Section 4 describes about the proposed approach and algorithm. In section 5, we have presented results obtained from simulations and compare these results with the least loaded algorithm. Finally, we have concluded the work.

2 Related Work

Load balancing in a distributed file system plays a crucial role to enhance the performance and response time. Various load balancing approaches have been worked out till now which can be categorized into static and dynamic in nature [3]. In [4] Yamamoto *et al.* proposed a distributed approach to balance the load using replication of original data. Authors have proposed two replication methods 1) In the path random replication method, replicas stored in the peers along the path of requesting to peer. 2) In the path adaptive replication method replicas stored only in the peers according to their probability of replication. This paper does not consider the physical capability of servers while selecting the server for replication. In [5] Rao *et al.* have presented a general framework for load balancing in distributed environment, named as HiGLOB. Here, authors have used two main components 1) *histogram manager* - generates a histogram to maintain a global information regarding to the distribution of the load in the system, and 2) *load-balancing manager* - reallocates the load whenever the node becomes overloaded or under loaded. However, there is overhead associated while constructing and maintaining the histograms. In [6] Zeng *et al.* have proposed a load re-balancing algorithm to work out the problem of load balancing among all chunk servers in the distributed file system. They have also ensured the reliability of the system where one chunk of a file and two duplicate copies are allocated in three different chunk servers at a time. In this algorithm authors have used the master server periodically for checking of chunk servers and to distinguish which chunk server is over-loaded and which is not. However, this master server has become a single-point of failure for the load balancer. In [7] Fan *et al.* have developed a new adaptive feedback load balancing algorithm (AFLBA) for the Hadoop distributed file system (HDFS) which uses two modes: 1) disk utilization rate model and 2) service blocking rate model. The proposed algorithm uses the storage utilization of each data node and probability of blocking client request of each data node. Since this algorithm is not distributed so it creates a performance bottleneck issue for the name node in the HDFS. Hasio *et al.* [1] and Chung *et al.* [8] have presented a distributed approach for load balancing in a distributed file system to minimize the performance bottleneck issue for the name node. They have used CHORD protocol to create an overlay of server nodes. Here, authors have considered the storage capacity of server as a load. A threshold value has been used to classify the under loaded and overloaded node in the system. Proposed algorithm is random in nature and does not guarantee for the reliability of the distributed file system and also does not deal with the physical aspect of a server. In [9] Malekian and Abdullah have studied the high traffic volume of services like video conferencing and other real time applications. Authors have proposed a mathematical model to fine end to end delay through intermediate nodes in the network and improves the overall network resource utilization.

3 Problem Statement

Distributed file systems provide a common virtual file system interface to all users as in DFS storage servers are distributed geographically and because of this load distribution of clients requests to these servers become uneven. This problem can be illustrated clearly through Figure.1. Here, we have taken five storage servers S1, S2, S3, S4 and S5 with their respective service rate (S_r) present in the system. Service rate of a server signifies the number of requests processed by a server in a given time. Initially at time t=0, we assume that each server receives an approximately equal amount of requests as shown in Fig. 1(a). We have taken total 8 requests to illustrate the scenario of our problem statement. In the second case as shown in Fig. 1(b) after time t=2, each server process the clients requests as per its service rate and server S1 requests gets over much earlier than other servers and S1 becomes idle. Server S3 and S5 are fully loaded and takes their time to process all requests. From this scenario, we can say that distributed file system does not utilize each server efficiently. In

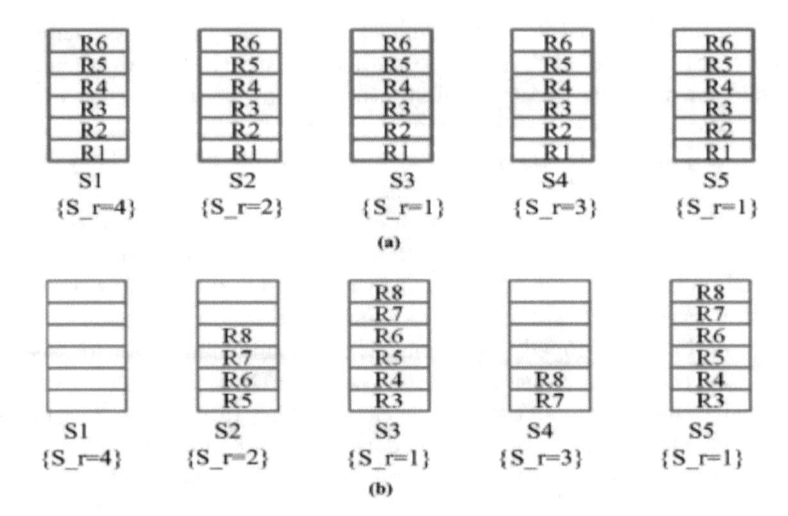

Fig. 1. Problem statement for load balancing (a) at time t=0, servers receive equal amount of client requests. (b) at time t=2, scenario of servers after processing the receive requests.

real-world situation, these requests are too large as compare to server service rate. So in order to increase the system performance some requests which are in queue must be migrated to the idle servers or least loaded server. Our aim is to avoid queue like situations and utilizing the capability of each server efficiently.

4 Proposed Approach

Here, we have proposed an algorithm that can balance the load of servers dynamically by considering its parallel processing capability and its request queuing capacity. Pro-posed approach takes two main parameters of a server 1) *Server request queue size* - buffer space to store the client requests to be handled by the server. 2) *Server service rate* (λ) - the number of CPUs available for processing the client request in a server. Modern servers are equipped with many features like multiple CPUs, large storage, high I/O capability etc. We have chosen the multiple CPUs feature as a main parameter for load balancing of our proposed approach. Following are the few assumptions that we have considered for our proposed approach:

- It is assumed that all the servers belong to same organization which can be geographically apart from each other. So each server maintains the replica of every server data.
- It is also assumed that all servers are strongly connected with each other through high bandwidth medium.
- Each server maintains global view which contains the information of its neighbors through master server.

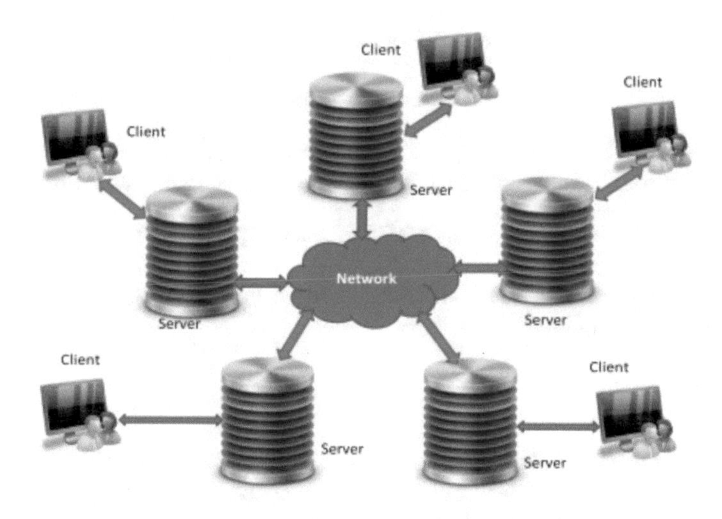

Fig. 2. Organization of distribute storage servers.

Fig.2 shows the general scenario of distributed storage servers. In fig.2, there could be N connected servers where N∈(1,2,3,....n) , in the system. Each server has properties like request queue, number of CPUs, storage capacity. Clients send their requests to the respective server. Many times the incoming request rate () gets increases to a particular server because of the series of client's requests to that data, stored within the server. In case, when a server gets too many requests than server buffers all them in their request queue and the size of request queue gets increases dynamically only upto its predefined threshold limit. Once, the request queue breaches the threshold limit than server is considered as overloaded server and triggers the load balancer. Load balancer classifies the least loaded server on the basis of their request queue and processing capacity. As soon as the least loaded server gets classified than overloaded server migrate its load to that server and balances the load. Various notations used in the proposed approach are represented as follows:
ρ - Current queue size of server.
λ - Service rate that is number of request processed simultaneously by a server.
$Q\_L_{threshold}$ - Threshold limit of server request queue.
$Q\_L_{current}$ - Current capacity of server request queue at time t.
ΔL_i - Additional load on server i.
F_j - Fitness value of neighbors of server i. Where $j \in (1,2,3,....n-1)$ We have considered the real world scenario where the server request queue size and service rate gets changes with respect to time t dynamically and represented as δ_ρ and δ_λ respectively.

$$\delta_\rho = \frac{\rho}{\delta_t} \qquad \text{and} \qquad \delta_\lambda = \frac{\lambda}{\delta_t} \tag{1}$$

Storage server is said to be overloaded if:

$$\delta_\rho > Q\_L_{threshold} \tag{2}$$

When server i where $i \in (1, 2, 3, ...n)$, is overloaded then it calculates the amount of extra load ΔL_i on that server which can be calculated as follow:

$$\Delta L_i = Q\_L_{current} - Q\_L_{threshold} \tag{3}$$

The condition when a load balancer module gets triggered on the overloaded server i is given below:

$$T(i) = \begin{cases} 1, & \Delta L > 0 \\ 0, & \Delta L \leq 0 \end{cases} \tag{4}$$

Once, the load balancer module is triggered, server i find the least loaded or idle server that can accommodate its load and adequately process the service requests. For this load balancer calculates the fitness value F_j that can be calculated using the following fitness function:

$$\Delta M_j = Q\_L_{threshold} - Q\_L_{current} \tag{5}$$

Here, δM_j is free request queue of server j. If δM_j is negative, then server j request queue is overloaded otherwise it is least loaded.

$$F_j = \alpha_1 \Delta M_j + \alpha_2 \lambda \tag{6}$$

Here, α_1 and α_2 are constants and may vary according to scenario such that:

$$\alpha_1 + \alpha_2 = 1 \tag{7}$$

For our proposed scenario, we have considered the value of α_1 and α_2 is 0.5 it is because both the parameters play the equal role in load balancing. In this way, load balancer calculates the fitness value for each neighbors of server i and select that server which has maximum fitness F_j value and migrate the ΔM_j amount of load to server j.

4.1 Proposed Algorithms

Proposed algorithms have been designed to balance the load of storage server in terms of client requests. Here, load balancer as shown in Fig. 3(a) continuously checks for the request queue size of server and tries to avoid the situation of overloading of any server by migrating the load to other idle or least loaded neighbor server in the system. Algorithms also checks and calculate the fitness value for the neighbor server to store them in a list shown in Fig. 3(b). Load balancer utilizes this list to select the server that has highest fitness value. Proposed algorithms also try to reduce the server response time by selecting the server with least CPU utilization. In this way, we utilize the server to increase the overall performance of the system.

Algorithm1. LBA(Servers, δp)

Input: Servers and δp

1. s← server; δp← current queue size;
2. $\Delta QL_{threshold}$ ← threshold limit of queue size;
3. **if**(δp < $\Delta QL_{threshold}$) **then**
4. // check server queue status
5. Add request to queue;
6. Process_request();
7. // processing the client request
8. **else**
9. //server is overloaded
10. S ← Find_server(server_neighbour_list L);
11. // find underloaded server
12. S ← migrate request;
13. Stop;

(a)

Algorithm 2. Find_server (server_neighbour_list L)

Input: server_neighbour_list L

1. **for** k=1 to L.size();
2. S1←L.get();
3. // get the server from neighbor list.
4. $F_k = \alpha_1 \Delta M_k + \alpha_2 \lambda$;
5. // calculate fitness value of server k.
6. temp_list t← F_k ;
7. // Add to temporary list.
8. **for** j=1 to t.size();
9. S2← max(F_j);
10. // select maximum fitness value.
11. **return** S2;

(b)

Fig. 3. Proposed algorithms (a) Load balancing algorithm. (b) Find neighbour server algorithm

5 Results

Performance analysis of proposed algorithm is done using simulations where we have created thousands of virtual client requests to be handled by 12 storage servers. All the servers work simultaneously with fixed number of CPU cores to process the client request quickly. Each server has a request queue to buffer the incoming client re-quests and storage capacity to store the data. For the given problem statement in section 3 where the load is unbalanced, it is assumed that half of storage servers get client requests and others remain idle. Our motive is to equally distribute the received client requests among the servers to avoid the scenario of overloading. In the simulation scenario numbers of storage servers are kept fixed with varying number of re-quests handling. We have also compared the obtained results with the least load balancing algorithm. Following table depicts the configuration parameter for our simulation environment. Fig.4 shows the

Table 1. Experimental parameters used for simulation environment

No. of Clients Requests	No. of Servers	No. of CPU cores available per server	Storage capacity of servers (GB)	Server queue length
800	12	7	500	15
1000	12	8	500	15
1200	12	9	500	20
1800	12	10	500	20
2400	12	11	500	20

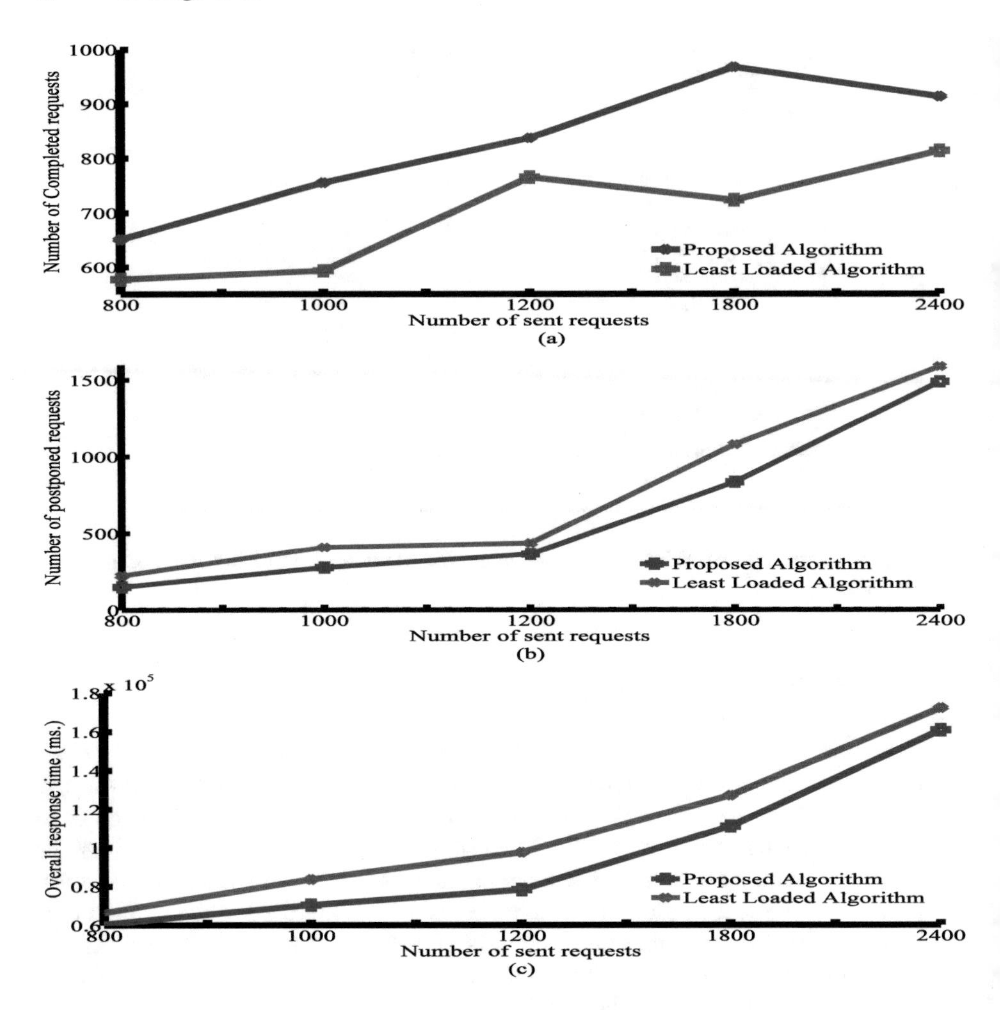

Fig. 4. Processed client requests (a) no. of sent requests vs. no. of completed requests. (b) no. of sent requests vs. no. of postponed requests. (c) overall response time

number of processed client requests by server in a given time. Here, Fig. 4(a) represents the graph between no. of sent requests vs. no. of completed request whereas Fig. 4(b) represents the graph between no. of sent requests vs. no. of postponed requests for the proposed and least load algorithms. In least loaded algorithm when any server get overloaded then load balancer selects the server of which request queue is least loaded without considering the CPU parameter. For the proposed algorithm we have considered the CPU parameter and from obtained results as shown in Fig. 4(a), Fig. 4(b) and Fig. 4.(c) that the proposed algorithm perform much better over the least load algorithm. In all set of client requests, proposed algorithm process more number of clients request with better overall response time as shown in Fig. 4(c).

6 Conclusion

In distributed file system, data is dispersed among different storage servers located geographically far away from each other. To provide the desired quality of service to the clients, performance of the distributed file system matters a lot. Response time is the major parameter that may affect the performance of the any distributed file system. Proposed algorithm claims to reduce the delayed requests and also reduces the overall system response time. It also considers the physical aspects of a server like available number of CPU cores in a server, request queue size or buffer to store the incoming client requests. Obtained result shows the improvements over previously worked least loaded algorithm and more number of client requests are processed by the system without delay and in case of overloading the load balance distribute the requests accordingly to neighbor servers.

References

1. H. C. Hsiao, H. Y. Chung, H. Shen, Y. C. Chao.: Load Rebalancing for Distributed File Systems in Clouds. In: IEEE Transaction on Parallel and Distributed System. vol. 24, Issue 5. (2013)
2. E. Levy, A. Silberschatz.: Distributed file systems: concepts and examples. ACM Computing Surveys (CSUR), vol. 22, Issue 4, pp. 321–374. (1990)
3. N. J. Kansal, I. Chana.: Existing Load Balancing Techniques In Cloud Computing: A Systemetic Review. Journal of Information System and Communication, vol. 3, Issues 1, pp. 87–91. (2012)
4. H. Yamamoto, D. Maruta, Y. Oie.: Replication Methods for Load Balancing on Distributed Storages in P2P Networks. In: Proc. of the 2005 Symposium on Applications and the Internet (SAINT05), pp. 264 - 271. (2005)
5. A. Rao, K. Lakshminarayanan, S. Surana, R. Karp, I. Stoica.: Load Balancing in Structured P2P Systems. In: Proc. of International Workshop Peer-to-Peer Systems (IPTPS). pp. 68–79. (2003)
6. W. Zeng, Y. Li, J. Wu, Q. Zhong, Q. Zhang.: Load rebalancing in Large-Scale Distributed File System. In: 1st International Conference on Information Science and Engineering (ICISE). pp. 265–269. (2009)
7. K. Fan, D. Zhang , H. Li, Y. Yang.: An Adaptive Feedback Load Balancing Algorithm in HDFS. In: IEEE 5th International Conference on Intelligent Networking and Collaborative Systems. pp. 23-29. (2013)
8. H. Y. Chung , C. Chang , H.C. Hsiao, Y.Chao.: The Load Rebalancing Problem in Distributed File Systems. In: IEEE International Conference on Cluster Computing (CLUSTER). pp. 117–125. (2012)
9. R. Malekian, A. H. Abdullah.: Traffic engineering based on effective envelope algorithm on novel resource reservation method over mobile internet protocol version 6. Int. Journal of Innovative Computing, Information and Control. vol.8, no.9, pp. 6445–6459. (2012)

Isotropic Magnetic Shielding of $Al(OH)_4^-$ in Aqueous Solution: A Hybrid Monte Carlo - Quantum Mechanical Computational Model

Bojana Koteska[1], Anastas Mishev[1], and Ljupco Pejov[2]

[1] Faculty of Computer Science and Engineering,
Rugjer Boskovikj 16, 1000 Skopje, Macedonia
{bojana.koteska,anastas.mishev}@finki.ukim.mk

[2] Faculty of Natural Sciences & Mathematics,
Institute of Chemistry, 1000 Skopje, Macedonia
ljupcop@iunona.pmf.ukim.edu.mk

Abstract. In the present work, we have addressed the issue of magnetic response properties of aqueous $Al(OH)_4^-$ ion. We develop and implement a hybrid statistical physics - quantum mechanical approach to compute the ^{27}Al NMR shielding tensor and the corresponding isotropic shielding. The complex hybrid approach has been implemented to account explicitly for the thermal motions of all ionic species along with the solvent (water) molecules under realistic conditions encountered during experimental measurements. In the developed approach, first, Metropolis Monte Carlo simulation (NPT ensemble) of water solution containing Al^{3+}, 4OH- ions, and 3000 water molecules in a cubic box, employing periodic boundary conditions is carried out. Subsequently, the MC "trajectories" are analyzed by time-series analytic methods (e.g. implementing the energy autocorrelation functions) so that out of a very large overall number of MC configurations that have been generated, only 100 representative ones are picked up, with negligible mutual statistical interdependence. NMR shielding tensors are subsequently computed for such chosen configurations at B3LYP/6-311++G(3*df*, 3*pd*) level of theory, using various approaches to include the environment of the "central" Al ion. In the simplest approach, all environment (within sufficiently large distance) is considered as being built up by point charges (accounted for explicitly or within the ASEC formalism). Further, the first solvation shell (consisting of 4 hydroxide ions) together with the central aluminum ions are described by a wavefunction, while the remaining solvent molecules are treated as point charges or the "bulk" solvent is considered to be a polarizable continuum. The convergence of isotropic shielding values with the environment description is analyzed and discussed.

Keywords: Metropolis Monte Carlo simulation, aluminum ion, quantum mechanical computation, magnetic response properties.

S. Loshkovska and S. Koceski (eds.), *ICT Innovations 2015*,
Advances in Intelligent Systems and Computing 399,
DOI: 10.1007/978-3-319-25733-4_5

1 Introduction

Aqueous aluminum(III) ionic species are of considerable interest in various areas of science and technology. These span a wide range of disciplines, from environmental science and geochemistry to biomedical sciences [17]. Having in mind that aluminum is the most abundant metal in Earth's crust, understanding of the structure and dynamics of its hydrated ionic species is of prime importance in relation to treatment of wastewaters and environmental processes. The essential chemistry of aqueous aluminum ionic species is, however, rich, complex and not yet completely understood. Exploring the structure and dynamics of complex molecular species in solution is far from a trivial task, even at the current stage of development of science. In contrast to the techniques aiming to target gas-phase species, in case of which the data interpretation has become almost a routine task, the situation in solution chemistry is characterized by much greater complexity. Even such essential and fundamental questions as the number of solvent molecules residing in the first coordination sphere around a central metal ion cannot be solved unambiguously yet by solely experimental techniques. Therefore, combining experimental data with theoretical models is of essential importance to be able to understand various issues in this area.

Among the variety of experimental methods which have been used to study the chemistry of aqueous complex ionic species, NMR spectroscopic techniques seem to be particularly important and suitable for a wide variety of purposes, especially in the con-text of biochemical significance of such species. This particular spectroscopic technique has proven to be crucial in elucidating the structural and dynamical characteristics of aqueous aluminum(III) ionic species [20]. ^{27}Al NMR studies have early indicated the existence of $Al(H_2O)_6^{3+}$ and $Al(OH)_4^-$ species in solution, along with Al_{13} polyoxo species, and at the same time it precluded the existence of $Al(OH)^{2+}$ and $Al(OH)_2^+$ ones [12]. In the context of better understanding the experimental observations related to the mentioned species, it is necessary to carry out theoretical simulations mimicking as closely as possible the in-liquid situation. However, from a theoretical viewpoint, this is not an easy task. Theoretical models of complex aqueous aluminum(III) species have so far mostly been based on finite-cluster species in gas phase, or embedded within the solvent treated as a polarizable continuum. The basic approach in this context is thus based on exploration of potential energy hypersurfaces of the mentioned species, either free or solvated with a finite (usually small) number of solvent molecules (the so-called "microsolvation" approach to theoretical solution chemistry). Considering the PESs, however, corresponds to complete neglect of the dynamical nature of the solution itself and the complex species embedded within at finite temperatures close to the ambient one (at which the actual experiments are being carried out). One way to circumvent this shortcoming is to develop a hybrid statistical mechanics-quantum mechanical approach to the problem. The basic idea of this approach is that the condensed phase system is initially simulated by statistical physics method (e.g. Monte Carlo or molecular dynamics), and then, subsequently to the statistical physics simulation, suitable chosen finite clusters are cut-out from the solution and treated in a more exact manner, e.g. by quantum mechanical methodology. A general common characteristic of all theoretical approaches to modeling of NMR properties is that the isotropic chemical shift values, the quantities that are actually measured in NMR experiments, are computed with respect to an internal standard. For example, the ^{27}Al chemical shift value of $Al(H_2O)_6^{3+}$ ionic species are computed with reference to the $Al(OH)_4^-$ ones. In order to model the in-liquid magnetic properties of $Al(H_2O)_6^{3+}$, thus, aside from the in-liquid isotropic shielding constant for these species, one has to compute the isotropic shielding constant for $Al(OH)_4^-$ s well.

In the present study, we explore some methodological and computational aspects related to computation of ^{27}Al isotropic shielding constant for $Al(OH)_4^-$ species. From a computational aspect, we implement and apply the map-reduce approach to some of the phases in the developed robust hybrid methodology.

2 Statistical Mechanics Simulations

The first phase of the implemented computational methodology for the present purpose involves generation of the structure of the liquid phase (aqueous solution that contains Al^{3+} and OH^- species in molar ratio 1:4). In the present study, to achieve this aim, we employ the Monte Carlo (MC) approach. MC simulations throughout the present study were carried out by the Metropolis algorithm, using the statistical mechanics code DICE [2]. We have chosen the isothermal-isobaric (NPT) ensemble, at $T = 298$ K, $P = 1$ atm to generate the structure of the liquid; experimental density of water of 0.9966 g cm^{-3} was taken to appropriately generate the cubic unit cell used further throughout the simulations. The cubic box actually included a single Al^{3+} ion, four hydroxide ions and 3000 water molecules; side length of the box was approximately 45 Å. Periodic boundary conditions were imposed, and the long-range corrections to the interaction energy were calculated for interacting atomic pairs between which the distance is larger than the cutoff radius defined as half of the unit cell length. The Lennard-Jones contribution to the interaction energy beyond this distance was estimated assuming uniform density distribution in the liquid (*i.e.* $g(r) \approx 1$). The electrostatic contribution was estimated by the reaction field method involving the dipolar interactions. To describe the intermolecular interactions, we have used interaction potentials consisting by a sum of Lennard-Jones $12-6$ site-site interaction energies plus Coulomb terms:

$$U_{ab} = \sum_i^a \sum_j^b 4\varepsilon_{ij} \left[\left(\frac{\sigma_{ij}}{r_{ij}} \right)^{12} - \left(\frac{\sigma_{ij}}{r_{ij}} \right)^6 \right] + \frac{q_i q_j}{4\pi\varepsilon_0 r_{ij}} \tag{1}$$

In (1) i and j are sites in interacting molecules a and b; r_{ij} is the interatomic distance between sites i and j, while all other terms have their usual meanings. To generate two-site Lennard-Jones parameters ε_{ij} and σ_{ij} from the single-site ones, the "geometric mean" combination rules were used:

$$\varepsilon_{ij} = \sqrt{\varepsilon_i \varepsilon_j} \tag{2}$$

$$\sigma_{ij} = \sqrt{\sigma_i \sigma_j} \tag{3}$$

To model water as a solvent, the SPC model potential parameters were used [1]. For Al^{3+} cation, the potential parameters from [10] were used, while the hydroxide ions were modeled by the so-called "simple-charge" potential parameters described in details in [18].

3 Calculations of Magnetic Shielding Tensors and Isotropic Shielding Constants

To compute the isotropic nuclear shielding constant for the studied aqueous species, we have first computed the nuclear magnetic shielding tensors; these are defined as mixed second derivative of the energy (E) with respect to the magnetic moment of the X-th nucleus ($\vec{m}_X$) and the external magnetic field ($\vec{B}$) [9]:

$$\sigma_X^{\alpha\beta} = \frac{\partial^2 E}{\partial \beta^\alpha \partial m_X^\beta} \tag{4}$$

In (4), the Greek superscripts denote the corresponding vector or tensor components. The average ^{27}Al isotropic shielding value was further computed as an average of the tensorial diagonal components:

$$\sigma_{\text{ iso}} = \frac{1}{3} Tr\sigma = \frac{1}{3}(\sigma_{11} + \sigma_{22} + \sigma_{33}) \tag{5}$$

In all calculations of magnetic response properties, it is of essential importance to achieve gauge invariance. In the present study, for this purpose we have used two approaches: the GIAO (gauge independent atomic orbital [8, 21, 19] and the CSGT (continuous set of gauge transformations) method [13, 14, 15]. All quantum mechanical calculations were carried out with the B3-LYP combination of exchange and correlation functionals, with the rather large 6-311++G(3*df*, 3*pd*) basis set for orbital expansion. All quantum-chemical calculations were performed with the GAUSSIAN09 series of codes [11].

4 Results and Discussion

Subsequently to MC simulations, we have analyzed the mutual interdependence of the generated structures by the energy autocorrelation function approach, as explained in the original works of Coutinho and Canuto [7, 5, 4, 6]. We have found out that configurations mutually separated by 3000 MC steps have negligible mutual statistical dependence. We could, therefore, safely pick up 100 such configurations and perform the time- and resource-consuming quantum chemical computations on these configurations only, and still get statistically significant results.

The simplest way to account for the in-liquid environment influence on the magnetic response properties of the "central ion" is to treat the environment as a set of point charges. This is, of course, a rather crude approximation to the real influence of the environment on the solute properties, but it allows us to discriminate between "classical" electrostatic and other, much more complex interactions that take place within the liquid. As a first approximation, therefore, we have first described the central Al^{3+} ion by a quantum mechanical wavefunction (at B3LYP/6-311++G(3*df*, 3*dp*) level, as explained before), while the first solvation shell (the four hydroxide ions) and the remaining solvent water molecules were treated as sets of point charges placed at the positions generated by MC simulations. The distribution histogram of the isotropic shielding constant values computed by the CSGT method for achieving gauge invariance is shown in Fig. 1 (the histogram computed by the GIAO method is very similar).

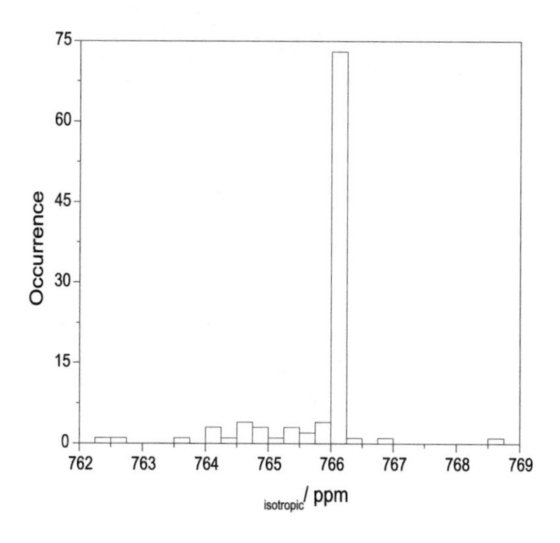

Fig. 1. The distribution histogram of the isotropic shielding constant values calculated approximating the complete in-liquid environment as a set of point charges (gauge invariance was achieved by the CSGT method.

As can be seen, the σ_{iso} values are characterized by rather narrow distribution, though certain values appearing at the distribution tails still exist, and correspond to unusual, rarely occurring configurations in the liquid. The average isotropic shielding values computed by the CSGT and GIAO approaches are 765.8643 and 756.8639 ppm respectively. Such close values strongly indicate that due to the large basis set used, both methods give basis-set-converged values of σ_{iso}. Instead of computing the shielding tensors for 100 independent configurations, however, one can use the ASEC approach developed by Coutinho and Canuto [3], which consists of superimposing the solvent atomic charges taken from M statistically uncorrelated MC-generated configurations, each scaled by

1/M. In the present study, we have carried out the ASEC computations implementing the map-reduce technique, as explained in details in our previous study [16]. The computed σ_{iso} values with the ASEC algorithm using both alternatives to achieve gauge invariance lead to essentially the same results as the averaged ones discussed before. This finding further supports the physical correctness of the ASEC methodology, aside from its simplicity.

A more realistic approach to the studied issue would be to treat at least the near-est neighbors of the central aluminum(III) ion (the first solvation shell) by a QM wave-function, while the remaining ("bulk") solvent molecules as sets of point charges. We have further therefore computed the magnetic shielding tensors and subsequently the isotropic shielding constants for 100 $Al(OH)_4$- species embedded in a sphere containing bulk water molecules treated as sets of point charges ranging up to distance equal to one half of the box length (for each particular snapshot from the NPT MC simulation). The $g_{Al\text{-}O(OH)}(r)$ RDF is shown in Fig. 2. Note that even each "internal" configuration of the $Al(OH)_4$- species is different in these 100 clusters, as the hydroxide ions were allowed to move during MC simulations. Distribution histograms of the σ_{iso} values computed by CSGT and GIAO methods are shown in Fig. 3 a) and b).

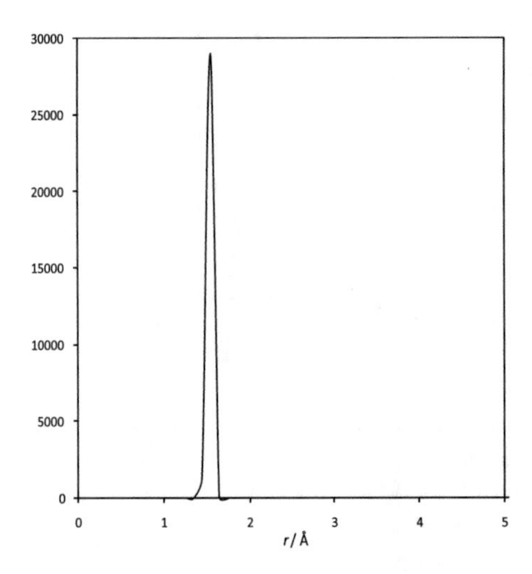

Fig. 2. The $g_{Al\text{-}O(OH)}(r)$ RDF computed from the equilibrated MC simulation.

As can be seen from Fig. 3, the distributions with the two methods for achievement of gauge invariance look very similar as well. The average isotropic shielding values computed by the CSGT and GIAO approaches with an "explicit first shell" approach (i.e. first shell explicitly described by a QM wavefunction) are 454.7421 and 455.0904 ppm respectively. Similarly to the discussion in the context of previous approach, we can say that the rather close average values obtained from the CSGT and GIAO approaches indicate well-converged values with the basis set size. At the same time, one should note the big difference between the values computed with the current approach, when the first solvation shell around Al(III) ion has been explicitly described by a wavefunction, and the previous, "fully electrostatic" method, where all neighbors within the liquid have been described as sets of point charges.

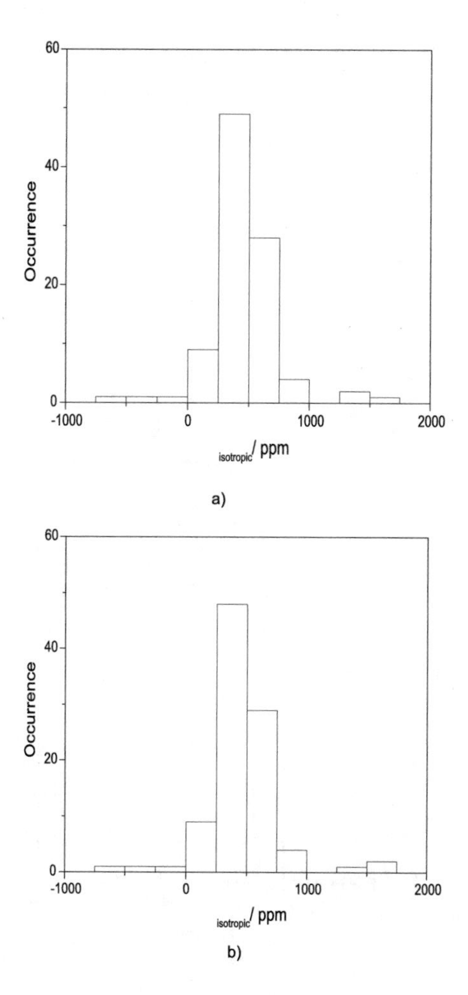

Fig. 3. Distribution histograms of the σ_{iso} values computed describing the first solvation shell with a QM wave-function and the bulk solvent molecules as sets of point charges with CSGT (a) and GIAO (b) methods for achievement of gauge invariance.

Simple description of the bulk solvent molecules by sets of point charges, however, aside from its simplicity, does not explicitly account for the possibility that the medium could actually be polarized by the strong electrostatic field generated by the central ion (as well as its nearest neighbors). We have therefore considered an alternative approach in the present study. Similarly as explained before, we have computed the magnetic shielding tensors and subsequently the isotropic shielding constants for 100 $Al(OH)_4$- species, described explicitly by a QM wavefunction, this time embedded in the "bulk" solvent, treated as a polarizable continuum.

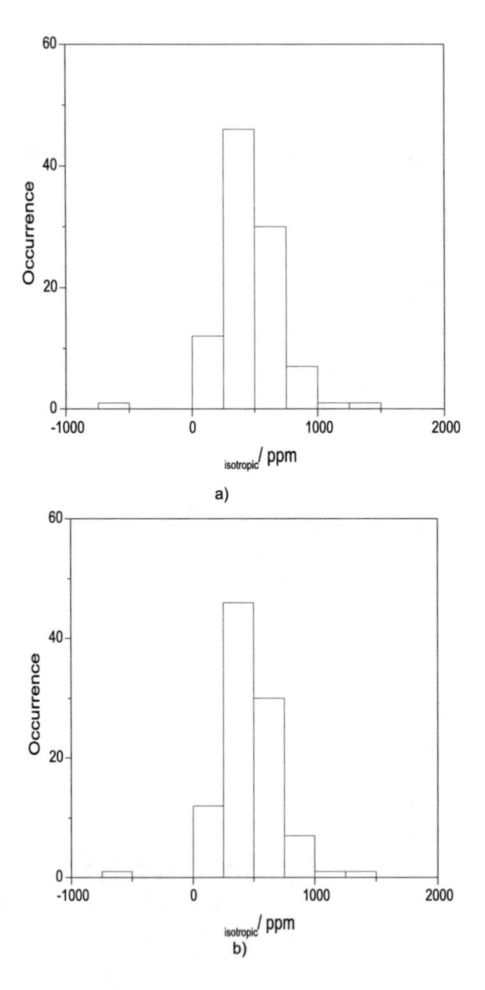

Fig. 4. Distribution histograms of the σ_{iso} values computed describing the first solvation shell with a QM wave-function and the bulk solvent as polarizable continuum with CSGT (a) and GIAO (b) methods for achievement of gauge invariance.

These calculations were performed within the polarizable continuum formalism of Tomasi and co-workers (PCM) using the integral equation formalism variant (IEFPCM)]. The distribution histograms of the σ_{iso} values computed by CSGT and GIAO methods are shown in Fig. 4 a) and b), respectively. The average isotropic shielding values computed by the CSGT and GIAO approaches within this, "explicit first shell + polarizable continuum" approach are 467.6568 and 469.0545 ppm respectively. These values differ by only about 10 ppm in comparison to those calculated by treating the bulk solvent as being built up by point charges. This indicates a good convergence of the currently presented results with the level of solvent description.

5 Conclusions and Further Work

In the present study, we have presented a robust, hybrid statistical mechanics quantum mechanical approach to rigorous treatment of the problem for computation of the isotropic shielding constant values of ionic species in liquid phases. Various possibilities of treating both nearest-neighbors of the central ionic species (Al(III) in the present case) and the "bulk" part of the solvent have been considered. Good convergence of the computed parameters both with the basis set size, the method of achievement of gauge invariance and the description of the bulk solvent has been obtained in case when the first solvation shell around the ion has been explicitly described by a QM wavefunction. We are further considering the convergence of the results by an explicit description of the second solvation sphere by a QM wavefunction as well. This work is in progress, with quite encouraging results.

References

1. Berendsen, H., Postma, J., Van Gunsteren, W., Hermans, J.: Intermolecular Forces, ed Pullman B (1981)
2. Coutinho, K., Canuto, S.: DICE: A Monte Carlo program for molecular liquid simulation. University of São Paulo, Brazil (1997)
3. Coutinho, K., Georg, H., Fonseca, T., Ludwig, V., Canuto, S.: An efficient statistically converged average configuration for solvent effects. Chemical physics letters 437(1), 148–152 (2007)
4. Coutinho, K., Canuto, S.: Solvent effects from a sequential Monte Carlo-quantum mechanical approach. Advances in quantum chemistry 28, 89–105 (1997)
5. Coutinho, K., Canuto, S.: Solvent effects in emission spectroscopy: A Monte Carlo quantum mechanics study of the $n \longleftarrow \pi^*$ shift of formaldehyde in water. The Journal of Chemical Physics 113(20), 9132–9139 (2000)
6. Coutinho, K., Canuto, S., Zerner, M.: A Monte Carlo-quantum mechanics study of the solvatochromic shifts of the lowest transition of benzene. The Journal of Chemical Physics 112(22), 9874–9880 (2000)
7. Coutinho, K., De Oliveira, M., Canuto, S.: Sampling configurations in Monte Carlo simulations for quantum mechanical studies of solvent effects. International journal of quantum chemistry 66(3), 249–253 (1998)
8. Ditchfield, R.: Self-consistent perturbation theory of diamagnetism: I. A gauge-invariant LCAO method for NMR chemical shifts. Molecular Physics 27(4), 789–807 (1974)

9. Dykstra, C.: Quantum chemistry and molecular spectroscopy. Prentice Hall PTR (1992)
10. Faro, T.M., Thim, G.P., Skaf, M.S.: A Lennard-Jones plus Coulomb potential for Al^3+ ions in aqueous solutions. The Journal of chemical physics 132(11), 114509 (2010)
11. Frisch, M., Trucks, G., Schlegel, H., Scuseria, G., Robb, M., Cheeseman, J., Scalmani, G., Barone, V., Mennucci, B., Petersson, G., et al.: Gaussian 09, revision A. 1. Gaussian Inc., Wallingford, CT (2009)
12. Haraguchi, H., Fujiwara, S.: Aluminum complexes in solution as studied by aluminum-27. Nuclear magnetic resonance. The Journal of Physical Chemistry 73(10), 3467–3473 (1969)
13. Keith, T., Bader, R.: Calculation of magnetic response properties using atoms in molecules. Chemical physics letters 194(1), 1–8 (1992)
14. Keith, T.A., Bader, R.F.: Calculation of magnetic response properties using a continuous set of gauge transformations. Chemical physics letters 210(1), 223–231 (1993)
15. Keith, T.A., Bader, R.F.: Topological analysis of magnetically induced molecular current distributions. The Journal of chemical physics 99(5), 3669–3682 (1993)
16. Koteska, B., Mishev, A., Pejov, L.: Magnetic Response Properties of Aqueous Aluminum (III) Ion: A Hybrid Statistical Physics Quantum Mechanical Approach Implementing the Map-Reduce Computational Technique. In: ICT Innovations 2014, pp. 33–43. Springer (2015)
17. Kubicki, J., Sykes, D., Apitz, S.: Ab initio calculation of aqueous aluminum and aluminum-carboxylate complex energetics and ^{27}Al NMR chemical shifts. The Journal of Physical Chemistry A 103(7), 903–915 (1999)
18. Mitev, P.D., Bopp, P.A., Petreska, J., Coutinho, K., Ågren, H., Pejov, L., Hermansson, K.: Different structures give similar vibrational spectra: The case of OH- in aqueous solution. The Journal of chemical physics 138(6), 064503 (2013)
19. Pulay, P., Hinton, J., Wolinski, K.: Efficient implementation of the GIAO method for magnetic properties: Theory and application. In: Nuclear magnetic shieldings and molecular structure, pp. 243–262. Springer (1993)
20. Qian, Z., Feng, H., He, L., Yang, W., Bi, S.: Assessment of the accuracy of theoretical methods for calculating ^{27}Al nuclear magnetic resonance shielding tensors of aquated aluminum species. The Journal of Physical Chemistry A 113(17), 5138–5143 (2009)
21. Wolinski, K., Hinton, J.F., Pulay, P.: Efficient implementation of the gauge-independent atomic orbital method for NMR chemical shift calculations. Journal of the American Chemical Society 112(23), 8251–8260 (1990)

GSolver: Artificial solver of word association game

Ercan Canhasi

Gjirafa, Inc.
Rr. Rexhep Mala, 28A, Prishtine, Kosovo
ercan@gjirafa.com
http://www.gjirafa.com

Abstract. In this paper we propose an artificial solver for a word association game. The possibility of a player to solve associations counts on the richness and deepness of players language and cultural qualifications. In order to provide answer(s) a human participant must accomplish a multiple memory search tasks for meanings of huge number of concepts and their frame of references. Hence the knowledge background (KB) of the proposed artificial solver is based on a large information repository formed by utilizing machine reading techniques for fact extraction from the web. As a KB we indirectly use the Albanian world-wide-web and the Gjirafa as a search engine. Complementary, the central processing unit (CPU) of the artificial solver is designed as a spreading activating network. The CPU treats provided hints and finds associations between them and concepts within the KB in order to incrementally compute and update a list of potential answers. Furthermore the CPU module is enriched by proposing a schema for finding the most promising solutions to be provided as the final answers. Experiments show that the accuracy of the system is as good as the average human player performance.

Keywords: Word Association Game, Search Engine, Knowledge Extraction, Spreading Activation Algorithm, Artificial Solver

1 Introduction

Games related to language can simply be categorized into two groups: word and language games. Since the word meanings are not very important in them, word games do not engage real language knowledge. An example of word game is Scrabble, in which two to four players take turn placing letters in a grid to form words defined in standard dictionaries. On the other side language games, where the word meanings have the central role, actively utilize natural language. Some of the well known language games are crosswords, "Who wants to be a millionaire?" and Jeopardy!^{TM}. Language games derive their complexity, impulse and motivation from the variety, richness and ambiguity of the natural language. This is also the main source of attraction for researchers from the fields of natural language processing and artificial intelligence.

S. Loshkovska and S. Koceski (eds.), *ICT Innovations 2015*,
Advances in Intelligent Systems and Computing 399,
DOI: 10.1007/978-3-319-25733-4_6

Table 1: Three different stages from WAG gameplay. Game starts with all of the fields covered(a). At a latter point of game some of hints will be reviled and/or some of associations will be solved (b). Eventually, the four sub-associations (Spiderman, theater, face, and carnival) and the final solution (mask) will get solved.

A1	B1
A2	B2
A3	B3
A4	B4
A	B
Final Solution	
C	D
C4	D4
C3	D3
C2	D2
C1	D1

a)

Movie	Antique
A2	B2
A3	Scene
A4	B4
A	Theater
Final Solution	
C	D
Makeup	D4
C3	Costumes
C2	D2
C1	D1

b)

Movie	Antique
Superhero	National
Saver	Scene
Spider	Puppets
Spiderman	Theater
MASK	
Face	Carnival
Makeup	Korca
Public	Costumes
Mimics	Rio de Janeiro
Honor	Dance

c)

In this paper we present the GSolver (Gjirafa Solver), which is a system designed to provide solutions for a word association game (WAG). WAG is a game usually played on TV quiz shows (Table.1), in which the player(s) first tries to solve one of the four supporting associations, and eventually uses obtained hints (solutions of the supporting associations) to solve the main association. The solution of each association set is a concept with a semantic relation to the terms in the corresponding column. Hints are usually indirectly related to each other, rarely totally unrelated to each other, but each of them is strongly related to the solution. At the beginning of the game all concepts are hidden (Table.1a). In sequel of the game, players reveal clues, reciprocally and one by one (Table.1b). After the announcement of each hidden term, the active player gets the right to guess a solution of any minor or major association. As an example, we showed in Table.1c that the solution for the column B is the word *theater* given the hints *antique* and *scene*. The semantic relation is rather obvious, *antique theater* and *theater scene*.

The main contribution of this paper is threefold: 1) paper presents an artificial intelligence system that attempts to solve the word association game for Albanian language 2) in it we also present new algorithm which improves the ranking of candidate solutions computed by the spreading algorithm 3) experiments show that the accuracy of the system is as good as the average human player performance.

The paper is organized as follows. Related work are summarized in Section 2. Section 3 describes the general system architecture and details about the KB and CPU of our artificial solver. Section 4 reports experiments carried out in order to evaluate the system. Conclusions and future works are in Section 5.

2 Related Work

Artificial Intelligence (AI) research and computer games are mutually beneficial combination [1]. AI in games has been one of the most successful and visible result of Artificial Intelligence research.

Language related games are the main focus of this work. One of the first works reported in literature of this kind is Proverb [2]. It has reached human-like performance on American crosswords. A method to implement a virtual player for the "Who Wants to Be a Millionaire?" (WWM) game has been proposed in [3]. The authors exploit the huge amount of knowledge available on the Web through Google and use advanced natural language processing methods to redefine the questions as a various questions. The system reaches an accuracy of 0.75. The another virtual player for the WWM game is recently presented in [4]. The main differences of the latter is that they adopt selected sources of information available on the Web, such as Wikipedia and DBpedia, rather than the whole Web, in an attempt to improve reliability of the answers. Furthermore they adopt a question answering methods instead of a search engine in order to improve the process of selecting the most reliable passages. Yet another very successful and famous artificial game solver is the IMB's Watson. In February 2011 the IBM Watson supercomputer, based on technology from the DeepQA project [5], has defeated two champions of the Jeopardy! TV quiz.

A very similar work to ours is one presented in [6]. They tackle the language game known as Guillotine, a game broadcast by the Italian National TV company. It involves a single player, who is given a set of five words (clues), each linked in some way to a specific word that represents the unique solution of the game. The authors in [6] presented an artificial player for the Guillotine game. The idea behind their implementation is to define a knowledge infusion process which adopts NLP techniques to build a knowledge base extracting information mainly from Wikipedia. For reasoning mechanism they adopted spreading activation algorithm that retrieves the most appropriate pieces of knowledge useful to find possible solutions. Although there are many similarities, our system differs in few essential senses: 1) WAG as a game differs in many ways 2) KB modeling is alike from many aspects 3) our KB is completely based on Albanian web where the results are obtained by search engine and last but not least 4) we perform on a minor language, i.e. the Albanian language, which makes the problem even harder and first of its kind.

3 The System Architecture

The visual summary of the GSolver is given in Fig.1. As it can be seen from the figure for each given hint system incrementally follows next steps to generate the potential solutions: *a)* The hints are provided as search queries to Gjirafa. *b)* Search results are used to update the spreading activation network . *c)* Spreading activation algorithm is used to calculate the list of potential answers *d)* Few different methods are combined and applied for better re-ranking.

The rest of the section reports the fundamental modules of GSolver.

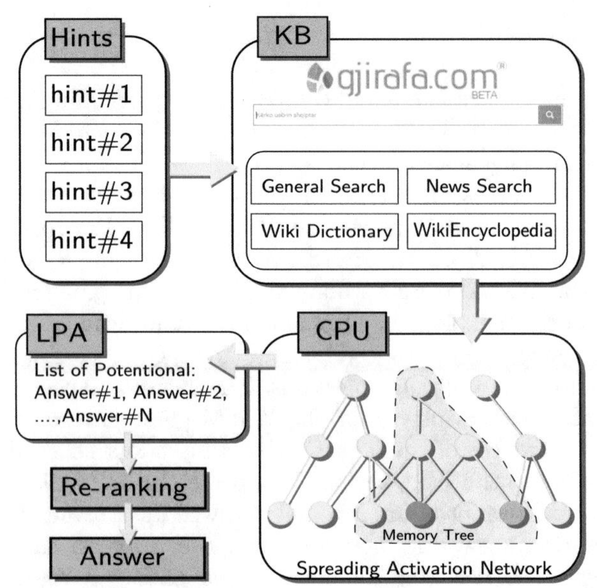

Fig. 1: The complete method realized by GSolver in order to discover the answer for the word associations game. It starts from provided clues and extracts relevant information elements via Gjirafa, that are used by the deductive logic apparatus (SAN) for producing the list of possible answers. Finally, the very short list of answers is formed by intelligent re-ranking algorithms.

3.1 GSolver's knowledge background

A comprehensive information repository should be formed for representing the lexical and semantic background knowledge of the GSolver. The KB used in this work is built by extracting information from textual sources on the Albanian web. In doing so we used: (1) the Albanian world-wide-web as the database, (2) Gjirafa as a search engine and, (3) few basic machine reading methods for knowledge extraction [7].

Gjirafa [8] is a powerful search engine specialized in Albanian language, able to utilize standard natural language pre-processing tasks [9] such are the tokenization, stop word removal, lemmatization, simple named entity recognition, and tf-idf scoring. We gather documents containing the given hints from Gjirafa by simply providing given hint words as search queries.

After an extensive analysis of the correlation between the hints obtained during the game play and the intermediary/final solutions, the following knowledge sources have been processed to build the knowledge background of the system:

1. general web search results: the word representing the solution is contained in the text of the search results, where some additional preprocessing has been applied.
2. vertical news search results: the word representing the solution is contained in the title or in the content of the news.

3. Dictionary: the Gjirafa search results filtered by sq.wiktionary.org domain: the word representing the solution is contained in the description of a lemma or in some example phrases using that lemma.
4. Encyclopedia: the Gjirafa search results filtered to sq.wikipedia.org as for the dictionary, the description of an article contains the solution, but in this case it is necessary to process a more detailed description of information.

Although all of the above mentioned types of sources convey textual information they have different inner structure, therefore an important problem was to standardize representation of the information they store, which is discussed in next few paragraphs.

Since the CPU of GSolver is implemented as a activation spreading network then the GSolver's KB should be represented as an interconnected network of nodes (elementary information trees, EITs) in order to be suitable for processing by CPU. Each EIT would represent elementary unit of information.

EIT is in fact two level N-ary tree, where: (1) the root contains reference to hint (or query); (2) the middle level nodes represent the source of information; and (3) leafs denote the words (concepts) retrieved by root (Figure 1). Since EITs' leafs can originate from different sources it is obvious that by the end of information retrieval process one should obtain a tree ready for further processing (Figure.1). What we have done by modeling the search results from different sources provided by Gjirafa is in fact creation of the systems KB. The next step is to develop an algorithm for retrieving the most convenient bit of knowledge related with the hints. Since the KB modeling is inspired by human-like system then the memory retrieval mechanism should simulate the cognitive mechanism of a human being in the most reliable manner.

3.2 GSolver's central processing unit

Spreading activation network [10,11] corresponds adequately to the graph theory of semantic memory. The plain spreading activation model is a semantic memory graph on which basic processing methods are applied. The graph consists of nodes interconnected by links. Links may be labeled and/or weighted and usually have directions, Furthermore the links can be either activatory (links with positive weight) or inhibitory (links with negative weight). The processing is initiated by labeling a set of source nodes with activation weights and proceeds by iteratively propagating that activation to other nodes linked to the source nodes. For each iteration, a termination condition is checked in order to end the search process over the network.

Given a spreading activation graph of nodes $n_1, ..., n_m$, each node has an assigned activation value at iteration t, $A_i(t)$. Since only some nodes should be able to spread their activation values over SAN, let F be a firing threshold determiner for nodes which tells whether a node is fired. At each iteration, every node propagates its activation to its neighbors as a function of its current activation value and the weights of the edges that connect it with its neighbors. The spreading strategy is described in the following:

Step 1 - Initialization: Iteration $t = 1$. The SAN is initialized by setting all activation values $A_i(t) = 0$, with exception of the clue nodes whose activation value is set to 1.

Step 2 - Marking: Each node n_i with activation value $A_i(t) \geq F$ is marked as $fired$.

Step 3 - Firing: For each fired node n_i, its output value is computed as a function of its activation level: $O_i(t) = (A_i(t)/t)(1 - D)$; Parameter D is the decay value which is experimentally set to $D = 0.2$;

Step 4 - Spreading: For each link connecting the $fired$ node n_i to the target nodes n_j, recalculate $A_j(t+1) = A_j(t) + w_{ij}O_i(t)$. Notice that, in order to avoid loops, once a node has been fired it cannot be fired again.

Step 5 - Termination check: $t = t + 1$ if $t < maxpulses \wedge fired(t)$ then go to *Step* 2 otherwise *End*. Here $fired(t) = true$ if there is at least one node fired at time t.

The final result of the spreading activation process is the activation level for each node in the SAN at termination time.

3.3 Re-ranking

Selecting the word with highest activation level is the most straightforward methods for proposing the unique solution to WAG game. Unfortunately, as it is reported in experiments (Figure 2), the accuracy gained by that scheme is too low. On the other hand, we observed that in very high percent of attempted games the solution was found among the first 30 nodes (words), when nodes are ordered descending by their activation levels. This observation shows that even though the activation level of nodes can support defining an order among candidate solutions, more advanced ordering scheme is required to select better ordering or even unique answer among them. Here we propose an enhancement for ranking obtained after activation value ordering. Briefly the idea is to additionally rank possible answers based on: (1) the number of hints ($nHint$) to which they are connected with; (2) the number of memory nodes (nMN) through which they are connected to clues; (3) the number of different type of memory nodes ($nDMN$) through which they are connected to clues in the original spreading activation network.

Given the new ranking parameters the final rank for nodes is calculated as follows:

$$w_i = al_i + (1 - (1/nHints_i)) + (1 - (1/nMN_i)) + (1 - (1/nDMN_i)) \quad (1)$$

4 Experimental Results

The goal of experiments was to measure the number of games solved by the system. A association game or the sub-association sets are considered as a solved depending on whether the solution occurs in the list of potential answers (LPS) produced by the system. LPS is produced either from 1) the nodes with the highest activation levels at termination time of SNA or 2) the nodes with the highest score calculated by re-ranking method summarized in Equation 1.

We used an artifact dataset in our experiments which was manually collected by monitoring the TV show [1]. Dataset includes 100 games, i.e the 20 sets of four sub-associations and one final association. Dataset also includes sets of four clues, attempted by human players during the TV show, along with their correct answers.

As for the performance measures, Precision (P) provides the accuracy of the method. It is computed as the ratio between solved games for which the method finds the correct answer, and attempted games: $P = \frac{SG}{N}$, here SG is the number of solved games, and N is the total number of attempted games. Another measure, P_{LPS}, estimates the accuracy whether a LPS is provided instead of a unique answer. It is computed as: $P_{LPS} = \frac{SG_{LPS}}{N}$,

We denote a game p-solved if the correct answer occurs among the top-p ranked possible solutions in LPS list. By changing the values of the variable p, we could measure the precision of our method in broader context, i.e when it does not provide a unique answer.

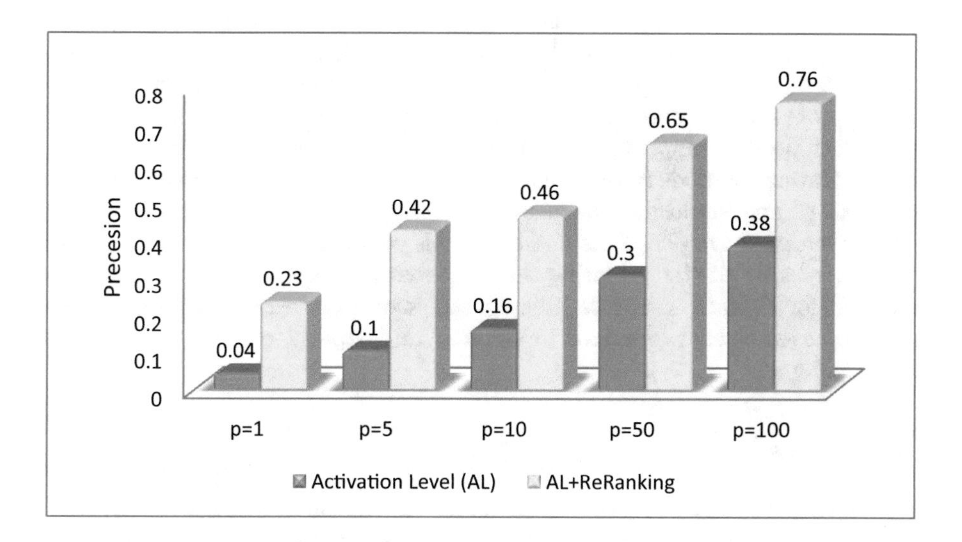

Fig. 2: Precision obtained by the activation level only ranking (AL) and activation level enriched by re-ranking (AL+ReRanking).

Detailed results of accuracy are visually reported in Figure 2. Looking strictly for unique answer results in relatively lower accuracy of the system (1st column in Figure 2). More promising results are reached by permitting longer lists of possible solutions (the rest of the columns in Figure 2). Employing advanced

[1] http://www.kohavision.net/ballina/oxygen/

re-ranking methods significantly boosts the accuracy of the pure SAN based method.

5 Conclusion and Future work

In this work we propose an artificial solver for a language game in which the objective is to guess a hidden association word semantically related to given hints. The essential objective was to formulate a knowledge base of the game solver. This was realized by extracting information from textual sources on the Albanian web and synthesizing them into a semantic network of elementary information trees. We designed the CPU of the artificial solver as a spreading activation algorithm capable of retrieving relevant words to the given hints. Experiments showed that the system accuracy is as good as the human player accuracy. Nevertheless there is room for improvements: 1) during the game play human players can guess the solution(s) at each turn and if the guess is not correct game continues with other player. The wrong guess can be included in our artificial solvers memory as a negative hint which is also our next step in systems development; 2) WAG as it is played on TV show has a couple of properties which for sake of simplicity has been completely ignored in this work. We plan to adapt GSolver to real gameplay by modeling the dynamics of game by means of game theory; 3) As an alternative to spreading activation based central processing we plan to employ the random walks based methodology [12]; 4) another possible improvement can be reached by integrating the document summarization methods in producing the summaries relevant to given hints and use them as additional knowledge source [13,14]. The presented method has a great potential for applications outside of this particular example. For instance, we have recently started adopting the presented approach to a query expansion task [15] for Gjirafa search results. In this scenario, terms from the original query are used as clues provided to GSolver, and expansion terms are keywords proposed by it.

References

1. Millington, I., Funge, J.: Artificial intelligence for games. CRC Press, (2009)
2. Littman, M.L., Keim, G. A., Shazeer, N.: A probabilistic approach to solving crossword puzzles. Artificial Intelligence 134:1, 23–55 (2002)
3. Lam, S.K., Pennock, D.M. Cosley,D., Lawrence, S.: : 1 Billion Pages= 1 Million Dollars? mining the web to play "who wants to be a millionaire?". In: Proceedings of the Nineteenth conference on Uncertainty in Artificial Intelligence, pp. 337–345. Morgan Kaufmann Publishers Inc.,(2002)
4. Molino, P., Lops, P., Semeraro, G., de Gemmis, M., and Basile, P.: Playing with knowledge: A virtual player for Who Wants to Be a Millionaire? that leverages question answering techniques. Artificial Intelligence 222, 157–181 (2015)
5. Ferrucci, D., Brown, E., Chu-Carroll, J., and Fan, J., Gondek, D., Kalyanpur, A., Lally, A., and Murdock, J. W., Nyberg, E., and Prager, J., at all.:o Building Watson: An overview of the DeepQA project. AI magazine 31:3, 59–79 (2010)

6. Semeraro, G., de Gemmis, M., Lops, P., Basile, P.: An artificial player for a language game. IEEE Intelligent Systems 27:5, 36–43 (2012)
7. Etzioni, O., Banko, M., Cafarella, M. J.: Machine Reading. In: AAAI, pp. 1517–1519. ACM, New York (2006)
8. Gjirafa Inc., Search Engine for Albanian Web `http://www.gjirafa.com`
9. Collobert, R., Weston, J., Bottou, L., Karlen, M., Kavukcuoglu, K., Kuksa, P.: Natural language processing (almost) from scratch. The Journal of Machine Learning Research 12, 2493–2537 (2011)
10. Collins, A., M., Loftus, E., F: A spreading-activation theory of semantic processing. Psychological review 82:6, 407 (1975)
11. Anderson, J., R.: A spreading activation theory of memory. Journal of verbal learning and verbal behavior 22:3, 261–295 (1983)
12. Sabetghadam, S., Lupu, M., and Rauber, A.: Which One to Choose: Random Walks or Spreading Activation?. Multidisciplinary Information Retrieval, pp.112–119. Springer, (2014)
13. Canhasi E., Kononenko, I.: Multi-document summarization via archetypal analysis of the content-graph joint model, Knowledge and Information Systems 41:3, 821-842. Springer (2014)
14. Canhasi E., Kononenko, I.: Weighted archetypal analysis of the multi-element graph for query-focused multi-document summarization, Expert Systems with Applications 41:2, 535-543. Elsevier (2014)
15. Billerbeck, B.: Efficient query expansion, Ph.D. thesis, RMIT University, Melbourne, Australia. (2005)

GIS Flood Prediction Models of "Kriva Reka" River

Darko Georgievski[1], Kosta Mitreski[2], Andreja Naumoski[2], Danco Davcev[2]

[1] GDI-GISDATA, Skopje, Macedonia
darko.georgievski@gdi.net
[2] Ss. Cyril and Methodius University in Skopje, Faculty of Computers Science and Engineering, Rugjer Boshkovikj 16, P.O. Box 393, 1000 Skopje, Macedonia
{kosta.mitreski,andreja.naumoski,danco.davcev}@finki.ukim.mk

Abstract. Floods are a natural phenomenon that can cause damage on town building, villages and farmlands, by increasing the water level of nearby river or river systems. The work in this paper aims to present the GIS flood prediction model for the "Kriva Reka" River. By providing early warning about the heavy rain from the national meteorological institute, in combination with our GIS flood prediction model, it will be possible to reduce the damage caused by the floods. The model contains analysis of the terrain data, the hydro-meteorological data, and visualizing the geographic river map of flooded areas. We provided the GIS prediction model with the necessary terrain data and hydro-meteorological data for a 5 years period. The visual results from the GIS model show critical areas, where in period of heavy rain, they are potential disaster zones. In the future, we plan to upgrade the GIS system to be available for the citizens via mobile platform, so we can increase the public awareness of such events and help public evacuation.

Keywords: Environmental Modeling, Forecasting, GIS Flood Prediction Model, Visualization, Meteorological Data

1 Introduction

Floods represent a natural phenomenon which causing damage to the human inhabitants and this can significantly be reduced. These types of weather events that are reinforced by the climate change in the recent years usually occur in spring or autumn. In spring, the increased air temperature and intensified melting of the snow, leads to increasing the level of the water in the main river and the catchment overall. This intense rainfall occurring in spring and autumn can easily cause problems with river catchments and their spill. The main problem is the huge sudden rainfall precipitation that is affecting the raise of the water level in the river catchment. Beside rainfall, the flood can be caused by natural landslides, formation of ice barriers of waterways, a collection of trees of a particular section of the river and etc. These water level fluctuations of the river can dramatically change the flow and the course of the river movement. This is the main cause of damage on its surroundings. Flood damage estimated at hundreds of millions of euros, which is a huge blow to the economies of

S. Loshkovska and S. Koceski (eds.), *ICT Innovations 2015*,
Advances in Intelligent Systems and Computing 399,
DOI: 10.1007/978-3-319-25733-4_7

the affected areas. But, more important are the areas where the river passes through the city or small villages and damages the inhabitant's homes.

1.1 GIS River Flood Prediction Related Work

The development of modern information technologies encourage the development of advanced technologies for management of spatial data and their integration with various disciplines into a single intelligent information system entitled GIS (Geographic Information System). GIS is a technology that combines geographic data, i.e. the location of natural and man-made objects on the ground, with various types of information (descriptive elements, attributes space) in a single database for the entire studied area. By connecting the GIS with the hydro-meteorological data, the experts could easily carry out specific spatial queries, and perform certain analyses and thus formulate a good prediction of future events.

River environmental modelling is a very active research area, several similar research studies exist using the HEC-RAS (River Analysis System) and GIS to model river floods [1, 2, 3]. In [4] a similar system is made using HEC-RAS and GIS in semi-arid regions of Iran. In this paper, we build our model to take computed water surface profiles generated from the HEC-RAS hydraulic model and draw floodplain regions in ArcView. Furthermore, in [5], the authors have used SCS Model have to simulate flood hydrograph for each sub-basin using HEC-HMS (Hydrologic Modelling System). Their research article further discusses how coupling of Digital Elevation Model (DEM) and flood hydrographs (integrated GIS – SCS hydrological model) could help to identify even the most critical flood prone areas within a sub-catchment. Parallel to the research studies done by instituters and researchers, the ESRI, one of the biggest GIS Company has create a lot of tools for generating hydrology and hydraulic modelling. On the ESRI blog site [6], they have described how to make floodplain delineation using only Arc Hydro models. Using the "Stream WSE from point WSE Measurements" and "Flood from stream WSE" models user is able to get precise map of the flooded area both as a raster of water depth in the floodplain and as a polygon feature of the flooded area [6]. Furthermore, in [7] authors has developed a flood hazard maps, from which the flood prone areas within the district would be identified. They have performed rainfall using HEC-HMS, hydraulic modelling using HEC-RAS, and prepared flood hazard maps [7]. There are a lot of other systems for floodplain which are not used ArcGIS, HEC-HMS and HEC-RAS like Imprints, WeSenseIt, and UrbanFlood. These are EU-funded projects that have developed unique forecasting and alert systems to warn communities of impending floods supported by European research and innovation [8]. To best of our knowledge this a first GIS flood prediction model for "Kriva Reka" catchment for prediction of critical region made by rainoff heavy rainfall. The typical approach for flood modelling in the process of flood prediction model in GIS consists from three modules.

1.2 GIS Modular Design of River Prediction Modelling

The first module is the ArcHydro tool to pre-process the terrain data [9]. The ArcHydro part of the GIS, evaluates data from the terrain model using the DEM (Digital Evaluation model). Additionally the Sink, Flow Direction and Flow Accumulation tools are used to estimate the flow and to calculate the accumulated water in the river catchment. Even more, the Stream definition and Stream segmentation are used to calculate the possible location of the water flow. Overall, all this analysis of the data can be used for hydrological model, or it can be used for flood prediction modelling caused in heavy rainfall scenario. The second module is the HEC-HMS (HEC - Hydrologic Model System) [10], which is used to deal with the hydrology of the river catchment. Hydrology is a process of understanding the amount of water, i.e. for a given storm, snowmelt, it calculates how much water will cover. So by working with hydraulics quantity of surface water and surface, it is possible to determinate how deep parts will be flooded. And the last, third module, is the HEC-RAS (HEC - River Analysis System) [11], is responsible for the hydraulic of the river system. Hydraulics is important part of the modelling process, because within its calculation, contains a time parameter which can help to calculate and thus alarm any possible flooding for a certain period of time. In this paper we present the complete description of the GIS flood prediction model for critical areas along the "Kriva Reka" River. Because the river is connected with several small rivers in the catchment, when it heavy rains falls, the accumulated water can cause damage to the buildings that are near the river shores. So, urgently is needed to build an early warning flood management system that will alert the town inhabitants when heavy rain is approaching. The heavy rain forecast is a part of the weather prediction on a national level.

The rest of the paper is organized as follows: Section 2 presents the site and data description used in this research, while the Section 3 presents the three parts of the GIS flood prediction management system (ArcHydro, HEC-HMS and HEC-RAS). In Section 4 we present the model, while the Section 5 concludes our paper and give direction for future work.

2 Site and Data Description

The subject of study of this paper is the catchment river area in the upper Krivo-Palanechkiot region around meter-station "Kriva Palanka", that is located in the Northwest part of Republic of Macedonia, (see Figure 1). The largest river in Krivo-Palanechkiot region is the "Kriva Reka" River.

The catchment of this river has an irregular rectangular shape and its stretching east to west-southwest. The total height difference in the "Kriva Reka" River catchment is 1958 meters. The highest point is Rouen on 2252 m and the lowest is the estuary Pchinja River at 294 m above sea level. The average elevation of the catchment is 862 m, and the average slope (fall) has a value of 17.1 degrees. Most of the "Kriva Reka" catchment is built of rocks with low permeability, making it hard surface runoff and infiltration of storm water low. It is one of the causes of pronounced erosion in the area and production of major erosive sediment. Low forest cover in the watershed can cause significant surface runoff during rainfall, because most of the waterways have stormy and quite variable flow in the year [12].

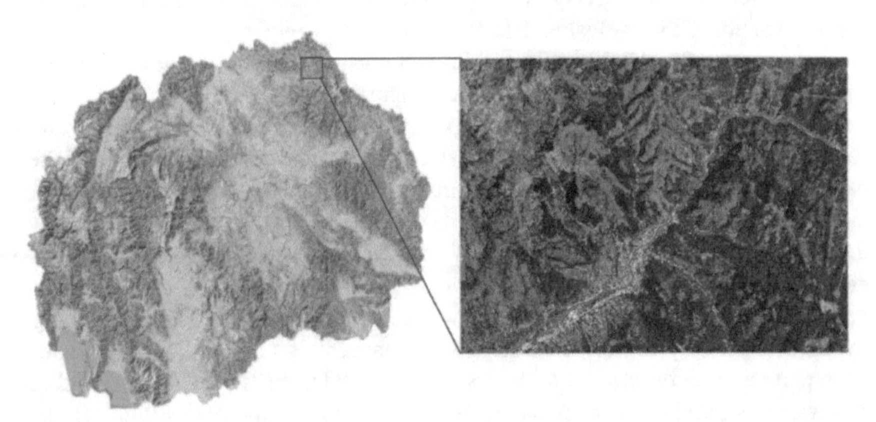

Fig. 1. Location of the studied area Republic of Macedonia (left) and the river catchment of the Kriva Reka River (right)

It should be noted that the catchment area of the "Kriva Reka" River tributaries that descend from Osogovski Mountains have high amounts of precipitation. Therefore, their waterways usually have water throughout the year, and maximum flows are during spring (May, June) and autumn (November) [12]. The right tributaries descend from the mountains Billy (Chupino Brdo, 1703 m), German (blue Chapter, 1390 m) and Kozjak (Backgammon, 1355 m). Because of small slope on the southern direction of these mountains, these tributaries are longer, have elongated catchment areas less medium height, lower average annual amount of precipitation and low forested. So, they have quite variable flow, and in the summer of the year is usually dry. In fact most of these waterways have stormy flow [13].

Beside the terrain data for the "Kriva Palanka" region, the model also contained data from the water level station "Kriva Palanka" for both, maximum and minimum levels of the "Kriva Reka" River over the 5 years period. Furthermore, the hydrometeorological data for this research was obtained from the Hydro-meteorological Institute of Macedonia. The data contains measurement of rainfall for this region in period from 2008 to 2012 (5 years).

3 The GIS Flood Prediction Management System

The GIS flood prediction model uses the terrain model of the "Kriva Palanka" region, the water level station - "Kriva Palanka", the amount of precipitation measured and types of land in "Kriva Palanka" region.

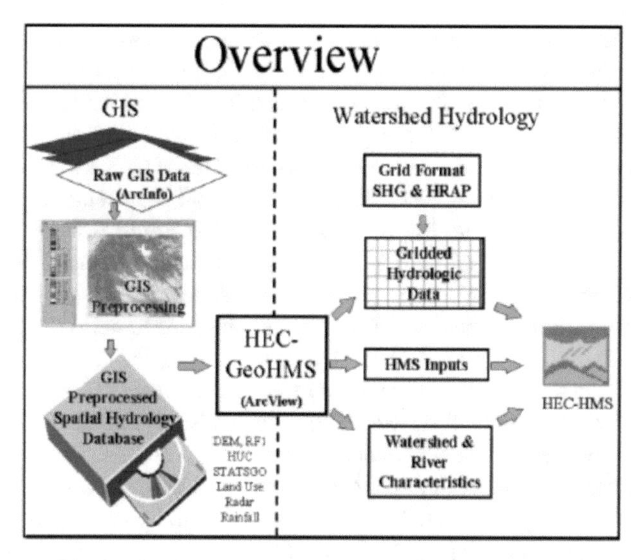

Fig. 2. The GIS flood prediction management system

This data is processed through GIS flood prediction model in six steps, by different tools as follows:

- Field pre-processing or terrain data (ArcHydro)
- Preparation of data for the HEC-HMS (HEC-GeoHMS) [14]
- Meteorological simulation for GIS flood prediction model (HEC-HMS)
- Preparation of data for the HEC-RAS (HEC-GeoRAS) [15]
- River analysis for GIS flood prediction model (HEC-RAS)
- Visualization of results of HEC-RAS (post-processing)

The ArcHydro tool pre-processes the terrain data using DEM. The AGREE method is used to process the data in DEM [16]. Field pre-processing is the milestone and the first step in the collection of field data to create GIS analysis. Its role is to define accurate hydrological DEM, including its elements, as well as a set of layers that will increase the visualization of the model, when used for hydro analysis. After pre-processing the terrain data, we continue generating a model to simulate the display of flooded areas during heavy rains. This is performed by simulation using HEC-HMS and then the HEC-RAS software tool [17].

HEC-HMS is a system for simulation of a complete hydrological process of drainage system. But, before data is processed by the HEC-HMS, we prepare the data by using the HEC-GeoHMS tool. The HEC-GeoHMS is a graphical tool that includes integration of data and graphical interface. Through a graphical interface that contains menus, toolbars and buttons, the user can perform analysis of field models, initialization river areas and sub-areas, as well as preparation for input into HEC-HMS hydrologic system. The next system is HEC-RAS (River Analysis System), which performs hydraulic calculations of the flow of water through rivers and canals, alteration in water analysis and water quality. The software is one-dimensional. The HEC-

GeoRAS represents the river in geographic form (flow, shore, river bed and cuts) is input data in HEC-RAS, which in turn, merges the hydrological and hydraulic data. The result of the HEC-RAS after analysed is displayed using the HEC-GeoRAS tool.

HEC-GeoRAS is a tool for pre-processing and post-processing of the data and serve as input in output of the HEC-RAS system. HEC-GeoRAS pre-processing task is a preparation of data using HEC-RAS and its visual representation. The main task of using this tool is to create RAS following layers: Central flow of the river, River bank (banks), Riverbed (flow path) and Cross sections lines. The HEC-RAS processing of the data is made by Steady Flow Analysis tool, which calculates the amount of water through the river. This is done, in order to build the basic model, on which we analyse the influence of the hydro-meteorological data. This basic model, assumes peaceful flow of the river without frequent changes of speed and level of the river. Once the analysis of the river in the HEC-RAS system is finished, the HEC-GeoRAS post-processing module hands the result.

Post-processing task of the GIS flood prediction model made by HEC-GeoRAS tool, displays the output model. This post-processing task, consist from several tasks to perform mapping and displaying the results. HEC-GeoRAS work with XML files (RAS GIS Export File), so using the "Import RAS SDF File" function we have converted the file .sdf obtained from HEC-RAS to XML file readable by the HEC-GeoRAS tools. This XML file serves as an input parameter to generate the analysis using the Layer Setup Tool. To obtain the area of interest, the Layer Setup Tool generate raster surface water in TIN format, needed for presentation of the model. Later, using the TIN format, more analysis can be done by cross sections lines and altitude values of the water surface using the "Water Surface TIN" tool in HEC-GeoRAS tool. The interoperability issue is also important step toward building the GIS prediction management system. The collected data from several stations is processed by the preprocessing module of the GIS model and then is imputed into the HEC-HMS. Later, the provided information regarding the terrain is passed to the GIS map module that can be local or web based system. Finally, the end user can build different scenarios regarding the provided models.

4 GIS Flood Prediction Model

In order to display the results from the GIS flood prediction model we use the "Floodplain Delineation Using Rasters" tool to take the output from the HEC-GeoRAS tool. First, we select a profile that we want to display of the flooded regions. This converts the TIN surface water raster to GRID based Rasterization Cell Size. Then we compare the heights of the water with GRID DEM raster. If the height of the water surface scatter is greater than the amount of raster DEM, that amount is added to the new raster to flooded areas. Once the HEC-RAS generates raster with flooded areas, the map is converted to vector form. Then the user can determine the boundary of a flooded area. The result from the GIS flood modelling is given on Figure 3.

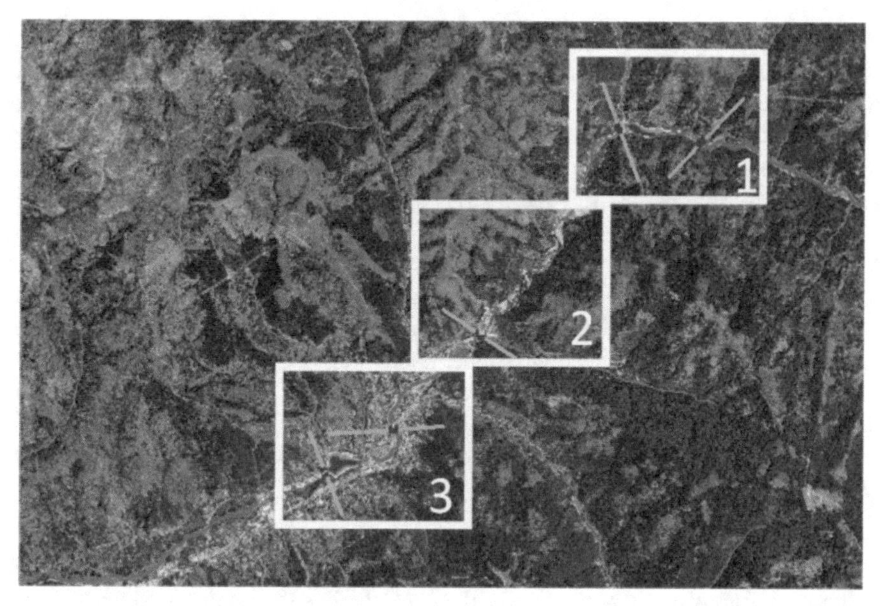

Fig. 3. GIS flood prediction model. The red spots represent critical places and horizontal lines represent possible dam building sites

As it can be seen from the model map, the red areas are possible areas where the flooding may occur, during a heavy rain for the "Kriva Reka" River. The upper regions of the "Kriva Reka" River, marker as 1, have a potentially of accumulating large amount of water, and then combined with the water accumulated by the region marked as 2, represent a flood threat for the "Kriva Palanka" inhabitants. The region 3, as it can be seen from the Fig. 3, flows through the center of the town and accumulates a large amount of water that needs to be handle properly by the authorities during heavy rainfall. Horizontal green lines represent possible dam building sites in order to prevent flooding of important sites.

5 Conclusion

This paper presents a GIS flood prediction model that can be used to help in prevention and reduction of the damage caused by heavy precipitation and thus major floods. The GIS flood prediction model, which acts are early warning system, shows the critical areas that can be found to be dangerous during heavy rainfall near by the "Kriva Reka" River. We used terrain model for "Kriva Palanka" region for the river catchment and the types of land to build the model. Additionally, we provided data for the amount of hydro-meteorological data and water level inside of the river catchment, for 5 years' time period. This data was process using the ArcHydro, HEC-HMS and HEC – RAS tools.

Using the ArcHydro Tool we have analysed the terrain data in DEM, and then we use to build hydrological model. This model, obtained by the HEC-HMS tool, based on the processed field data, hydro-meteorological data (amount of rainfall) and the types of land, integrates calculations on the amount of water received from precipitation. Then these results were analysed and visualized by the HEC-RAS tool, particularly HEC-GeoRAS tool. The result of the river analysis tool produced a geographic map, which serves as an early warning system of "Kriva Reka" River area. Using the model we have shown the flooded regions that would occur when heavy rain falls in "Kriva Reka" river catchment. Using the GIS flood prediction model, we have created a realistic picture of the flooding that can occur in this river catchment. The models can be evaluated on the basis of new collected meteorological and hydrological data for certain period of time. This is done primarily by measuring the level of the "Kriva Reka" river system and collecting weather data for this region.

As future work, we plan to develop mobile version as an early warning system of the GIS flood prediction model, which will alert the inhabitants of the "Kriva Reka" River, and even more to provide a realistic scenarios for wider region around the "Kriva Reka" river catchment.

Acknowledgement. This work was partially financed by the Faculty of Computer Science and Engineering at the Ss. Cyril and Methodius University in Skopje.

References

1. Mioc, D., Nickerson, B.G., Anton, F., Fraser, D., MacGillivray, E., Morton, A., Tang, P., Arp J.P., & Liang, G.: Web-GIS Application for Flood Prediction and Monitoring. FRIAR 2012, Dubrovnik, Croatia, (2012)
2. Lorenzo A., Thielen. J., Pappenberger, F.: Ensemble hydro-meteorological simulation for flash flood early detection in southern Switzerland, Journal of Hydrology, doi:10.1016/j.jhydrol.2011.12.038, (2012)
3. Chan, Y., Mori, M.: The Construction of a Flood Monitoring System with Alert Distribution Using Google Earth and 3D GIS, Journal of Disaster Research 8(3), pp. 512-518, (2013)
4. Salajegheh, A. Bakhshaei, M., Chavoshi , S. Keshtkar, A.R. Najafi H. M.: Floodplain mapping using HEC-RAS and GIS in semi-arid regions of Iran, DESERT 14, accessed at 22-10-(2014), `http://jdesert.ut.ac.ir/pdf_21750_021ed7dbef413559fbe3139a99956e66.html`
5. Bhankaurally M., Y., Nowbuth M. D., Bhavana U.: Flood Hazard Mapping by Integrated GIS SCS Model, International Journal of Geomatics and Geosciences Volume 1, No 3, (2010), `http://www.ipublishing.co.in/jggsvol1no12010/EIJGGS2017.pdf` accessed at 22-10-2014,
6. Bergeron P.: Floodplain delineation using only Arc Hydro models, 2014, accessed at 22-10-(2014), `http://blogs.esri.com/esri/arcgis/2014/03/17/floodplain-delineation-using-only-arc-hydro-models/`

7. Okirya M., Rugumayo, A., Ovcharovichova J.: Application of HEC HMS/RAS and GIS Tools in Flood - Modeling: A Case Study For River Sironko – Uganda, Global Institute for Research and Education, (2012), http://www.gifre.org/admin/papers/gjedt/1221-19-31.pdf
8. European Commission - IP/14/950.: Faster, more accurate flood warnings through EU research, 28-08-2014, accessed at 22-10-2014, http://europa.eu/rapid/press-release_IP-14-950_en.htm?locale=en
9. David, R.M.: ArcHydro: GIS for Water Resources, (2002)
10. US Army Corps of Engineers.: HEC-HMS (Hydrologic Engineering Centre – Hydrologic Modelling System), 2010, http://www.hec.usace.army.mil/software/hec-hms/documentation/HEC-HMS_Users_Manual_3.5.pdf
11. US Army Corps of Engineers.: HEC-RAS (Hydrologic Engineering Centre – River Analysis System), (2010), http://www.hec.usace.army.mil/software/hec-ras/documentation/HEC-RAS_4.1_Users_Manual.pdf
12. Milevski, I.: The hydrographic features of the Kriva Reka River IGEO portal, accessed at 5.5.2014, http://www.igeografija.mk/Portal/?p=668
13. Dimitrovska, O., Milevski, I., "Билтен за Физичка географија, Квалитет на површинските води во сливот на Крива Река (Bulletin for Physical Geography, Quality of surface waters in the catchment of the "Kriva Reka" River)", Skopje, Republic of Macedonia, (2005)
14. US Army Corps of Engineers.: HEC-GeoHMS, Hydrologic Engineering Center - Geospatial Hydrologic Modeling Extension, (2010), http://www.hec.usace.army.mil/software/hec-geohms/documentation/HEC-GeoHMS_Users_Manual_5.0.pdf
15. US Army Corps of Engineers.: HEC-GeoRAS, Hydrologic Engineering Center GIS Tools for Support of HEC-RAS using ArcGIS, (2011), http://www.hec.usace.army.mil/software/hec-georas/documentation/HEC-GeoRAS_43_Users_Manual.pdf
16. Hellweger, F.L.: AGREE - DEM Surface Reconditioning System, (1999)
17. Maidment, D.R., Djokic, D.: Hydrologic and hydraulic modelling support with geographic information systems, ESRI Press, Redlands, (2000)

Collaborative Cloud Computing Application for Water Resources Based on Open Source Software

Blagoj Delipetrev[1], Aleksandra Stojanova[1], Ana Ljubotenska[1], Mirjana Kocaleva[1] Marjan Delipetrev[2] and Vladimir Manevski[2]

[1] Goce Delcev University, Faculty of Computer Science, Krste Misirkov 10-A, 2000 Shtip, Macedonia
{blagoj.delipetrev,aleksandra.stojanova,ana.ljubotenska, mirjana.kocaleva}@ugd.edu.mk
[2] Goce Delcev University, Faculty of Natural and Technical Sciences Krste Misirkov 10-A, 2000 Shtip, Macedonia
{marjan.delipetrev@ugd.edu.mk,vladimir.manevski@ugd.edu.mk}

Abstract. This article presents research and development of a collaboration cloud computing application for water resources based on open source software. The cloud computing application is using a hybrid deployment model of public – private cloud, running on two separate virtual machines (VMs). The first one (VM_1) is running on Amazon web services (AWS) and the second one (VM_2) is running on a Xen cloud platform. The collaborative cloud application has four web services for 1) data infrastructure (DI), 2) support for water resources modelling (WRM), 3) user management and 4) water resources optimization (WRO). The most important characteristic of the cloud application is a real-time geo-collaboration platform supporting multiple users. This research shows the capability to scale and distribute the cloud application between several VMs. The cloud application was successfully tested in the Zletovica case study in a distributed computer environment with concurred multiple access.

Keywords: Cloud computing, Web GIS, Geospatial software.

1 Introduction

Presently, most of the existing software for water resources is desktop-based, designed to work on a single computer. The classical desktop software has many limitations starting from computer processing power, storage, data and model portability. Another very important limitation is sharing or collaborating on joint data and models between multiple users. The desktop software is limited in availability and accessibility that are often restricted by physical location. Which brings us the main research question of this article: Is it possible to develop water resources application that alleviates most of the issues mentioned before and provides seamlessly endless storage, computing power, is available everywhere and anytime, supports multi-user platform, diminish data portability issue, and it is flexible for upgrading the existing and connecting/adding other software components.

S. Loshkovska and S. Koceski (eds.), *ICT Innovations 2015*,
Advances in Intelligent Systems and Computing 399,
DOI: 10.1007/978-3-319-25733-4_8

The only feasible solution lies in the web and cloud. There are various examples of web applications [1,2], cloud web services [3,4], and mobile applications [5] in the water domain. Previously mentioned systems often need “only a web browser” to be used, hiding all implementation details from the end-users. These advantages can significantly increase the number and the diversity of users.

The cloud application for water resources is continuation of previous research [6] and has four web services for 1) data infrastructure (DI), 2) support for water resources modelling (WRM), 3) user management, and 4) water resources optimization (WRO). The cloud application enhancements are the following:

- The application is distributed / deployed on two VMs. The VM_1 is running as a micro instance of Amazon web services (AWS) public cloud, and the VM_2 is running on a Xen cloud platform at the University Goce Delcev in the Republic of Macedonia.
- The web service for support of WRM that runs on VM_1, and the DI web service that runs on VM_2, are communicating with WFS-T (Web Feature Service - Transactional) XML messages over the internet, demonstrating distributed computer environment.
- Additional optimization algorithms nested stochastic dynamic programming (nSDP) and nested reinforcement learning (nRL) are included into the web service WRO.
- Clear demonstration of hybrid cloud is presented, where VM_1 is part of the AWS public cloud, and VM_2 is running in the private cloud. The advantage of this distributed computer environment is that the data security and protection can reside in the private cloud (VM_2), while the web services are in the public cloud (VM_1).
- The cloud application was tested with several students, demonstrating that it can work in distributed environment supporting concurrent multiple users.

The presented research is a geospatial web application [7] that was built using OGC standards (WMS, WFS-T), OpenLayers library, Geoserver, PostgreSQL and PostGIS. The application provides geo-collaboration platform [8] where multiple users in real time can jointly model water resources using web geospatial tools provided by OpenLayers library.

The cloud computing application was tested with data from the Zletovica river basin located in the north-eastern part of the Republic of Macedonia. The test is a proof of concept that this application can be a foundation for a modern collaboration cloud based solution. The application url www.delipetrov.com/his/ provides video presentation and explanation of the system components, guides how to use the services etc.

2 Design and Implementation

The cloud application has four web services:

1. DI.
2. Support of WRM.
3. User management.
4. WRO.

Figure 1 shows the web services and the data communication links represented by arrows. The communication between the web service for support of WRM and the DI web service is asynchronous, or on demand.

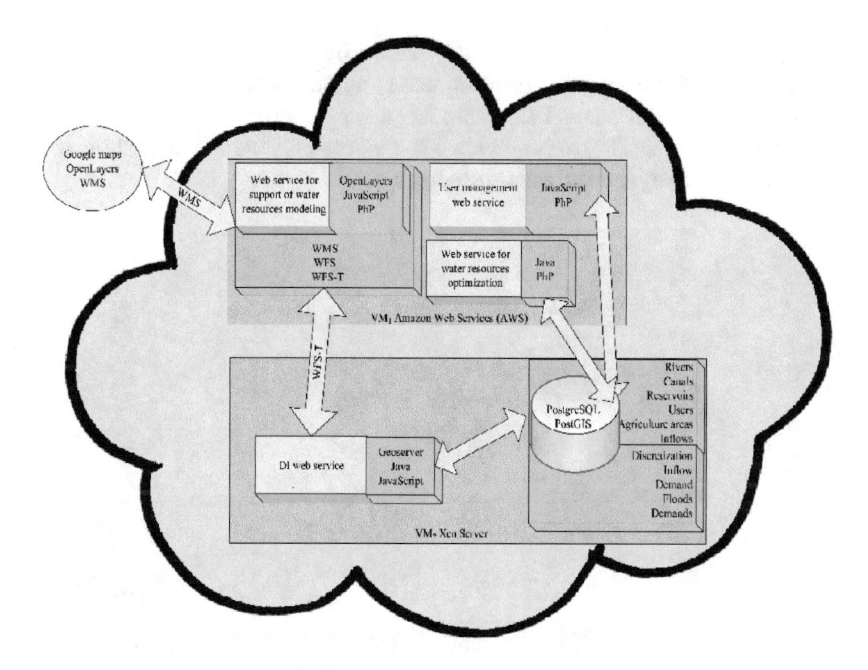

Fig. 1. Design of the cloud computing application for water resources

2.1 DI Web Service

The DI web service is composed of two geospatial software components 1) HMak database created in PostgreSQL and PostGIS and 2) GeoServer. The HMak stores six geospatial vector layers: rivers, canals, reservoirs, users, inflows, and agriculture land, and their attribute tables, that are used by the web service for support of WRM. Additionally, the HMak stores around 40 time series data tables used by the WRO web service. The DI web service is running on VM_2 that resides in the private cloud on the Xen cloud platform.

The PostgreSQL, PostGIS and GeoServer are exceedingly used for creating spatial data infrastructure (SDI). More importantly, GeoServer is as a middle tier application that connects the HMak database on one side and provides OGC web services (WFS-T) on the other side. In our cloud application, GeoServer provides WFS-T interface for the web service for supporting WRM.

2.2 Web Service for Support of WRM

The web service for support of WRM is intended to provide a web interface accessible by multiple users simultaneously for creating and editing geospatial water resources elements just like in classical desktop applications, as shown in Fig. 2. The six geospatial layers are designed to allow only a specific type of geospatial data for each layer, e.g. points for reservoirs, inflows, users; polylines for rivers and canals; and polygons for agricultural areas. Each layer has a simple attribute table, only used in demonstration purposes. The web service for support of WRM is a customized geospatial web service designed for water resources.

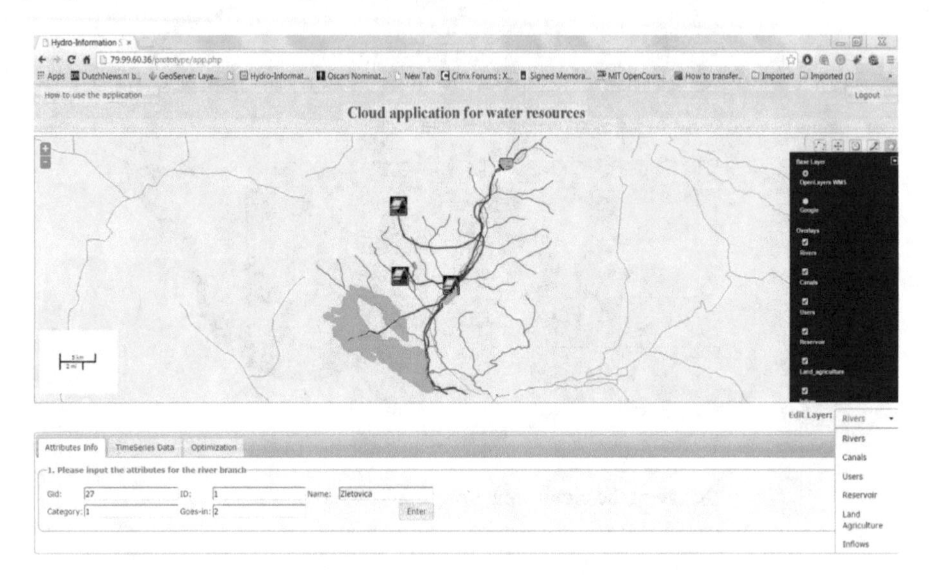

Fig. 2. Cloud application interface

The web service for support of WRM is developed using PHP, Ajax, JavaScript and the most importantly OpenLayer library. The OpenLayer library supports OGC standards (WMS, WFS-T). The web service for support WRM is using WMS to connect to two different basic map providers: 1) Google Maps and 2) OpenLayers WMS. This allows users to select the background map from the menu as shown in Fig. 2, where OpenLayers WMS is used.

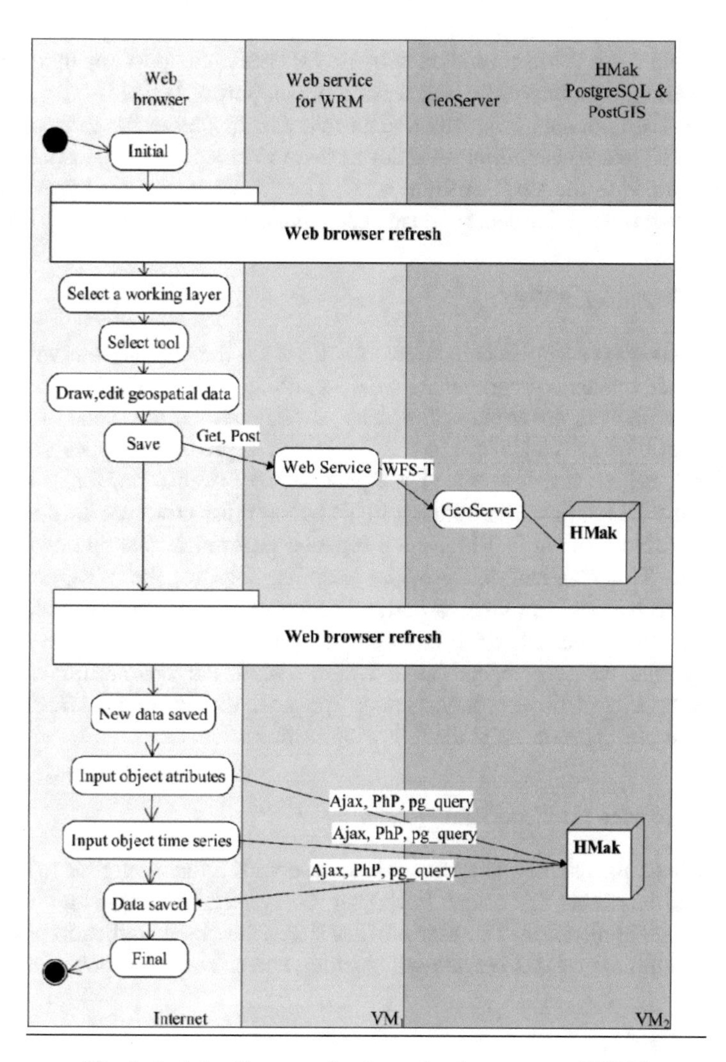

Fig. 3. Activity diagram of web service for support of WRM

The OpenLayer library creates WFS-T communication between the web service for support of the WRM user interface, running on VM_1, and the geospatial data stored in HMak where the GeoServer acts like a middle tier, running on VM_2. The user can change or create geospatial objects from the web service for support of WRM interface, which creates a WFS-T message on VM_1 that it sent to the GeoServer at VM_2. The GeoServer translates the WFS-T message and correspondingly changes the geospatial object data that is stored in HMak. When user makes a browser refresh, it makes a WFS-T request to the GeoServer to read the geospatial data from the HMak database stored in VM_2 and sent it back to the user interface in VM_1. The WFS-T communication is based on XML messages. The browser request also generates a WMS request that fetches the data from WMS providers with HTTP "get" and "post" methods. Figure 3 shows the activity diagram of the web service for support of WRM.

The WFS-T do not support attribute data that accompany every geospatial object. Additional PHP and Ajax scripts were developed to work with the geospatial data attributes, presented in the tab "Attribute info". The "Time series data" provides a possibility to attach time series data for each geospatial object that is stored in HMak.

2.3 Web Service for WRO

The WRO web service implements three algorithms for the optimal reservoir operation named 1) nested dynamic programming (nDP) 2) nSDP and 3) nRL. Further discussion about the algorithms is not explained, because it goes beyond the scope of this article. The three algorithms are coded in Java and, developed as prototype applications.

The WRO web service has several components: PHP web form, for data input and upload into the HMak database, running in VM_1, the three prototype Java applications for each algorithm residing in VM_2, and a separate page for results presentations again in VM_1. The "Optimization" tab provides user interface for the WRO web service. There are three buttons, and when a button is pressed, the appropriate prototype algorithm application is selected which connects to HMak database, reads the data, calculates the optimal reservoir operation and stores back the optimization results into HMak. The JavaScript library "highcharts.js" presents the results (A screenshot of an example result plot is presented in the following section).

2.4 Web Service for User Management

The web service for users' management is simple with a main purpose to control the cloud application access and usage. Each user receives its own login and password to access the cloud application. The user profiles stores the cloud application usage time. Further development of this service will include users' computer power and storage usage.

3 Results and Tests

Important milestone is the deployment of the cloud application between the two VMs running on separate physical servers. The VM_1 is a micro instance on the AWS, and the VM_2 is running on the Xen cloud platform. The VM_1 has 8 GB HDD, 1 GB RAM and Ubuntu 13 as an operating system. The VM_2 has 30 GB HDD, 1 GB RAM and Fedora 16 as an operating system. The VM_2 is running on a physical server IBM x3400 M3 with four-core Intel Xeon E5620 2.40 GHz with 12 MB of cache per processor. The AWS management console and the Citrix XenCenter, respectively manage the VM_1 and VM_2.

Figure 4 shows the hydro system Zletovica model created by the web service for WRM. The hydro system contains the reservoir Knezevo, river network, canal network, towns as users and agricultural areas. The towns and reservoir titles are added additionally and are not part of the web service.

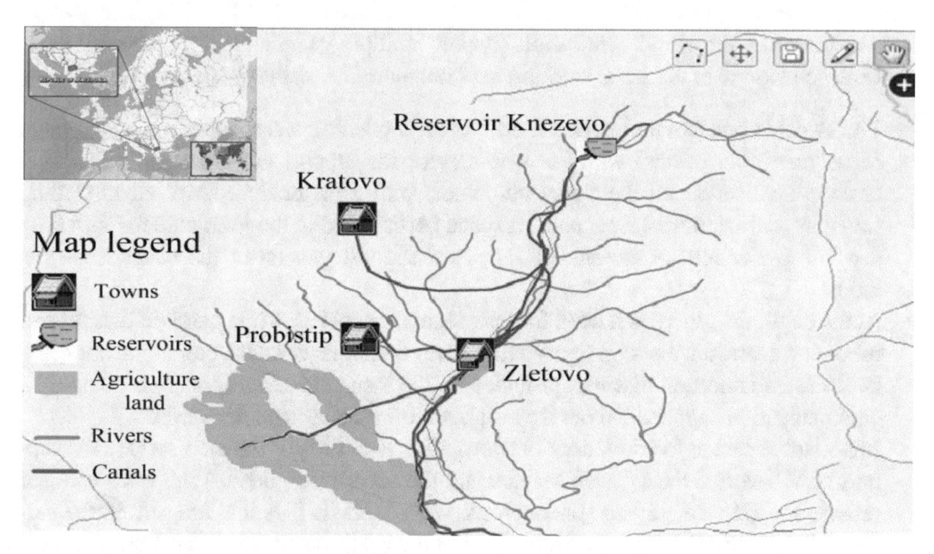

Fig. 4. Water resource model of the Zletovica river basin

The WRO web service was tested using data from an existing study of the Zletovica river basin. Three nDP, nSDP and nRL datasets were uploaded as CSV files into the HMak using the WRO web service interface. Figure 5 presents the optimal reservoir storage results of the three algorithms.

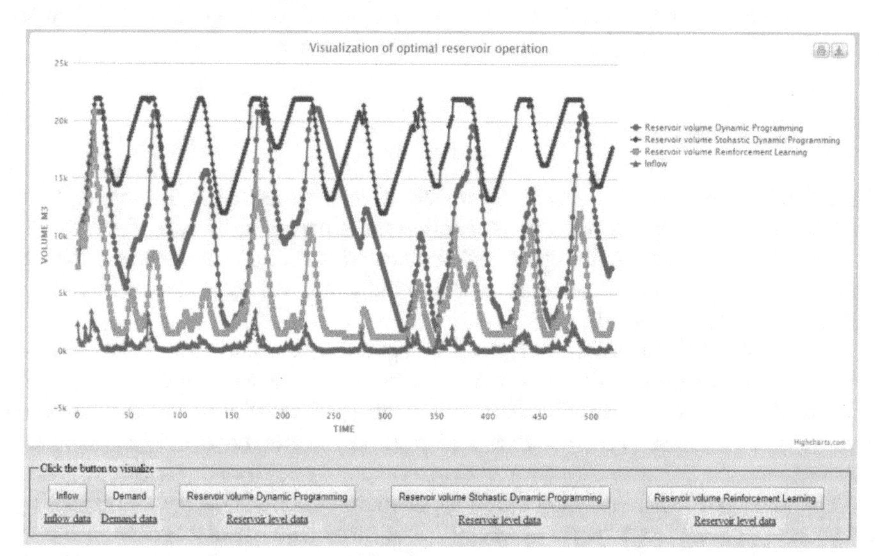

Fig. 5. Optimal reservoir operation graph from the three algorithms displayed by the WRO web service

The collaborative cloud application was tested by six students for several hours. The students were using the cloud application from their homes and communicated try instant messaging services (Skype and Facebook) between each other. The idea was to demonstrate that the cloud application support multiple geographically dispersed users that can collaborate in the same working environment. The main conclusions are:

- The students collaborated between each other modelling water resources (often from Macedonia river basins) with the web service for support of WRM just as using a desktop application on their own computer, with a major difference, all modelling was done online, in real time and everyone participated in the joint model. With only a web browser refresh the student can view the last version of the water resources model.
- At the beginning, there is a need for task separation, because it is possible that several users can work on the same geospatial data, therefore constant communication between users is needed that was provided by the instant messaging services. The miscommunication happened in our first trials, and was corrected afterwards.
- Important aspect is the data flow between VM_1 and VM_2 or the web service for support of WRM and the DI web service and the users' computers. With each browser refresh, the DI web service generates six WFS-T XML files that are sent to the service for support of WRM and from there to the user's computer. Even a small model can generate substantial network traffics. Our simple example of the Zletovica river basin model at each refresh generated around 3 MB. This can be an important issue if a complex model with many geospatial objects is considered.

The test general conclusion is that the cloud application is stable and functional, supporting multiple users with increased workload. The system performance measures, the workload and other characteristics were not performed, although both AWS management console and Citrix Xen control provides that information.

4 Discussion

The cloud computing paradigm "Only a web browser is needed to use the application" is accomplished. The presented cloud application is deployed on two VMs working in a distributed computer environment and demonstrates all cloud advantages, such as diminishing concerns about working platform, software versions, data portability, and other implementation details.

Further, the NIST definition of cloud computing is considered to evaluate the presented cloud application. The first two essential characteristics of the cloud application are "on-demand self- service" and "broad network access." The cloud application is available and accessible all the time and from anywhere and it only requires a web browser.

The third and fourth essential cloud application characteristics are the capability for "resources pooling" and "rapid elasticity." The basic adjustment concerning the workload can be performed by increasing the current VMs computational power. The VMs workload can be monitored over the AWS console and XenCenter and adjust appropriately. The cloud application components, standards and programming languages are interoperable and can be deployed on an unlimited number of servers and connect appropriately. The issues concerning scalability and resource pooling can be resolved by creating many data repositories similar to the HMak deployed on a number of VMs, storing large quantities of geospatial and other types of data. Additionally, several GeoServer instances can connect to multiple data repositories, creating a giant DI system. The web service for support of WRM can connect to the multiple GeoServer instances, and finally the web services can be replicated into several VMs. Depending on the number of users, the workload, the storage capacities, the processing power, the number of servers available, etc., the optimal cloud application environment can be adapted.

The last essential characteristic of cloud computing is "measured service" which is rudimentary supported by measuring the time of each user's usage of the system. This satisfies cloud computing criteria, but needs to be vastly improved (e.g. with measuring processing power consumption, storage capacity utilization, etc.).

Concerning service models, the presented cloud computing application belong to software as a service (SaaS). Users with a web browser access the cloud application and do not care about underlying cloud infrastructure. The current deployment model is hybrid of public - private cloud because the VM_1 is running in public cloud AWS, and VM_2 is on private Xen cloud platform.

The most valuable feature of the cloud application is its real time collaboration platform capabilities. Multiple users using only a web browser can work jointly with the web services and collaborate in the same working environment in real time. An example is the web service for supporting WRM is when a user saves the current work. After that moment all other distributed users with just refreshing the web browser window can see the change (new/modified rivers, users etc.). All of the data and models are stored in HMak and users do not have to be concerned about hardware and software support infrastructure.

Another important concern about implementing cloud solutions is the data protection and safety. Often many companies and organization dismiss implementation of cloud computing solutions just because their data will be stored somewhere on the internet raising major concerns about its safety and protection. This prototype cloud application makes an elegant solution where services are residing in the public cloud, while the data is stored in the private cloud. If for instance an attack happens on the cloud application, to protect the data the private cloud VM_2 can be disconnected from the public cloud AWS VM_1. Another key point is that data resides inside the institution, and only the web services are "outsourced." This concept can be applied in many organizations where the data needs to be stored internally.

5 Conclusion

This research demonstrates that there are available open source software and technologies to create complex and robust collaborative cloud application for water resources. The application is a SaaS hybrid cloud solution running on two virtual machines VMs, from which VM_1 runs on AWS and VM_2 on Xen cloud platform. The cloud application was tested in the Zletovica case study and with real-time multiple users that collaborated and jointly modelled and optimized water resources.

The cloud application can be further improved by connecting to new data repositories. Additional modelling, optimization and other decision support services have already been envisioned and can be added to the existing platform, so that it can evolve into a fully cloud based water resources modelling system.

References

1. Choi, J., Engel, B., Farnsworth, R.: Web-based GIS and spatial decision support system for watershed management. Journal of Hydroinformatics, 7, 165-174 (2005)
2. Horak, J., Orlik, A., Stromsky, J.: Web services for distributed and interoperable hydro-information systems. Hydrology and Earth System Sciences. 12(2), 635-644 (2008)
3. Burger, C. M., Kollet, S., Schumacher, J., Bosel, D.: Introduction of a web service for cloud computing with the integrated hydrologic simulation platform ParFlow. Computers and Geosciences. 48, 334-336 (2012)
4. Quiroga, V. M., Popescu, I., Solomatine, D., Bociort, L.: Cloud and cluster computing in uncertainty analysis of integrated flood models. Journal of Hydroinformatics, 15(1), 55-70 (2013)
5. Jonoski, A., Alfonso, L., Almoradie, A., Popescu, I., van Andel, S. J., Vojinovic, Z.: Mobile phone applications in the water domain. Environmental Engineering & Management Journal (EEMJ), 11 (5) (2012)
6. Delipetrev, B., Jonoski, A., Solomatine, D. P.: Development of a web application for water resources based on open source software. Computers & Geosciences, 62, 35-42 (2014)
7. Scharl, A.: Towards the geospatial web: Media platforms for managing geotagged knowledge repositories. Springer (2007)
8. MacEachren, A. M., Brewer, I.: Developing a conceptual framework for visually-enabled geocollaboration. International Journal of Geographical Information Science, 18(1), 1-34 (2004)

Rule - Based Model for Medical Knowledge Presentation and Reasoning in Clinical Decision Support Systems

Liljana Aleksovska – Stojkovska[1], Suzana Loshkovska[2], Deska Dimitrievska[3]

[1] MAK-System Corp., Chicago, IL, USA
Liljana.A.Stojkovska@gmail.com
[2] University „Ss. Cyril and Methodius – Skopje“, Faculty of Computer Science and Engineering, Skopje, Republic of Macedonia,
suzana.loshkovska@finki.ukim.mk
[3] University „Ss. Cyril and Methodius – Skopje“, Medical Faculty, Skopje, Republic of Macedonia
deskadimitrievska@gmail.com

Abstract. The two core components of every clinical decision support system (CDSS), which are crucial for the success of the system, are the knowledge base and the reasoning engine. A great deal of the artificial intelligence research has focused on determining the appropriate knowledge representation and reasoning methods to achieve high performance of the knowledge-based systems. This paper proposes a model for medical knowledge presentation and reasoning, which is used in a clinical decision support system for managing asthma in school-aged children. To promote customization and flexibility, we use rules for formal presentation and reasoning of the general asthma related knowledge and the individual patient specific facts. This paper contributes to the medical informatics research by demonstrating an effective method for knowledge presentation and reasoning in a specific clinical decision support system, which can also be applied to similar systems.

Keywords: Asthma, Clinical Decision Support System (CDSS), Knowledge base, Knowledge representation, Peak Expiratory Flow (PEF), Reasoning engine, Rules

1 Introduction

The knowledge base and the reasoning engine are intelligence-embedded components of the Clinical Decision Support System (CDSS) [1], which play a major role in determining the success of the system. The performance of the system greatly depends on the quality of its knowledge base and the effectiveness of the methods to process that knowledge [2]. Constructing a good knowledge base, which consists of collecting the relevant medical knowledge from the specific domain, its systematization and technical formalization in a form that is human understandable but computer-interpretable, deserves special attention in the development process of CDSS [3,4,5]. There are many different methods for representation and reasoning of

S. Loshkovska and S. Koceski (eds.), *ICT Innovations 2015*,
Advances in Intelligent Systems and Computing 399,
DOI: 10.1007/978-3-319-25733-4_9

the medical knowledge, which have different strengths and weaknesses and can be more or less suitable for different systems. This paper focuses on the rule-based method and elaborates a model for knowledge presentation and reasoning within a specific CDSS – PedAst, designed for managing asthma in school aged children.

The paper is organized as follows. Section 2 reviews the related research and specific systems designed to support the decision making in the area of Pulmonology. Section 3 provides a high – level overview of the PedAst system describing the main components. Section 4 provides general information about the rule-based method for knowledge presentation and reasoning. Section 5 demonstrates how the rule-based method is applied in the PedAst system. The last Section 6 discusses the advantages and disadvantages of the model and provides directions for future enhancements.

2 A Review of the State-of-the-Art

Computerized decision support systems in the clinical area of Pulmonology originate since the mid 1970's. Among the first systems was PUFF - designed to become a practical assistant to the pulmonary physiologists by performing interpretation of the pulmonary function tests [6]. Its knowledge base was built with production rules. The system's performance was satisfactory enough to be used daily in clinical service.

The need for enhancements of the production rule formalism in PUFF, such as ability to represent prototypical patterns of disease, motivated the creation of a prototype-directed system, called CENTAUR [7], which combines the advantages of production rules and frames (prototypes). Similar to PUFF, CENTAUR produces pulmonary diagnosis based on interpretation of pulmonary function test results.

A more recent system, designed for detecting asthma and COPD patients based on the analysis of patient's data captured during the routine visits of the primary care physicians is Asthma Critic [8]. The knowledge base of the system was mainly derived from the Asthma and COPD guidelines of the Dutch College of General Practitioners and took years of iterative process to construct it.

Modern electronic systems for asthma care vary from small tools for patient self-care, to integrated systems that are designed to assist clinicians in the process of clinical decision making [9]. Asthma Tracker is a small piece of software for Blackberries and provides a list of common asthma medications and common asthma triggers to avoid. Stop Asthma is a rule-based decision support system for pediatric asthma management [9]. A new system called RespDoc introduced in [10], is designed for managing childhood asthma based on measurements of Fraction of Exhaled Nitric Oxide (FeNO). The knowledge base of RespDoc consists of algorithms for determining the degree of airway inflammation, based on the FeNo measurements. The system described in [11] was developed for asthma management in an emergency department, using an algorithm based on Bayesian Network to detect asthma in the patients.

The various systems presented in this review are all valuable and useful in covering certain aspects of asthma control, but there is a need for an integrated system that will allow continuous control of the patient's asthma conditions in all settings.

Our goal is to build a prototype of such system that will fulfill this need, while at the same time will contribute to the medical artificial intelligence in general.

3 Overview of the PedAst System

PedAst (Pediatric Asthma) is a web-based modular system, including modules that cover the patient in home, school and healthcare settings to ensure continuous monitoring and control of the asthma conditions. The system is intended to provide the following functions [12]:

- Capturing, storing and real-time access to the patient's medical information;
- Opening communication channels between the parents, school nurse and doctor;
- Supporting the decision making process by assessing the patient's state based on the entered data and generating alerts, reminders and recommendations;
- Recognition of patterns in the collected data set and identification of factors that improve or worsen the patient asthma conditions.

The system is composed of the following modules [12]:

Patient Module – designed for the child's guardians for capturing and storing data related to the asthma conditions.

School Module – designed for the school nurse for following the patient's treatment plan and providing communication channel with the doctor's office.

Doctor Module – designed for the medical practitioners for following the patient's condition and providing support when making diagnosis and prescribing medications.

Core Module – composed of knowledge base, reasoning machine and patient's electronic medical records. This is the brain of the system, which applies the medical knowledge on the patient individual data and generates specific decisions.

Admin Module – designed to provide means for configuring and customizing the system. A main part of this module is the Knowledge Base Editor, designed for knowledge experts to build and update the knowledge base.

The PedAst system is expected to promote increased self-care and improve clinical management of the asthma conditions, allowing the patients to live a near-normal life. The system is expected to increase independency of the patient's guardians or even the older patients themselves in monitoring and controlling the asthma conditions, which would be very helpful in today's dynamic world.

4 Rule-Based Methods for Knowledge Representation and Reasoning

The rest of this paper will focus on presenting the knowledge base of the PedAst system, as a structural collection of medical knowledge and the Reasoning Engine as a collection of algorithms that apply the medical facts to the patient data to generate specific conclusions. Among the different methods for knowledge presentation, we chose the rules method as most suitable for our system. This section will describe the method and provide a rationale why we decided to use rules.

In general, the knowledge can be: declarative and procedural [5]. The declarative knowledge is expressed through propositions and sentences. Propositions are statements that can either be true or false, which can be connected by Boolean operators „and“, „or“ and „and not“ to form sentence. Procedural knowledge is more task oriented and provides more explicit information about what action can be taken or what conclusion can be drawn from the declarative knowledge. For example: „The patient’s blood pressure is 190/120 mm Hg“ is declarative statement. The statement „IF the patient’s blood pressure is > 180/110 mm Hg, THEN the patient may suffer stroke“ is procedural knowledge.

Rules are statements in the form „IF <condition> THEN <action>“ [5]. The <condition> part is a logical expression that consists of a single statement or multiple statements linked with Boolean operators. The idea is for a given variable and a bound, to check if the value of the variable is within or outside of the bounds and take action based on the result. The <action> part represents a specific task that should be performed, given that the logical condition is true. For example: „IF the patient’s temperature is above 38 C, THEN give him/her a fever reducing medicine“. In clinical settings, these rules are widely used for alerts and reminders, such as alert to the doctor if the patient’s blood pressure drops bellow certain value.

At run-time, the reasoning engine determines which rules will be applied by searching for patterns in the rules that match patterns in the entered data [13,14]. This is a cyclic process that consists of three sequential steps:

- Match rules (find all of the rules that are triggered by the current context);
- Select rules (prioritizes the rules to determine the execution order);
- Execute rules (it often leads to new facts or goals being added to the knowledge base which will trigger the cycle to repeat until no new rules can be matched).

During the reasoning process, the rules can be chained together until conclusion is reached. The two basic reasoning strategies are [13,14]: forward-chaining, (data-driven), where the reasoning starts with known data or facts and looks for rules which apply to the facts until a goal is reached or backward-chaining (goal-driven), where the reasoning starts with a goal and look for rules which apply to that goal until a conclusion is reached.

The rule-based method is fairly simple and straight forward. It allows an explicit presentation of the knowledge in a clear unambiguous way that matches the human’s natural reasoning process. This approach supports inspecting and debugging of the medical knowledge to ensure completeness and accuracy of the statements, which ultimately impacts the system's reliability. Writing programming code to process these rules is relatively easy as there is a conceptual relationship between the if-then knowledge rules and if-then statements in a programming language. The rules support modularity (small pieces of information can be represented into independent statements) and flexibility (certain rules can be easily modified without major impact on the rest of the knowledge base). This is very important for our system with a dynamic knowledge base, which is constantly being updated with new discoveries. In addition, some of the rules are patient – specific and need to be modified at each doctor's visit (i.e new therapy prescribed, the ranges of the PEF zones are changed due to the child growth etc.). As such, the rules are quite popular and have been widely used in many CDSSs, such as: MYCIN, PUFF, ONCOIN, UML and Arden.

5 Applying the Rule-Based Model in PedAst

5.1 The Knowledge Base in PedAst

The knowledge base in the PedAst system consists of general knowledge applicable to all patients and individual knowledge base specific to patient individuals [15]. The need for keeping separate knowledge bases for the general patient population and for individual patients is that asthma, as many other chronic diseases, has some general characteristics applicable to the majority of the patients, but also there are many patient-specific facts that determine the management of the disease [15]. For example, it is a well known rule that if the patient experiences recurrent symptoms of wheezing, shortness of breath, chest tightness and cough, then it has to be further evaluated for asthma [15]. However, different asthma patients have different symptoms at different time of the day or year, may respond differently to irritants, such as allergens and exercise, and may react differently to specific medication [15]. While the general rules and common facts about the disease can be extracted from the medical books, web-sites and human experts, the patient's specific rules must be based on the findings relevant to that patient. These rules are usually set by a medical professional after careful examination of the patient and reviewing the patient's history or the rules can be generated automatically by the system by identifying patterns in the entered data.

The PedAst system uses relational database for encoding, storing and maintaining the medical knowledge. This is not a new concept - incorporation of production rule systems in relational database has been regularly used in active databases as well as in inductive and deductive databases, which involve rules for learning purpose [16]. However, the common characteristic for all these methods is that they include programming of the rules on a database server level, while our approach is based on treating the rules as data instead of programming code. This approach promotes flexibility, efficiency and robustness [16].

The declarative knowledge in PedAst consists of entities definitions and entities relationships, while the procedural knowledge is expressed in form of rules. Each of the entities, such as tests, results, allergies, medications, diagnosis, etc. is represented with a dedicated table, where the entities are defined with a unique code, descriptive name and attributes. The relationships among the different entities are represented with foreign key constraints. The rules are statements in the following form:

IF <condition> THEN <action> (1)

The condition on the left side represents a logical statement, which when satisfied will trigger the action on the right side. The logical condition is generally checking if the value of a given variable is within a defined range. The action part can have various outputs, such as making conclusion, generating an alert or updating the patient's medical record with a new fact. An example of a simple rule would be: "If the Peak Expiratory Flow (PEF) is in the RED zone, then generate an alert message for the patient to go straight to the emergency room". The rules are also presented in tabular form and stored in dedicated tables.

5.2 Formal Representation of the General Knowledge

A typical example from the general knowledge base in the PedAst system is the management of test results. Within the system, there is a tight connection between the following three entities: <TEST>, <RESULT> and <ACTION>. For each of these entities there is a separate table designed to define the entities and their relationships.

The <TEST> is identified with a test code, descriptive name, mnemonic (a short symbolic name), result type (numeric or alpha-numeric), minimum and maximum result. The <RESULT> is identified with a test code (as link to the corresponding test), result code, value, result type and action code, as a link to the action to be triggered when this result is received. The <ACTION> is defined with an action code and description. Each test can have many results, which represents 1:N relationship between <TEST> and <RESULT>. The result can trigger follow-up actions, i.e repeat the test, order another test, alert the doctor, etc. On the other side, the same action can be applicable to different results. This implies M:N relationship between <RESULT> and <ACTION>. Figure 1 represents the entity-relationship diagram for test results.

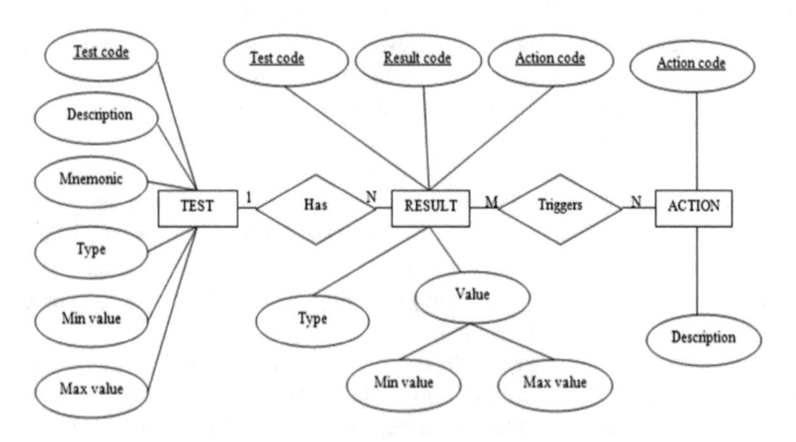

Fig. 1. Entity-relationship diagram for test results

The triple (TEST, RESULT, ACTION) is a base for generating the following rule:

$$\text{IF} \quad \langle \text{TEST}=\text{T} \quad \text{AND} \quad \text{RESULT}>=\text{R}_{min} \quad \text{AND} \quad \text{RESULT}<= \text{R}_{max}\rangle \quad \text{THEN ACTION} \tag{2}$$

where RESULT refers to the currently entered result for the given TEST, while R_{min} and R_{max} define a range associated to a specific action.

In some cases there could be correlation of the entered test result with a previous result and based on that correlation, a specific action can be triggered, such as confirm a diagnosis or perform additional tests. Another table is designed for that purpose, with the fields: test1, result1, test2, result2, action. The rule would look like this:

IF <TEST1=T_1 AND RESULT1=R_1 AND TEST2=T_2 AND RESULT2=R_2> THEN ACTION (3)

The same concept is applied to other area of the general knowledge base, such as medications, immunizations, allergies etc.

This method of configuring the knowledge, promotes system flexibility. The facts and rules that constitute the general knowledge base are actually parameters of the PedAst system as they dictate the system's functionality and allow customization of the system, without the need to modify the source code. They are entered and updated by a knowledge expert through the Admin Module, which provides an editor, as a collection of various forms with user-friendly graphic interface.

5.3 Reasoning of the General Knowledge

The reasoning engine in the PedAst system mostly uses forward-chaining (data-driven) reasoning. In case of the given example about entering test results, the reasoning consists of searching through the rules specific to test results and finding the corresponding combination (TEST, RESULT) that matches the entered data and executing the corresponding action. The ACTION will trigger execution of the corresponding piece of code, which can have various outcomes depending on the condition. For example, the following actions are possible depending on the result for the test 4420 – Immunoglobulin E (IgE), with expected results within the range 0 – 350.

- CONDITION 1: Test result within the normal range (0 – 150) =>
  ACTION 1: None
- CONDITION 2: Test result within the defined range, but higher than normal (151 – 350) =>
  ACTION 2: MSG_DR "Send message to the doctor"
- CONDITION 3: Test result out of the defined range (351 – 1000) =>
  ACTION 3: RPT_TST "Repeat the test" (suspected error and the test needs to be repeated).

The reasoning process is represented with the block – diagram as shown in Fig. 2.

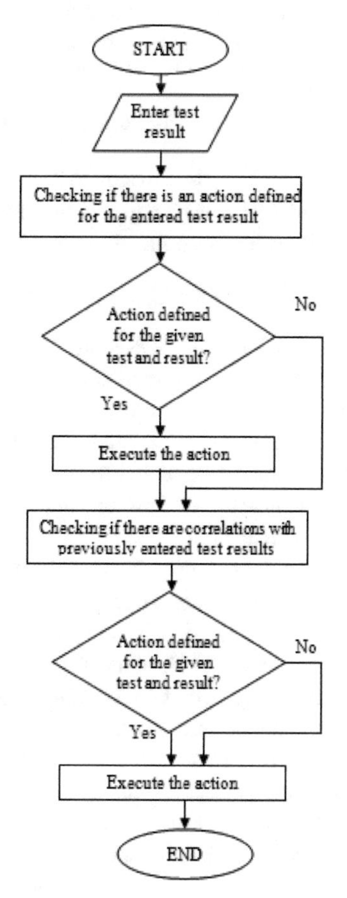

Fig. 2. Reasoning process at results entry

5.4 Formal Representation of the Individual Knowledge

The same rule-based tabular concept can be applied for modeling the patient individual knowledge, which consists of facts specific to an individual patient. As an example, we will use the Asthma Care Plan, which is a set of patient-specific rules, prescribed by a medical professional after careful examination of the patient's conditions. These are the two main components of the Asthma Care Plan:

- Peak Expiratory Flow (PEF) measurements, which include: best PEF and PEF ranges that define the green zone, yellow zone and red zone.
- Medications, which include rescue medication and dose to be taken in emergency cases, control medication and dose to be taken on regular basis for controlling the asthma symptoms and allergy medication and dose to be administered for controlling the allergies.

This information is stored in a table, where each of the facts represents a separate column and for each patient there is a separate record. The main entities in the individual Asthma Plan are: <PEF>, <MEDICATION> and <ACTION>. The formal representation of the rule for controlling the PEF reading would look like this:

$$\text{IF} <\text{PEF} >= P_{min} \text{ AND PEF} <= P_{max} > \text{THEN} <\text{ACTION}> \tag{4}$$

where PEF is the current PEF reading, while P_{min} and P_{max} are the ranges of the particular PEF zone (i.e green, yellow and red), as determined by the patient's doctor.

The rule for controlling the correct medication and dose would look like this:

$$\text{IF} <\text{MED} <> M_{pres} \text{ OR DOSE} <> D_{pres}> \text{THEN} <\text{ACTION}> \tag{5}$$

where MED and DOSE are referring to the medication currently taken, while Mpres and Dpres are the medication and dose prescribed by the doctor (i.e rescue, control, and allergy medication).

The Asthma Care plan can be revised by the patient's doctor at any time and the rules can be updated (i.e depending on the current condition, the doctor may change the medication or dose or change the PEF ranges because of the patient growth). This is a typical area of the system's knowledge base, where the rules as a method can support the required frequent updates, due to their modularity.

It is interesting to mention that some of the patient's individual rules can be automatically generated by the system by identifying patterns in the patient's data collected over time [15]. This is possible by implementing data mining algorithms, which would analyze certain variables and their change over a period of time. For example, some patients may experience worsening of their asthma symptoms during the Spring season, due to severe pollen allergies, which would be reflected on the PEF readings and the captured asthma symptoms. Background jobs can be scheduled to monitor and analyze these variables and based on the conclusions, the system can send a recommendation to the doctor to adjust the patient's therapy for the Spring season.

5.5 Reasoning of the Individual Knowledge

The asthma log is populated on a regular basis by the patient's guardians and the school nurse, where information such as patient's current symptoms, PEF readings and administered medication is entered. At this point, the entered data is matched against the prescribed asthma care plan and the appropriate action is executed. A typical scenario would be entering the current PEF reading, which initiates a search through the patient's asthma care plan to find in which of the three ranges (green, yellow or red) the current PEF reading belongs. Once the corresponding range is identified, the action associated to that range is triggered. In this situation, the rule defined with (4) in the previous text is applied. For example, if the entered PEF is in the yellow zone, which is 50% do 80% of the patient's best peak flow, the patient is advised to take bronchodilator, such as Albuterol every 4 to 6 hours. Also, the administered medication is compared with the prescribed medication and if discrepancy is noted, the system prints a warning message (rule #5).

6 Conclusion

This paper demonstrated how the rule-based methods are used for representing and processing the medical knowledge in a specific decision support system – PedAst, for managing asthma in school-age children. The model uses relational database for encoding and storing the rules. This approach enforces separation of the knowledge from the code, which provides many advantages in the process of implementing and managing knowledge [16]:

- Flexibility - changes in the rules can be made easily, without modifying the code;
- Usability - the rules can be configured by knowledge experts, with basic computer skills, through some interface for rules editing, while changing a portion of code requires programming skills and a development environment;
- Portability/Re-usability – the database with rules can be easily exported and used for other systems/platforms;
- Modularity/Extendibility -- the knowledge is organized in small and discrete units, which allows extensions to the knowledge base without forcing substantial revisions, so the knowledge base can be built incrementally;
- Explicitness – the rules allow presentation of the knowledge in a clear and precise way, which promotes a straight-forward programming.

The disadvantages of the rule-based technique are that in some cases it can be difficult for the experts to transfer their knowledge into distinct rules and many rules may be needed for the system to be effective.

Limitation of the current version of the PedAst system is that it does not have capability to handle uncertainty in the knowledge facts. Adding probability of the rules, based on which the rules can be prioritized, is considered as a direction for future enhancement of the system to reflect the uncertainty and the fuzziness of the real medical knowledge domain.

References

1. Kuan-Liang, K., Chiou-Shann, F.: A Rule-Based Clinical Decision Model to Support Interpretation of Multiple Data in Health Examinations. Journal of Medical Systems; Vol. 35 Issue 6, p1359 (2011)
2. Purcell, G.P, BMJ 330, 740 (2005).
3. Aleksovska – Stojkovska, L., Loskovska, S.: Clinical Decision Support Systems - Medical Knowledge Acquisition and Representation Methods. In: IEEE International Conference on Electro/Information Technology, pp. 1-6, Normal, IL, (2010)
4. Gamberger, D., Prcela, M., Jovic, A., Smuc, T., Parati, G., Valentini, M. at all.: Medical knowledge representation within Heartfaid platform. In: Proc. of Biostec Int. Joint Conference on Biomedical Engineering Systems and Technologies, pp.205-217 (2008)
5. Verlaene, K., Joosen, W., Verbaeten, P.: Arriclides: An Architecture Integrating Clinical Decision Support Models. In: 40th Annual Hawaii International Conference on System Sciences (HICSS'07), pp.135c, (2007)
6. Aikins, J. S., Kunz, J. C., Shortliffc, E. H., and Fallat, K. J.: PUFF: An Expert System for Interpretation of Pulmonary Function Data. In: Comput Biomed Res. 16(3), pp.199-208, (1983)
7. Aikins, J. S.: Prototypes and Production Rules: an Approach to Knowledge Representation for Hypothesis Formation. In: Proc 6th IJCAI, pp.1-3 (1979)
8. Kuilboera, M. M., AM van Wijka, M., Mossevelda, M., van der Doesb, E. at all.: Feasibility of AsthmaCritic, a decision-support system for asthma and COPD which generates patient-specific feedback on routinely recorded data in general practice. In: Family Practice Vol. 19, No. 5, pp. 442-447 © Oxford University Press (2002)
9. Bi, J., Abraham, A.: An intelligent web-based decision support tool for enhancing asthma guideline adherence. In: ACM IHI'12, pp. 51-60, Miami, Florida (2012)
10. Rigopoulou, A. V., Anthracopoulos, M.B., Katsardis, C. V., Lymberopoulos, D.K.: RespDoc: A new Clinical Decision Support System for Childhood Asthma Management based on Fraction of Exhaled Nitric Oxide (FeNO) Measurements. In: 35th Annual International Conference of the IEEE EMBS, pp. 1306-1309, Osaka, Japan (2013)
11. Dexheimer, J., Arnold, D., Abramo, T., Aronsky, D.: Development of an Asthma Management System in a Pediatric Emergency Department. In Proc of AMIA Annual Symposium, pp. 142-146 (2009)
12. Aleksovska – Stojkovska, L., Loskovska, S.: Architectural and data model of clinical decision support system for managing asthma in school-aged children. In: IEEE EIT, pp.1-6, Mankato, MN (2011)
13. Dennis Merritt, G.: Best Practices for Rule-Based Application Development, http://msdn.microsoft.com/en-s/library/aa480020.aspx, (Accessed: 15 Jan 2014)
14. Kong, G., Xu, D. L., Yang, J.B. The University of Manchester: Clinical Decision Support Systems: A Review On Knowledge Representation And Inference Under Uncertainties. In: International Journal of Computational Intelligence Systems,vol.1, no.2, pp. 159-167 (2008)
15. Aleksovska – Stojkovska, L., Loskovska, S.: Data Mining in Clinical Decision Support Systems. In: DEIT 2011, Indonesia, Recent Progress in Data Engineering and Internet Technology Lecture Notes in Electrical Engineering Volume 156, 2013, pp 287-293
16. Sawar, M.J., Abdullah, U., Ahmed.: Enhanced Design of a Rule Based Engine Implemented using Structured Query Language. In: Proceedings of the World Congress on Engineering, Vol I WCE 2010, London, U.K. (2010)

ECG for Everybody: Mobile Based Telemedical Healthcare System

Stevan Jokic[1,2], Ivan Jokic[1], Srdjan Krco[2], Vlado Delic[1]

[1] Faculty of technical sciences, University of Novi Sad, Serbia
{stevan.jokic,ivan.jokic,vdelic}@uns.ac.rs
[2] DunavNET, Serbia
srdjan.krco@dunavnet.eu

Abstract. This paper presents telemedical system entitled ECG for Everybody. ECG for Everybody is a mobile based telemedical healthcare system for heart health check. Key components of the system are: ECG device, mobile applications, web platform with portals. Main system capabilities are ECG acquisition, ECG signal review service by physician involved in the system, automated analysis, transmission to the server for permanent storing and remote access.

An attractive mobile application is implemented for heartbeat detection and HRV analysis using only mobile device camera (PPG approach, no additional device is needed).

ECG signal simulator (generator) is designed as a mobile application and a very affordable device which transforms signal generated by the mobile app to the standard ECG signal which can be further captured by any ECG device. ECG simulator is an affordable solution for various ECG signal visualizations on the real ECG equipment. Education and equipment testing are just some of possible applications of the ECG signal simulator solution.

Keywords: ECG, telemedicine, healthcare, ECG signal HRV analysis, Photplethysmgram, ECG simulator.

1 Introduction

Technology progress of the mobile telecommunication systems provides a fertile field for design and implementation of new modern healthcare telemedical systems. Mobile telecommunication improvements are attractive for design and implementation of modern healthcare telemedical systems for two reasons. The first reason is introduction of a high speed Internet access protocols and the second is mobile device performance improvements. Broadband Internet access in mobile phones very obviously improves client-server healthcare telemedical systems by providing faster and cheaper access to the system services. Progress in mobile devices is manifested through increased portion of smart phones and improved mobile devices hardware. Hardware capabilities like x GHz CPU, x GB of RAM memory, in a very recent history were reserved only for a PC configurations, nowadays, mobile devices are equipped with a

S. Loshkovska and S. Koceski (eds.), *ICT Innovations 2015*,
Advances in Intelligent Systems and Computing 399,
DOI: 10.1007/978-3-319-25733-4_10

high speed dual/quad core processors, GB of RAM, increased screen resolution and diagonal size. All this mobile device improvements allow implementation of high autonomy healthcare systems, closed in the mobile device of the user.

Typical client-server architecture design paradigm use mobile devices for data collection and transmission only. Data processing and analysis are executed on the server side and online available to physician. For an end user of the telemedical system this approach is not efficient. Much more attractive approach is to implement data analysis in the mobile device and provide results in nearly real-time to the end user. This is a new approach in design and implementation of healthcare telemedical systems which is more attractive to end users. A whole healthcare process can be initiated and closed on the user side. This approach is user centric, thus any user may initiate a whole healthcare process from his mobile device and receive analysis results anywhere he is and at any time.

Benefits of usage telemedical systems are well presented [1]. Telemedical systems offer remote patient's health monitoring, eliminates distance barrier of rural areas. Several papers offer exhaustive review of existing telemedical systems [1, 2], as well as review of earlier reviews [3]. Reviewers usually analyse a large set of papers from the appropriate topics and focus on the one hundred of the implied papers [3]. Common conclusions of existing telemedical systems from reviews are:

- Technical efficiency in remote patient monitoring
- Easy to use, for physicians, as well as for end users without medical knowledge
- Economic feasibility of the TS building
- Reduced staying in the hospitals
- Improved patient compliance, satisfaction and quality of life.

Further conclusions from the review paper are that a lot of presented papers consider only technical aspects of system design without long term user evaluation. Thus, paper [2] present results of 66 papers which consider end user acceptance. Results are not as positive as we may expect and they show that 56% of the users (nearly half only) emphasize benefits of the telemedical system, 36% of them report big flaws of telemedical systems and 8% of them give advantage to the traditional treatment.

2 Motivation

Cardiovascular diseases cause more than 15 million deaths in the world each year, according to the World Health Organization (WHO; Geneva). From the CDC site [4] America's Heart Disease Burden facts are:

- About 600,000 people die of heart disease in the United States every year–that's 1 in every 4 deaths.
- Heart disease is the leading cause of death for both men and women. More than half of the deaths due to heart disease in 2009 were in men.
- Coronary heart disease is the most common type of heart disease, killing nearly 380,000 people annually.

- Every year about 720,000 Americans have a heart attack. Of these, 515,000 are a first heart attack and 205,000 happen in people who have already had a heart attack.
- Coronary heart disease alone costs the United States $108.9 [4] / $312.6 [5] billion each year. This total includes the cost of health care services, medications, and lost productivity.
- Cardiovascular disease (CVD) causes more than half of all deaths across the European Region.
- CVD causes 46 times the number of deaths and 11 times the disease burden caused by AIDS, tuberculosis and malaria combined in Europe.
- 80% of premature heart disease and stroke is preventable.

In a 2005 survey, most respondents—92%—recognized chest pain as a symptom of a heart attack. Only 27% were aware of all major symptoms and knew to call 911 when someone was having a heart attack. About 47% of sudden cardiac deaths occur outside a hospital. This suggests that many people with heart disease don't act on early warning signs. Recognizing the warning signs and symptoms of a heart attack is key to prevent fatal outcome.

Early Action is Key to reduce this numbers [4], but conventional healthcare is not adapted to the new societal expectations where everything is expected to happen in a speed of click. Thus, our system is user centric and we are using mobile device of the user as a proxy for the healthcare access which user can very easy initiate and receive results nearly real time. Our system can significantly reduce coronary heart disease fact numbers and increase life quality by early detection of symptoms associated with a serious heart disease.

3 Related Work

Technology for telemedicine exists, for more than 10 years, however, there is not too many widely accepted commercial telemedicine systems. In this section are presented similar healthcare solutions.

One of the competitors is AiveCore [6]. They are oriented to the ECG device manufacturing. Device requires mobile device and is manufactured especially for specific cases of iPhones and Android phones. They provide and service for physician review of the arrived signals. Our benefits are that our ECG device can operate and with electrodes and thus it is applicable for long period monitoring, AiveCore has only contacts for fingers. AiveCore also cannot be applied for different ECG channels like our device which has additional contacts. Price of the AiveCore is about 160$ which is much higher than our device. Our web portal is not closed only for our devices, we offers and forum based portal for users with recordings from other devices.

ECG Check [7] is a solution very similar to the AiveCore, but less complete and without server platform. It's oriented for custom made telemedicine systems.

Another available solution is eMotion ECG Mobile [8]. This solution requires mobile device and does not provides web services for their users. Device operates only using standard electrodes and captures single channel of the ECG. This solution is less complete than ECG for Everybody and device is less flexible for end users and more expensive.

ECG pen [9] is elegant ECG device which looks like a big pen and can capture one channel of the ECG. Signals are recorded on the device and by USB they should be downloaded to the PC and displayed. We think that this couldn't be more complicated and less efficient to the end users. There is no web service support for ECG pen customers.

Life Watch V [10] is Android based mobile phone with embedded healthcare sensors and applications. Users have call service 24/7, from the site it looks like physician support is optional and this facility should be implemented by third party.

Summarized comparison is presented on the following table. Comparison covers several aspects like: does device manufacturer provides and services or platform for their users, is it possible to capture ECG without electrodes, ECG channel number, price. Without platform, devices are not suitable for individual usage or for small institutions which cannot develop their own platform.

CorScience [11] provides ECG device which uses electrodes to capture ECG and does not provides a mobile software or platform to end users.

One of big advantages of existing devices is certifications which they claim. Some of mentioned ECG devices are FDA certified. ECG for Everybody offers open, comprehensive, affordable approach for ECG healthcare. Recordings from another devices can be uploaded in the ECG for Everybody system. Very affordable ECG device solution is provided without competition in the price.

Table 1. ECG device/system comparison

	Platform	Physician	Use Electrodes	Without electrodes	Channel number	Certified	Price [€]
Aive-Core	YES	YES	NO	YES	1	YES	200
ECG Check	NO	NO	YES	NO	1	NO	103
eMotion ECG Mobile	NO	NO	YES	NO	1	NO	890
ECG pen	NO	NO	NO	YES	1	NO	200
Life Watch	NO	YES	NO	YES	1	YES	400
COR-Science Bluetooth ECG	NO	NO	YES	NO	2	YES	1000
ECG for Everybody	**YES**	**YES**	**YES**	**YES**	**1/2**	**NO**	**< 10 manufacturing**

4 Project Details

In this section are presented details about architecture and ECG for Everybody [12] system components.

4.1 Affordable ECG Device

The first version of the ECG devices uses three electrodes and sends baseband ECG signal to the mobile device audio interface. QRS complex morphology can be detected using this approach but significant part of the signal is reduced by audio filters in the mobile device.

Later versions of the ECG device introduce ECG capturing from thumbs (without standard ECG electrodes placing on the body of user) and signal adaptation for the standard audio interface like it is available, for example, on the mobile devices. By using the signal adaptation for the standard audio interface in the mobile application it is possible to receive ECG signal with all significant spectral components. ECG capturing without placing is very attractive for end users and fast heart health checks. Of course, signal captured using electrodes is less prone to artifacts caused by user movements, but with carefully usage signal from thumbs can achieve the same quality as signal captured with standard electrode placed on the user chest. On the Fig. 1 is displayed ECG device usage. ECG capturing with electrodes is useful for a long period of recording.

The latest version of the ECG device offer capturing of six channels of the standard ECG: I, II, III, aVR, aVF, aVL channels. This ECG device version can operate also without electrodes by placing thumbs on the appropriate contacts and by contacting point above right leg of the user.

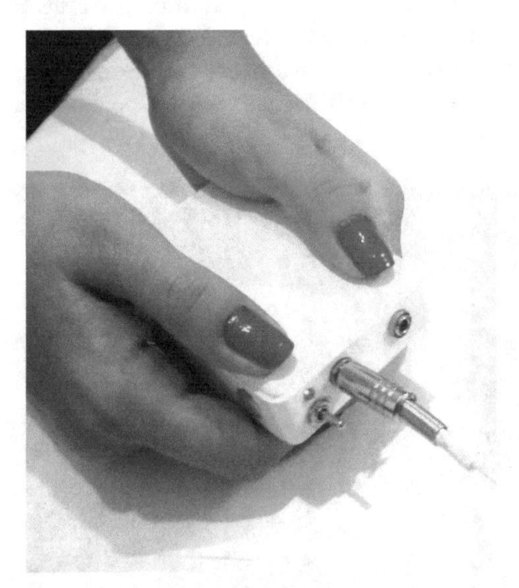

Fig. 1. Affordable ECG device in usage

Typical approach for ECG device building them from scratch as a small computers, use AD converter, controller, memory, communication module (usually for communication with mobile device), different memories etc. This approach provides a complex device which acts as a small computer able to capture ECG signal. If we look in the mobile device hardware we will see that there are all components mentioned in the hypothetic ECG device design approach. A lot of resources available in the mobile device are not used or doubled by the typical ECG device design approach. Mobile device resources like CPU, memories etc. are much better than in the embedded solutions.

Our solution uses mobile devices resources as much as it is possible starting from the audio input interface. ECG for everybody device has analog interface for ECG capturing and module for signal adaptation to the standard audio interfaces. ECG signal is adapted for the audio interface and starting from AD conversion, further processing is performed by a very rich set of mobile device resources. ECG device autonomy on battery is much better than for hypothetically ECG devices with the standard a small computer based architecture. ECG for Everybody ECG device uses only general-purpose components like a standard operational amplifiers, resistors and capacitors available in every electronic shop for a very low price. Manually manufacturing price of the proposed ECG device is less than 10$ including battery, connectors and the device case. Mass production can decrease this price and result with a final device price affordable for everybody.

4.2 Mobile Application

In the ECG for Everybody system mobile application performs ECG capturing, ECG displaying, real time ECG analysis, ECG recording locally on the user mobile device, user reporting about analysis results, interfaces for email messages sending with attached images of the ECG as well as interface for communication with physician involved in the system.

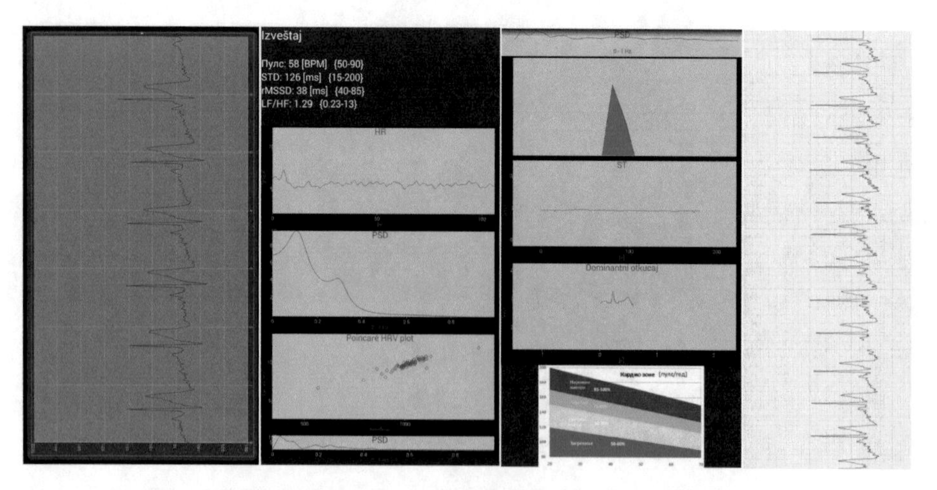

Fig. 2. ECG display in the mobile application and generated report

Real time ECG analysis covers pathologically heartbeats detection, ST level analysis and HRV analysis. Pathologically heartbeat analysis is based on use of efficient algorithms based on polynomial function modeling [13]. HRV analysis estimates several standard medical time domain statistical: heart rate, STD (Standard deviation), rMMSD (Root Mean Square of the Successive Differences), spectral: AR (Autoregressive model) estimation, LF/HF (Low/High Frequencies) ratio, FFT) and geometrical features (Poincaré plot, HRV histogram – HRV triangular index) [14].

User activity is analyzed by acceleration sensor processing. This information is useful in ECG signal analysis. During a signal processing, mobile application displays a various health messages and advices. On the Fig. 2 are displayed several screens from the mobile application like a real-time ECG signal displaying and analysis results. All screens and analysis results can be sent as images using email messages.

From the mobile application, ECG signals can be uploaded to the ECG for Everybody web platform described below.

4.3 Heartbeat Detection Using Only Mobile Device Camera

A very attractive solution for heart rate analysis which does not requires any additional device is designed using only camera and flash on the mobile device PPG (Photoplethysmogram). Mobile application detects heartbeats by detecting small changes in skin captured using in device embed camera. Heart rate analysis may covers heart rate estimation and stress level estimation as well as cardio training guide. Several methods for stress estimation using heart rate signal are well known in research papers. Cardio training becomes more and more popular and this application guides user through training by matching target heart rate to reach desirable cardio zone, for example instructions through the "fat burning training" for practically user. By using elastic strap mobile device can be placed on the user arm and thus provide flexibility to practicing, for example running and performs continuous HR analysis. For cardio training exercise timing is very important, thus application guide user through practicing. For example, if user target practicing is fat burning then application guide user to keep appropriate heart rate for specified period of time, regarding to achieve defined goal. Application explains that in the first period of the training body will use more glucose, thus it's important to user follow application guide for training.

4.4 ECG Simulator

ECG simulator is realized as a mobile application and appropriate hardware. Mobile application provides interface for signal selecting. Simulated ECG signal may originate from various ECG data bases. ECG signal is generated using audio interface of the mobile device and by designed hardware is adapted for standard ECG device inputs. Hardware for adaptation is embedded in the connectors of the cable, thus cable looks like standard 3.5mm cable. This ECG simulator is useful for testing but it can be used in medical school for attractive presentation of real ECG signals on the standard ECG equipment. In the real conditions it is hard to detect some pathologic sequences because they occur very rarely, thus ECG simulator can be used in education for manifesting of rarely ECG shapes in the more realistic environment like a real ECG devices.

4.5 Web Platform

Web platform is implemented as an application the Google App Engine cloud platform [15]. Several web services are available to end users. Facilities of web part of the ECG for Everybody system are records storing, remotely access, data visualization (ECG signal and reports), physician access to records which requires physician opinion, forum based portal with reviewed user messages by physician.

Access to the forum based web portal is open for all users and they can post their records, possible captured and by another device like scanned images. User comments are reviewed and rated by physicians. Aspects of this crowdsourcing web application as well as several use-cases in which are going to be used facilities of the proposed telemedicine system will be considered in FP7 project IoT Lab funded by EU [16]. IoT Lab is a European Research project which aims at researching the potential of crowdsourcing. ECG for Everybody can be part of IoT Lab use case scenarios to help researcher to collect data from users. Collected data may be useful for researcher to validate different algorithms for ECG analysis, for example stress estimation based on the HRV analysis. Students of medicine can be frequently forum users where they can find different signals and check their knowledge or learn something new.

5 System Users

Our users are institutional and individual. Our system provides physician ECG review service and commercial platform user account subscription. Existing health centers with our system improve communication with patients, provide recovery monitoring to their patients, shorter hospitalization with appropriate reporting. Medical workers during visiting patients can communicate with physician in the health center using our system. Medical institution does not need our ECG review service, but they do need a platform for user record managment.

Sport/fitness institutions, retirement homes use combination of platform subscription and individual heart health check requests. Sport institution can check health or relaxation of players and regarding to results plan further trainings. Sevilla player Antonio Puerta (22) died on 28 August as a result of a weakness of the right ventricle, which is a common cause of sudden cardiac arrest among athletes. This young lives loosing should be avoided [17]. Elderly people very often live alone and far from big medical centers, ECG for Everybode can increase elderly people safety feeling and reduce solicitude of their younger family members who are not able to be near their senior family members.

Regarding to a very affordable price of our components, our users may be and medical educative institutions. Affordable ECG device and ECG simulator provide opportunity to visualize various ECG signals in much more realistic situation than on the book image or on the blackboard.

We should keep in mind that heart diseases are the most dangerous for people around fifties, not for very old people, thus people in their fifties years are our target group for individual users. This is positive because they are still very good tech equipped and able to use our services. Modern trend of living does not leave time for opportunely health checking, thus this segment of users need a fast, attractive to use heart health check service.

First users of the system are medical high school and technically university.

6 Conclusion

In this work is presented complete telemedical solution for fast, easy to use, affordable heart health check. System architecture is user centric, thus user initiate a whole process of heart health check and receive analysis using mobile device. Heart health analysis results may be obtained by automatic analysis implemented already in the user's mobile device, or by our physician involved in the system or using crowdsourcing mechanisms available on the platform. User may initiate processing at anytime from anywhere he or she is. ECG for Everybody provides affordable, attractive to use ECG device, mobile application for ECG capturing, real-time processing and visualization as well as a web platform for ECG records storage and remote access.

Further work will be oriented to the implementation finalization and platform improvement through collaboration with research/industry institutions, applying to different open calls for projects and startup accelerators as well as popularization of the platform concept and developed solutions in the system ECG for Everybody.

Our mission is to save hearts for nice fillings, to save our loved ones.

Acknowledgements. The presented study is performed as part of the project "Development of dialogue systems for Serbian and other South Slavic languages" (TR32035), funded by the Ministry of Education, Science and Technological development of the Republic of Serbia.

References

1. Martínez, A., Everss, E., Rojo-Álvarez, J.L., Pascual Figal, D., García-Alberola, A.: A systematic review of the literature on home monitoring for patients with heart failure. J Telemed Telecare Jul 01, 2006; 12: 234-241 (2006)
2. Hailey, D., Roine, R., Ohinmaa, A.: Systematic review of evidence for the benefits of telemedicine. J Telemed Telecare Mar 01, 2002; 8: 1-7.G. Young, "Book style with paper title and editor," in Plastics, 2nd ed. vol. 3, J. Peters, Ed. New York: McGraw-Hill, 1964, pp. 1–9 (2002)
3. Ekeland AG, Bowes A, Flottorp S.:Effectiveness of telemedicine: a systematic review of reviews. International Journal of Medical Informatics, 2010 Nov;79(11):736-71 (2010)
4. Centers for Disease Control and Prevention, http://www.cdc.gov/heartdisease/facts.htm, http://www.cdc.gov/dhdsp/data_statistics/fact_sheets/fs_heart_disease.htm
5. Heart-Healthy and Delicious, http://millionhearts.hhs.gov/abouthds/cost-consequences.html
6. AliveCor http://www.alivecor.com/home
7. ECG Check http://www.ecgcheck.com/
8. eMotion ECG http://www.megaemg.com/products/emotion-ecg/
9. HeartCheck™ PEN, http://www.theheartcheck.com/Portable-EKG-Machines.htm

10. Life Watch V, http://www.lifewatchv.com/
11. Cor Science, BT3/6: Bluetooth resting ECG device, http://www.corscience.de/en/medical-engineering/products/ecg/bluetooth-ecg-device.html
12. ECG for Everybody, http://ecg4everybody.com
13. Jokic, S., Delic, V., Peric, Z., Krco, S., Sakac, D.: Efficient ECG Modeling using Polynomial Function. Electronics and Electrical Engineering, No. 4(110). – P. 121-124. Kaunas: Technologija (2011)
14. Task Force of the European Society of Cardiology and the North American Society of Pacing and Electrophysiology . 1996. Heart rate variability: standards of measurement, physiological interpretation and clinical use. Circulation 93:1043–1065.
15. Google App Engine, https://appengine.google.com/
16. IoT Lab, European Research project, www.iotlab.eu/
17. Fifa to discuss cardiac testing , http://news.bbc.co.uk/sport2/hi/football/africa/6972281.stm

Proposal and Experimental Evaluation of Fall Detection Solution Based on Wearable and Depth Data Fusion

Samuele Gasparrini[1], Enea Cippitelli[1], Ennio Gambi[1], Susanna Spinsante[1], Jonas Wåhslén[2], Ibrahim Orhan[2], and Thomas Lindh[2]

[1] Dipartimento di Ingegneria dell'Informazione, Universita' Politecnica delle Marche, Ancona, Italy I-60131
{s.gasparrini, e.cippitelli, e.gambi, s.spinsante}@univpm.it

[2] School of Technology and Health, KTH, Royal Institute of Technology, Stockholm, Sweden SE-100 44
{jonas.wahslen, ibrahim.orhan, thomas.lindh}@sth.kth.se

Abstract. Fall injury issues represent a serious problem for elderly in our society. These people want to live in their home as long as possible and technology can improve their security and independence. In this work we study the joint use of a camera based system and wearable devices, in the so called data fusion approach, to design a fall detection solution. The synchronization issues between the heterogeneous data provided by the devices are properly treated, and three different fall detection algorithms are implemented. Experimental results are also provided, to compare the proposed solutions.

Keywords: Depth camera, Inertial sensor, Data fusion, Synchronization, Fall detection

1 Introduction

Fall is defined by the World Health Organization as *an event which results in a person coming to rest inadvertently on the ground or floor or other lower level* [1]. This problem affects particularly the aged population and, as stated in [2], approximately 28-35% of people aged 65 and over fall each year, increasing to 32-42% for those aged over 70 years of age. These numbers are confirmed also in EU28 and EEA countries, where approximately 100000 older people die from injury due to a fall each year [3].

The direct consequences correlated to a fall could be: superficial cuts, broken or fractured bones, and abrasions or tissue damage. Also the "*long-lie*" condition, defined as involuntarily remaining on the ground for an hour or more, following a fall, represents a serious risk for the health. As stated in [4], half of elderly people who experience a "*long-lie*" die within 6 months. Taking into account all these aspects, a reliable and secure system to monitor an elderly during his daily life is strongly recommended. It must ensure an adequate robustness against false alarms, and be unobtrusive at the same time.

S. Loshkovska and S. Koceski (eds.), *ICT Innovations 2015*,
Advances in Intelligent Systems and Computing 399,
DOI: 10.1007/978-3-319-25733-4_11

In the literature, the initially proposed solutions tried to use wearable devices to solve this task. In [4], tri-axial accelerometers are placed on the trunk and the thigh of 10 volunteers that perform ADLs (Activities of Daily Living), and simulate falls. Kangas *et al.* [5] used a tri-axial accelerometer attached to belt at the waist, involving also elderly people in the ADLs subset of a test campaign. An alternative research approach uses cameras as a source of information to detect risky activity [6].

Recently, the availability of cheap depth sensors, has enabled an improvement of the robustness in camera based approaches for fall detection solutions. In particular, the Kinect sensor, i.e. the RGB-Depth device used in this publication, has been adopted in different implementations, as presented in [7] and [8].

In the last years, thanks to the growth of computational resources, the combination of the previous solutions became possible and this led to an improvement of the performance. These solutions exploit an approach defined as "data fusion", and examples of joint use of Kinect sensor and wearable devices are visible in [9]. The synchronization issues between Kinect and wearable devices, to the best of our knowledge, is not totally covered in the literature. In view of this fact, we use the synchronization approach described in [10] to design fall detection systems that exploit heterogeneous data provided by different sensors. It is also worth noting that we started creating a database of ADLs and falls, containing visual and acceleration data, that can be exploited to compare different solutions [11].

The remaining part of this paper is organized as follows. In Section 2 the synchronization approach is presented. Section 3 describes the proposed fall detection solution. Experimental results are discussed in Sect. 4, while Sect. 5 is dedicated to concluding remarks.

2 Synchronization

The synchronization issue between a wearable inertial device and a vision based device, namely the Microsoft Kinect sensor, has been addressed in [10]. In this work, the transmission and exposure times of the frames captured by Kinect are exploited to synchronize the RGB-D sensor with two inertial measurement units (IMU) from Shimmer Research. Figure 1 shows the devices involved in the synchronization process. An ad-hoc acquisition software allows to simultaneously capture data from Kinect, connected via USB cable, and from the accelerometers, linked via Bluetooth to the same PC, running the acquisition software. The same software applies a timestamp when each packet, or frame, arrives at the PC, using the $QueryPerformanceCounter$ and $QueryPerformanceFrequency$ C++ functions. The synchronization is realized by exploiting these timestamps, taking into account the transmission times of Kinect frames and any possible delays caused by the Bluetooth protocol. Figure 2a shows, in red, the sequence of skeleton samples provided by the Kinect sensor, while the green and blue lines represent the packets sent by the accelerometers. As visible, the number of packets received from each accelerometer, is much greater than the number of frames captured by Kinect, because the sampling rate of the Shimmer is 10 ms

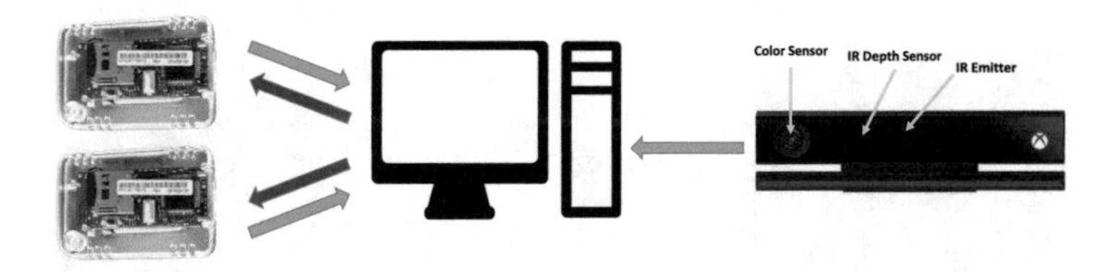

Fig. 1. The synchronization scheme involves a Microsoft Kinect v2 sensor, connected via USB cable to a PC, and two accelerometers onboard the wearable devices, linked via Bluetooth to the same PC.

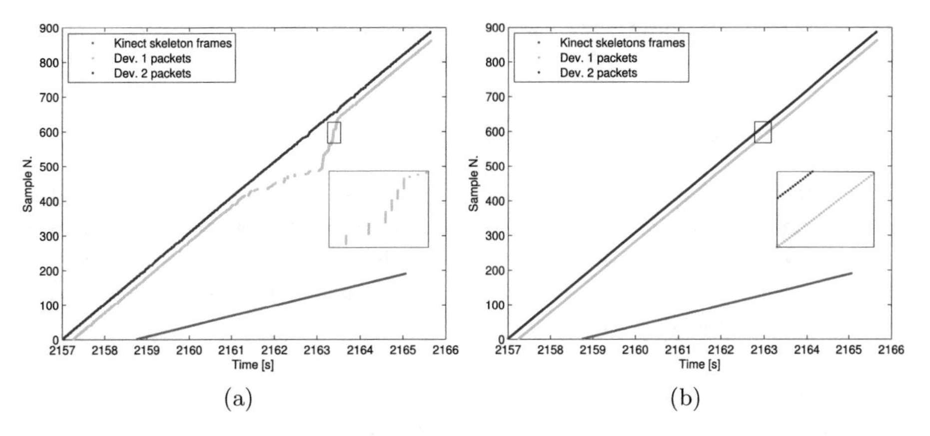

Fig. 2. (a) Raw sample time correlated to Skeleton frames and packets generated by the two accelerometers, (b) same curves after linearization.

while Kinect outputs data approximately every 33 ms. The rectangle contains a zoomed interval of the data sent by device 1. The nonlinear trend is caused by the behaviour of the Bluetooth protocol, as highlighted in [12]. Indeed, each packet arrives at the PC with a variable delay that must be corrected to enable the synchronization with the Kinect data. Using a linear regression algorithm, the wearable device curves are linearized, and the result is shown in Fig. 2b. The zoomed area shows that delays between subsequent packets have been corrected.

The aim of the synchronization process is to associate one acceleration sample to each Kinect skeleton frame. Thus, after having linearized the samples from accelerometers, the following operations have to be done:

- synchronization of skeleton and depth frames captured by Kinect, by using timestamps provided by the Microsoft SDK 2.0;
- compensation of the transmission time of the skeleton frame, which is the same as the depth frame;
- association of the closest in time acceleration sample to each skeleton frame.

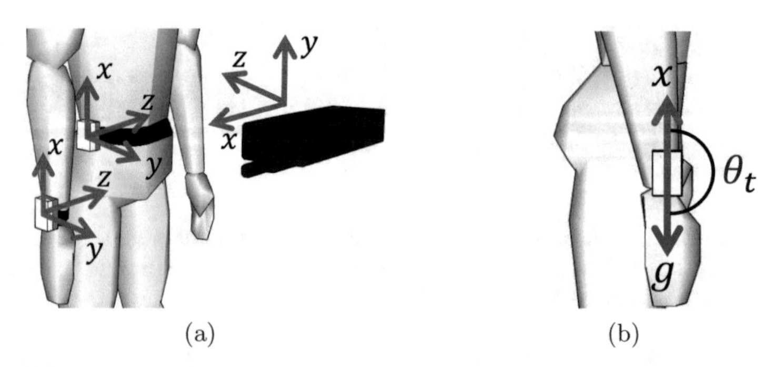

Fig. 3. (a) Setup of Kinect v2 and wearable sensors, (b) orientation of accelerometers angle.

3 Fall Detection

The synchronization algorithm summarized in the previous section is used to perform data fusion in a fall detection solution. It is possible to use Kinect and acceleration data to design reliable fall detection algorithms. The idea is to propose different algorithms, that can compute different parameters, and to evaluate their performances.

3.1 System Setup

The system setup includes two IMUs, mounted on the wrist and on the waist of the subject, and a Microsoft Kinect v2 sensor, placed as shown in Fig. 3a. A Shimmer device is placed to the right side of the body, constrained to the waist using a belt, since Kepski *et al.* [9] recommend to place the sensor to the trunk or lower back. Another accelerometer is placed to the right wrist, to simulate a smartwatch. The Kinect sensor monitors the test area, and it is positioned at about 1.5 meters from the floor and 2 meters from the person.

3.2 Acceleration Data Processing

The Shimmer device integrates the Freescale MMA7260QT accelerometer that provides 3-axis raw accelerations data. They are converted into gravity accelerations (X, Y, Z), taking into account possible biases. The acceleration magnitude is:

$$M_{acc} = \sqrt{X^2 + Y^2 + Z^2} \quad (1)$$

and the angle θ_t between the X axis and the g vector (Fig. 3b) is given, as defined in [13], by:

$$\theta_t = atan2\left(\sqrt{Z^2 + Y^2}, X\right) \quad (2)$$

Normally when the person is standing, with arms parallel to the body, the angle θ_t measured by both the accelerometers is equal to 180°. When the person lies down on the floor this angle should be 90°.

3.3 Algorithms 1 and 2

The first implemented solution for fall detection exploits acceleration data from the wrist IMU and skeleton information from Kinect. In particular, the following information are considered:

- variation in the skeleton joint position;
- M_{acc} of the wrist accelerometer;
- θ_t angle of the wrist accelerometer, after the extraction of the gravity acceleration component.

The first parameter, i.e. the variation of a skeleton joint position, considers the socalled $SPINE\_BASE$ joint (J_{SPB}), located at the base of the spine [14]. As visible in Fig. 3a, the y axis represents the vertical component in the reference system of the skeleton, and it can be monitored to evaluate any movements referable to a fall. In the first captured frames, the reference y value of the J_{SPB} joint is evaluated and then, if the difference between the actual value and the reference one exceeds a threshold of 50 cm, an irregular activity is detected. The second considered parameter is the magnitude of acceleration, revealed by the wrist IMU. In this case, an acceleration peak greater than 3g, as suggested by [15], has to be found within a time interval of two seconds, centered in the time instant where the irregular activity of the skeleton has been identified. The third parameter is represented by the orientation of the sensor. In order to detect a fall, the angle θ_t should be around 90° for a not negligible amount of time. In the proposed implementation, a threshold value of 90°, with a guard interval of 20°, for at least 0.5 s, has been considered. If all the parameters satisfy the chosen conditions, the action is classified as a fall.

The Algorithm 2 computes the same parameters as Algorithm 1, but it exploits data from the accelerometer placed on the waist of the subject.

3.4 Algorithm 3

The third implemented solution avoids using the orientation of the accelerometers, and exploits the following parameters:

- variation in the skeleton joint position;
- distance of the J_{SPB} joint from the floor;
- M_{acc} of the waist accelerometer.

The parameter that indicates an irregular activity is the remarkable change in the y component of the J_{SPB} joint. Then, the distance of that joint from the floor is also evaluated. The floor is modeled as a plane, which is automatically detected from the first available skeleton frame. Given the plane equation:

$$ax + by + cz + d = 0 \tag{3}$$

the constant term d is computed using the following equation:

$$d = -(ax_0 + by_0 + cz_0) \tag{4}$$

where $v_n = [a, b, c]$ is the orthogonal vector to the plane, and $P_0 = [x_0, y_0, z_0]$ is one point in the plane. In the proposed approach, the vector v_n is evaluated as the vector that models the spine, assuming that the subject is standing at the beginning of the test, that is when the plane is identified. Considering two vectors that identify two joints of the spine, namely the J_{SPB} and the $SPINE\_MID$ joint (J_{SPM}), the following equation is used to find the normal to the floor vector:

$$v_n = \frac{J_{SPM} - J_{SPB}}{||J_{SPM} - J_{SPB}||} \tag{5}$$

while the point belonging to the plane is one of the ankle joints of the subject. The distance $dist_{SPB}$ between the SPB joint and the floor is evaluated using the following equation:

$$dist_{SPB} = \frac{|v_n \cdot J_{SPB} + d|}{||v_n||} \tag{6}$$

When the distance $dist_{SPB}$ decreases below a threshold value (20 cm), the algorithm evaluates the time instant and selects a time window of 2 seconds in the M_{acc} trajectory, looking for an acceleration peak greater than 3g. If also this peak is found, the action is classified as a fall.

4 Results and Discussion

The designed algorithms have been tested in a laboratory environment, on a database composed by 11 healthy volunteers. The people involved in the test are aged between 22 and 39, with different height (1.62-1.97 m) and build. The actions that populate the dataset are separated in two main groups: ADLs and Fall. Each activity is repeated three times by each subject involved. The whole database, containing 264 different actions for a total number of 46k skeleton and 230k acceleration values, is available at [11]. The proposed algorithms are implemented in MATLAB and they have a very low complexity. The time required to process a sequence of skeleton and acceleration data goes from 2 to 6 ms, having sequences with durations in the interval 2.5-15 s. The detailed experimental protocol is provided in Table 1. Results are evaluated, as defined in [5], in terms of sensitivity, specificity and accuracy.

Table 2 shows the accuracy evaluated over the entire dataset, for the three considered algorithms. Algorithm 1 is the less invasive one because it simply relies on the accelerometer placed on the wrists and, despite it shows a specificity of 98%, it is characterized by a sensitivity of 59%, which means that a quite poor set of falls are correctly detected. Looking at Table 2, it can be noticed that the most difficult fall to detect is the *side* one, featured by an accuracy of 48% while the highest accuracy (82%) is reached by the backward fall that ends up lying, labelled as *back*. The weakness of Algorithm 1 is represented by the orientation of the accelerometer. In fact, even if the person is fallen and he/she is lying, the arm could be not parallel to the floor, thus avoiding the detection of the

Table 1. Experimental protocol

Category	Activity	Description
ADL	*sit*	The subject sits on a chair
	grasp	The subject walks and grasps an object from the floor
	walk	The subject walks back and forth
	lay	The subject lies down on the mattress
Fall	*front*	The subject falls from the front and ends up lying
	back	The subject falls backward and ends up lying
	side	The subject falls to the side and ends up lying
	EUpSit	The subject falls backward and ends up sitting

fall. In order to have better performance, it is necessary to use the accelerometer placed on the waist, which provides a more reliable information about the orientation of the subject's body. The sensitivity and specificity of Algorithm 2 reach respectively the percentage of 79% and 100%; looking more in detail at Table 2, it gives an accuracy of 100% for each test, except the *EUpSit* fall test. In this specific case, the orientation of the accelerometer does not give values below the chosen threshold, because the torso remains perpendicular to the floor in almost all the tests. The correct detection of the *EUpSit* fall is attained using Algorithm 3 (specificity 99%, sensitivity 100% and accuracy 99%), that exploits the distance from the floor of the J_{SPB} joint instead of the accelerometers orientation. The variation of the y axis during an *EUpSit* fall is shown in Fig. 4a,

Table 2. Accuracy of the three fall detection algorithms for each activity

Category	Activity	Accuracy		
		Algorithm 1	Algorithm 2	Algorithm 3
ADL	*sit*	97%	100%	100%
	grasp	100%	100%	100%
	walk	100%	100%	100%
	lay	97%	100%	100%
Fall	*front*	54%	100%	97%
	back	82%	100%	100%
	side	48%	100%	100%
	EUpSit	52%	18%	100%
Average		79%	90%	99%

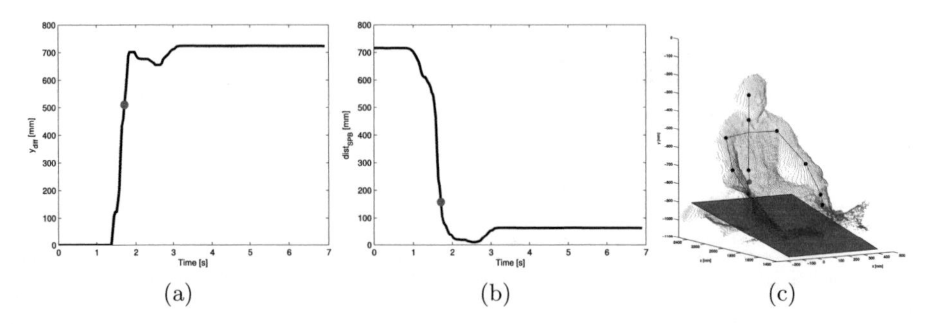

Fig. 4. (a) Variation of the y component of J_{SPB}, (b) its distance from the floor during an *EUpSit* fall, and (c) person's point cloud with highlighted in red J_{SPB} and detected plane.

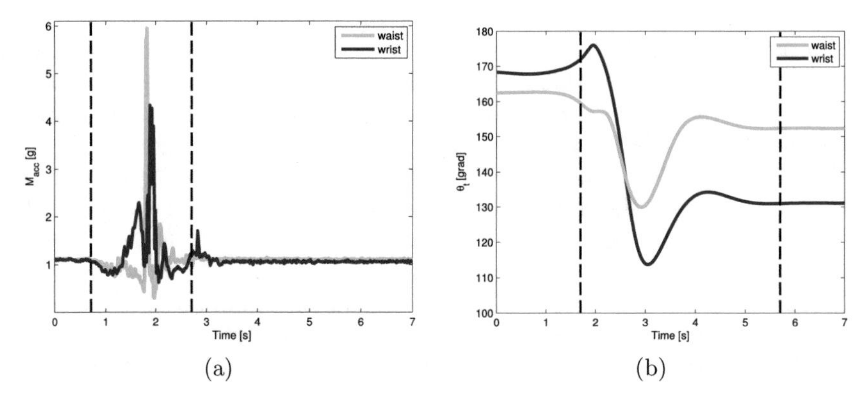

Fig. 5. (a) Acceleration and (b) orientation revealed during an *EUpSit* fall.

and the threshold that indicates an irregular activity is reached after about 1.8 seconds from the beginning of the action. The second parameter that Algorithm 3 checks is the distance between the floor and the J_{SPB} joint, revealing a value below the threshold of 20 cm, as shown in Fig. 4b. Figure 4c shows the subject's point cloud in the final phase of *EUpSit* fall, the red plane is used as a reference element to model the ground. The skeleton is superimposed to the person and J_{SPB} is highlighted by a red circle. Finally, the algorithm selects a time window of 2 seconds and searches an acceleration peak greater than 3 g in the waist accelerometer data, as depicted in Fig. 5a. Algorithm 1 and 2 fail to detect this fall because they consider accelerometer orientations that do not reveal an angle lower than the threshold of 110°, as can be noticed from Fig. 5b.

The most used features in fall detection solutions are extracted from accelerometers, gyroscope or pressure sensors, and include magnitude vectors and tilt angles [13]. Considering only these features, it is quite difficult to detect falls where the subject ends up sitting. In fact, only a few previously published works include this type of fall in the evaluated dataset. Kangas *et al.* [5] reach a *sensitivity* of 97.5% and a *specificity* of 100% considering a dataset of 6 falls and 4 ADLs. Despite the large number of performed tests, in all the considered falls the subjects end up lying, limiting the application scenarios. Pierleoni *et al.* [16], in their fall dataset, consider syncope and backward falls ending up sitting. The authors state very high performance in terms of *sensitivity* and *specificity*, with an average *accuracy* of 100%. Most of the approaches based on wearable devices try to detect the acceleration peak and evaluate the orientation of the device, to estimate the posture of the subject when he/she is on the ground. Thus, the wearable sensor must be positioned on the subject's body giving a special attention to its orientation. The solution proposed herein exploits a vision-based device and a wearable device, combining heterogeneous information by a data fusion approach. The orientation of the subject is not evaluated using the wearable device in Algorithm 3, but exploiting the information provided by the camera, which allows to identify the subject on the floor even if he/she is sitting. A drawback of the proposed solution is due to the fact that it is based on the skeleton data automatically extracted from Kinect Microsoft SDK from raw depth data. The joints estimation algorithm has been developed for gaming purposes, when the subject stands is in front of the sensor. Algorithm 3, tested on a database of 11 people performing the proposed experimental protocol, fails only in one *front* fall, as can be noticed from Table 2. In that case, Microsoft SDK is unable to estimate the skeleton when the subject is falling, and the fall is classified as an ADL because the conditions on the skeleton joints are not satisfied. This issue may be solved by integrating a barometric pressure sensor in the wearable device and using that data to evaluate the height of the waist from the floor.

5 Conclusion

This work proposed fall detection solutions exploiting skeleton data computed using the Microsoft Kinect sensor, joint acceleration data. By means of an ad-hoc synchronization algorithm, vision-based data and inertial data can be associated and used to design a simple and reliable fall detection solution. The wearable accelerometer device makes it easy to distinguish a fall from a "lying on the floor" condition because of the different acceleration vector magnitude, while the Kinect sensor is able to estimate the body orientation and the distance from the floor, enabling to identify a fall where the subject ends up sitting.

Future works will concern enriching the dataset, involving more people in the tests and considering different ADLs and falls, in order to allow a more extensive testing of the proposed algorithms.

Acknowledgments. This work was supported by STSM Grants from COST Action IC1303 *Algorithms, Architectures and Platforms for Enhanced Living Environments* (AAPELE).

References

1. World Health Organization,
http://www.who.int/violence_injury_prevention/other_injury/falls/en/
2. World Health Organization. Ageing and Life Course Unit: WHO global report on falls prevention in older age. World Health Organization, France (2008)
3. Safety for Seniors, www.eurosafe.eu.com/csi/eurosafe2006.nsf/wwwVwContent/l2safetyforseniors-seniornew.htm
4. Bourke, A.K., O'Brien, J.V., Lyons G.M.: Evaluation of a threshold-based tri-axial accelerometer fall detection algorithm. Gait Posture 26, 194-199 (2007)
5. Kangas, M., Vikman, I., Wiklander, J., Lindgren, P., Nyberg, L., Jämsä, T.: Sensitivity and specificity of fall detection in people aged 40 years and over. Gait Posture 29, 571-574 (2009)
6. Rougier, C., Meunier, J., St-Arnaud, A., Rousseau, J.: Robust Video Surveillance for Fall Detection Based on Human Shape Deformation. IEEE Trans. Circuits Syst. Video Technol. 21, 611-622 (2011)
7. Gasparrini, S., Cippitelli, E., Spinsante, S., Gambi, E.: A Depth-Based Fall Detection System Using a Kinect Sensor. Sensors 14, 2756-2775 (2014)
8. Marzahl, C., Penndorf, P., Bruder, I., Staemmler, M.: Unobtrusive Fall Detection Using 3D Images of a Gaming Console: Concept and First Results. In: Wichert, R., Eberhardt, B. Ambient Assisted Living. LNCS, pp. 135-146. Springer, Heidelberg (2012)
9. Kepski, M., Bogdan, K., Austvoll, I.: Fuzzy inference-based reliable fall detection using kinect and accelerometer. In: Rutkowski, L., Korytkowski, M., Scherer, R., Tadeusiewicz, R., Zadeh, L.A., Zurada J.M., Artificial Intelligence and Soft Computing. LNCS, vol. 7267, pp. 266-273. Springer, Heidelberg (2012)
10. Cippitelli, E., Gasparrini, S., Gambi, E., Spinsante, S., Wåhslén, J., Orhan, I., Lindh, T.: Time Synchronization and Data Fusion for RGB-Depth Cameras and Inertial Sensors in AAL Applications. To appear in: Proceedings of IEEE ICC 2015 - Workshop on ICT-enabled services and technologies for eHealth and Ambient Assisted Living, 8th June, London (2015)
11. TST Fall Detection Database,
http://www.tlc.dii.univpm.it/blog/databases4kinect
12. Wåhslén, J., Orhan, I., Lindh, T., Eriksson, M.: A Novel Approach to Multi-Sensor Data Synchronisation Using Mobile Phones. J. Auton. Adapt. Commun. Syst. 6, 289-303 (2013)
13. Pannurat, N., Thiemjarus, S., Nantajeewarawat, E.: Automatic Fall Monitoring: A Review. Sensors 14,12900-12936 (2014)
14. JointType Enumeration, https://msdn.microsoft.com/en-us/library/microsoft.kinect.jointtype.aspx
15. Kangas, M., Konttila, A., Lindgren, P., Winblad, I., Jämsä, T.: Comparison of low-complexity fall detection algorithms for body attached accelerometers. Gait Posture 28, 285-291 (2008)
16. Pierleoni, P., Belli, A., Palma, L., Pellegrini, M., Pernini, L., Valenti, S.: A High Reliability Wearable Device for Elderly Fall Detection. IEEE Sensors J. 15, 4544-4553 (2015)

Affordable Smart Living System Prototype Based on Generic Environmental Sensors

Zarko Popovski, Vladimir Trajkovik

Faculty of Computer Science and Engineering, University "Ss Cyril and Methodious", "Rugjer Boshkovikj" 16, P.O. Box 393, 1000 Skopje, Republic of Macedonia
zarko.popovski@gmail.com, trvlado@finki.ukim.mk

Abstract. This paper presents our approach to create affordable self aware Ambient Assisted Living System (AAL) that has possibilities to follow activities and take care of the humans who live in certain healthcare environment. The developed prototype is applicable to different healthcare environments, but in this paper we focus on nursing homes populated by elderly people or people with disabilities. The system presented in this paper is using environmental sensors and body sensors. Using this kind of AAL system can increase quality of life to its users.

Keywords: Body Sensor Networks, Environmental Sensors, Assisted Living, Health Care, Smart Home, Self Aware Systems

1 Introduction

User aware AAL systems are systems that collect and analyze data from different sensors in real time and can define context of behavior of end users. These systems could be very promising for helping elderly and people with disabilities. They can also provide information needed to decrease economic cost for medical centers. There are many companies that are interested for developing systems that analyze collected patient data and depending on patient Electronic Health Record (EHR) or Electronic Medical Record (EMR) take specific actions. These systems are mostly used in medical institutions because it's important to track patient activities in real-time. Also some variations on these systems could be used individually for making homes more secure and for smart house automation [7].

AAL systems are built by using many different sensors needed for collecting environmental data: temperature, humidity, light detection, sound detection, motion detection and body sensors for reading: heart rate, blood pressure, brain activity, oxygen capacity etc. Collected data from sensors is evaluated and analyzed by the system and different actions are executed according to some predefined rules. These actions can be used to provide help in medical treatments but also for early diagnosis and preventing illness. Social networks are also very good source for tracking patients' real time activities, which creates data that can be used in combi with data collected from the AAL sensors to generate more precise diagnosis. Analyzing social data [5] shared by

S. Loshkovska and S. Koceski (eds.), *ICT Innovations 2015*,
Advances in Intelligent Systems and Computing 399,
DOI: 10.1007/978-3-319-25733-4_12

users with similar health conditions is important because it can help for generating the recommendations [8] for improving patient health. Social networks can be also used for sending notifications to specific user or group of users.

Many medical centers give portable diagnostic devices to the patients allowing automatic generation of notification whenever some parameter from EHR is changed. Depending on the EMR, they can give instructions to the patient what actions to take. This provides possibility for real time monitoring of patient vital data regardless to his/her location. Real time tracking on patient's health status in combination with AAL system can find and notify the closest medical center for sending the ambulance vehicle or calling taxi for transportation in case of emergency. AAL system can inform doctors about patient's condition prior to his/her arrival in the medical institution.

AAL systems are usually implemented in medical centers. Implementation on AAL system for personal usage is expensive. Generic sensors can decrease cost of implementation of AAL systems. Generic sensors can be reconfigured automatically depending on the needs of end users. These sensors can be used for implementation on informative AAL systems where the main role would be notifying or recommending the end user for some actions.

In this paper we'll present prototype AAL system based on general AAL design architecture of generic sensors. [1]. It focuses on cheap, affordable implementation that includes easy reconfiguration, data persistency and availability of offered services. We investigate the total data throughput needed to establish such AAL system.

We use generic portable sensors and analyze collected data from the sensors. The prototype includes: custom developed data collection server that collects all data from the sensors, transforms and encrypts them for easy transferring trough the Internet and post processing, mobile devices for notifying the person for the current status, and cloud based backend where all data are collected, transformed by the data collection server, analyzed and processed.

The rest of the paper is organized in five sections, Section2, Related Work, analyzes similar systems to prototype AAL system developed in this paper, emphasizing their positive and negative features. Section 3 explains operational architecture and implementation of our prototype system for Ambient Assisted Living, based on the custom developed portable sensors. Section 4 presents deployment of prototype system and two simple test use case scenarios. Section 5 concludes the paper.

2 Related work

There are already many developed systems. We'll try to present some of them.

In [1], general architecture for AAL based on mobile, web and broadband technologies is presented. Mobile devices are used for collecting data from environmental and body network sensors, but also for easy accessing the specific person data. Raw data is preprocessed, filtered by noise, and then processed using healthcare algorithms that transforms raw data into relevant information which is distributed and used by different services. This model includes processing of data aggregated from social networks needed to give different recommendation to the user and medical centers along with data collected by sensors.

"Smart Monitor" [2] is another AAL system, but in contrast of previous one "Smart Monitor" is based on video capturing and face recognition. As AAL system can be used in different medical health care centers, also can be used for home automation. This system belongs to category intelligent AAL systems because it can be configured to react on specific object in movement but also can automatically detect, track and recognize object of interest and react on specific event without qualified employee for monitoring and reporting. Detection of configured object is based on video content analysis and sensor data analysis. The user can set specific action to be triggered after executing some event without need for qualified employee for monitoring. Actions that can be executed are: sending data to some service, or activating specific device. Thanks to the remote controller, this AAL system can be triggered remotely. Different environmental sensors and cheap cameras enable this system to monitor variety of activities in the premises where it is implemented.

Authors of [3] give a solution for intelligent part of AAL system, in their particular case solution for a kitchen. "Smart Kitchen" is part of AAL system based on different sensors used for collecting environmental data generated by different activities in the kitchen including control on different devices in the kitchen. The sensors communicate with the main system using different standards like, LAN, Wi-Fi, IR, RFID, or ZigBee. The system enables end users to start different devices in the kitchen, monitor temperature, or schedule different activities like starting washing machine, turning off the light and other energy saving activities.

Younghwan [4] describes proposes context aware smart system for covering bigger areas. This system is mainly used for preventing of crimes and different accidents. Its composed of the two parts sensor network and data processing server. Aggregator sensor network is built with cheap sensors connected and positioned on different locations. Different types of sensors used, generate very complex raw data. Data is processed by data fusion server that processes the data using different algorithms and generate data in different formats. Our prototype AAL system follows general principles of aggregation of data, processing the information and generation of recommendations. In addition, it focuses its specific use on elderly people or people with disabilities. The elderly people will benefit using proposed AAL system with constant notifications for powered devices, for taking medicaments if needed, automatically calling ambulance if needed, navigating to specific object, tracking user health status, activities or execute user specific predefined actions. The people with disabilities will also benefit from proposed AAL system with executing user specific action triggered by specific event detected by environmental sensors, navigating to specific object, recommending most suitable activity for specific period of day and more. Proposed AAL system is also focused on cheap, affordable implementation that includes easy reconfiguration, confidence and availability of the offered services.

3 Architecture of our Ambient Assisted Living Prototype based in generic sensors

Our system relies on network of sensors, both environmental and body sensors, connected using home Wi-Fi network and exposing their data to the Internet. In this phase of system development we have investigated only environmental generic sensors also we use standard body sensors. Each sensor in the network is independent, portable and wireless so it is relatively easy to make reconfiguration of the sensor network [6]. Quality of aggregated data relies on sensor activity, so it is very important that sensors are positioned on locations that will collect different data for future analysis. Each sensor module is developed with sensor component, microcontroller, Wi-Fi module, and controlling module that can be used for powering on / of different home electrical equipment which contain solid-state relay modules.

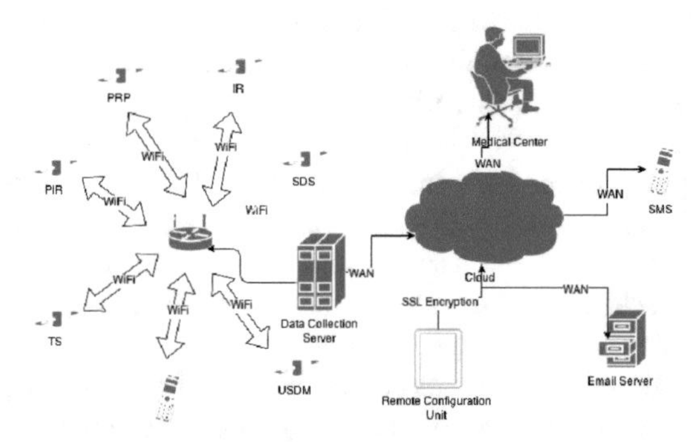

Fig. 1. General AAL implementation diagram

Every sensor is connected to Wi-Fi router, and then by LAN to Data Collection Server (DCS) for data processing [19,20,21,22]. Processed data is sent via the Internet to different application programming interfaces (APIs), like SMS message, email, medical center, etc. Depending on the API type, DCS is formatting data in specific format depending on requesting device (hardware or software). The AAL system can be also configured remotely by encrypted connection. The sensor network consists of different environmental portable sensors, wearable devices, and the mobile phone where wearable sensors are connected and used for collecting data. Aggregated data is filtered and analyzed [17,18] for executing action.Fig.2 presents implementation on generic portable sensor module used in this paper.

Each sensor module is based on same principles, which make them portable, safe and independent. The main idea is that the sensor will be connected to microcontroller that will aggregate data and do some basic data transformation before sending to the DCS. Depending on usage type, we can recognize two types of sensor modules. First type of sensor modules listens, collects and formats data using integrated MCU and then sands collected data to the DCS utilizing Wi-Fi network.

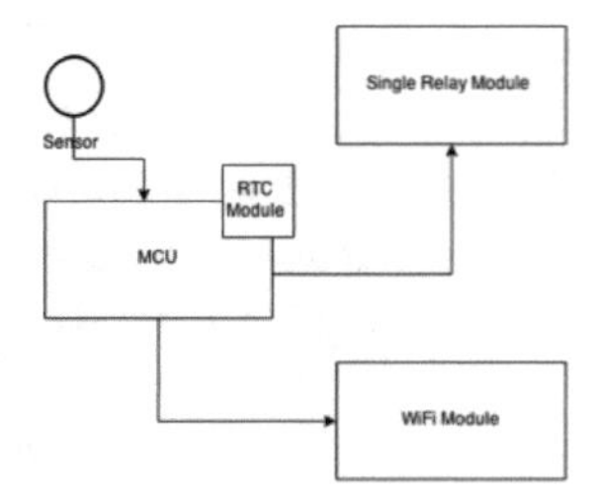

Fig. 2. Generic Portable Sensor Module Implementation

The second type of sensor modules has building relays that can be triggered by specific configured action. The sensor modules are independent because in case of when the Wi-Fi network is disconnected they can continue working with the latest good configuration thanks to integrated MCU unit. Once Wi-Fi become active they will reconnect and sync collected data, or if there is backup Wi-Fi connection they can reconnect to the alternative and then connect back to the primary Wi-Fi connection. Since these sensors modules are not only used for basic home automation, but also for developing complex AAL system they need to be precise. That is enabled by RTC module that can be configured and used by the MCU unit. The MCU is used for scheduling data collection periods using the RTC module, collecting data from the used sensor, formatting data and sending to the DCS using the Wi-Fi module. Prototype of generic sensor designed on proto-board is presented on Fig 3.

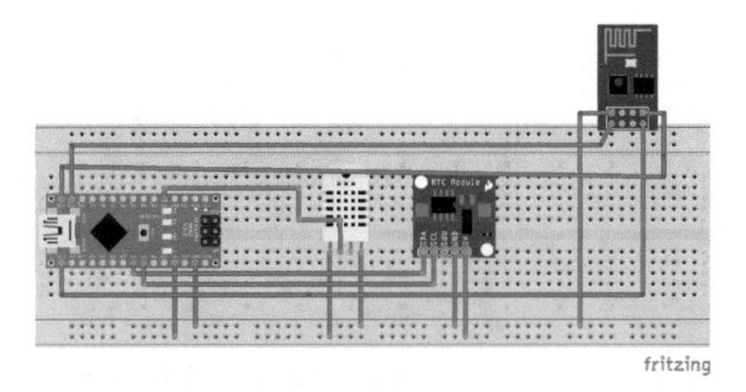

Fig. 3. Prototype from generic sensor

For better understanding how this generic sensor can be developed, basic sensor is developed on proto-board presented on Fig 4.

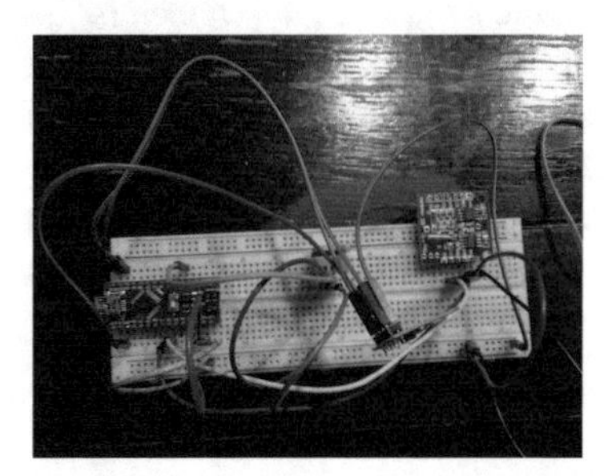

Fig. 4. Generic sensor developed on proto-board

Presented prototype is developed using Arduino Nano V3.0 MCU [22], DHT11 Temperature and Relative Humidity Sensor [24], AT24C32 I2C RTC DS1307 [23] and ESP8266 WiFi Module [25]. The MCU use analog I/O port for reading from temperature sensor and RTC module. Communication between the MCU and RTC module is based on I2C protocol. ESP8266 module [26] is used as Wi-Fi module because it can be used as independent Wi-Fi module which can control 2 independent devices using 2 onboard GPIO ports .In this case, communication between the MCU and ESP8255 Wi-Fi Module is based on serial communication. Figure 5 presents diagram for simple data gathering and processing algorithm. DCS is most important unit and is used for data collecting, analyzing and notifying. Initially all sensors are sending initialization code to the DCS, TS module collects the current temperature and tries to send to the DCS, DCS depending on different cases sends notification thru NS, but in case of when TS lost connection to DCS, it tries to find alternative DCS to send collected data. If alternative DCS is not active, TS module tries to use preconfigured Internet connection to send direct notification or in case of when the Internet connection is not available it will start with logging data that will be uploaded to the DCS once the system connects to the network.

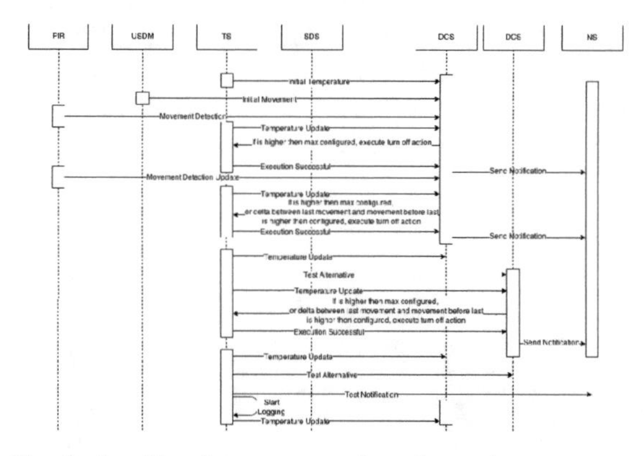

Fig. 5. Simple algorithm for movement detection and temperature changes

In some specific cases where the people who use this system suffer from some illness and health status needs to be monitored more often, we can use mobile body sensors for measuring temperature, heart rate, brain activity and gravity (using sensors from the mobile phone). So if some of the EMR parameters that these sensors can measure is changed, the person is notified and different methods for stabilizing those parameters are recommended based on the current health status [15,16]. These data are also sent to the DCS for analyzing and generating more relevant information for the person. Body sensor network built with different sensors for collecting bio data is presented on Fig.6.

Fig. 6. Body sensor network

Communication on body sensors with the mobile phone is based on Bluetooth Low Energy (BLE) modules, so when the sensors are queried, collected data is sent to the DCS using Wi-Fi network. Body sensor network of this type can be also configured to connect to the 3G networks for sending data to the medical centers. There are also sensors that monitor multiple body parameters and can track sleeping activity and quality of sleeping [21], respiration, snoring, and sleep cycles. They are usually positioned on the bad bellow the mattress. Collected data from the sensors are very complex and for that reason different standard methods are used for analyzing those data: machine learning [13,14], classifying, clustering [9,10,11,12]. Analyzed data can be used for early discovery of the illness based on data generated from daily activities, discovered symptoms and changed characteristics in EMR.

4 Prototype System Deployment

Portability of the sensors used in the project described in this paper gives possibility of different use case scenarios. The scenario that will be explained is developed for nursery home for elderly and people with disabilities. Figure 7describes 3D design of an example apartment with partial implementation of the system designed in this paper. It should be noted that envisaged deployment cost of the described installation will be in the range from1500 EUR to 2000 EUR depending of the number of the sensors used, access point and DCS configuration. That makes this system affordable for widespread usage.

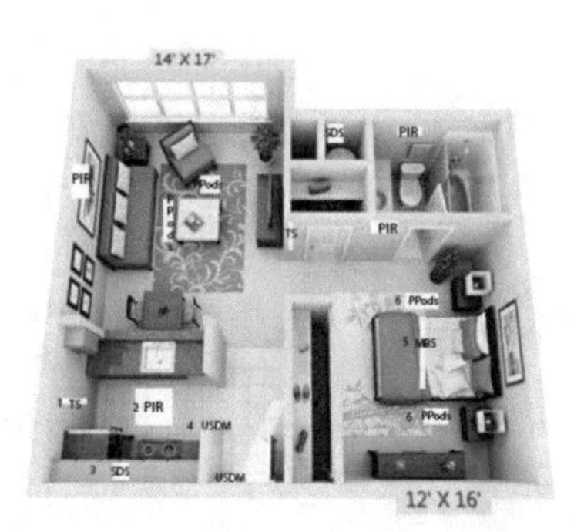

Fig. 7. Nursery Home (Smart Apartment)

Environmental sensors used in this case are motion sensors (PIR), pressure pods (PRP), ultra sound distance meters (USDM), temperature sensor (TS) and smoke detector sensor (SDS). As body sensors in this case we have used portable wearable device that includes following sensors: heart rate sensor (HRS), body temperature sensor (BTS), mobile body sensors (MBS). Any Bluetooth enabled smart phone can be used as wearable device data collection unit. Beside data collection unit, mobile device usually contain some sensors: accelerometer, compass, gyroscope. Most of the smart phones these days contain GPS sensor for precisely geo locating. Using mobile sensors, wearable sensors and pressure pods, the health care centers will actively monitor the person who lives in this apartment. The care takers will know when that person is resting, when he / she is in movement, whether the person has left some powered device or monitored premises in which case they can track him/her by the GPS sensor. Using HRS and BTS, the caretakers will actively know the patient health status. Sensors used in this scenario are sensors 5-MBSand 6-PPods. The next scenario is where the person who is using this AAL system is located in the bedroom and the all sensors in the room are working whole day and night. Depending on the daytime and the person activity, some sensors are sending data in shorter or longer periods. Figure 8 describes one possible daily routine scenario as line chart: The person wakes once at the night to go to the toilet and the next taking off is early at the morning in 06:00. Then the person has different activities. The person is making one nap in between 16:00 – 17:00 o'clock. Figure 9 displays data used for this line chart .Each request made by sensor modules is approximately 1kB of data. Depending on detected activities the sensors emit from 10 – 20 times hourly which means from 10kB – 20kB are spend per sensor module hourly. The sensor module works in two working modes depending on detected movements. Normal working mode at the night would be 10kB/h or with detected movements 15kB/h because we assume that during the night there are smaller number of user activities. During the day, due more user activities, the sensors are emitting data normal 15kB/h and 20kB/h with movements detected. Some sensors, because of their importance are emitting constantly during the day and during the night.

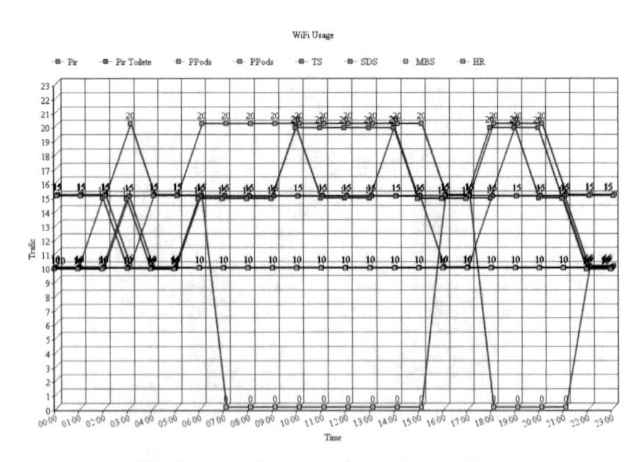

Fig. 8. Wi-Fi usage thru 24h period

From the chart presented on Fig. 8, one can conclude that with all working sensors, data transmitted to the central Wi-Fi router in normal mode and low active periods is approximately 90kb/h and in most active periods is 105kb/h which proves that this AAL network based on portable generic Wi-Fi sensors can work even on basic SOHO routers. Generic sensors used in this prototype have drawbacks. One is the security risk, which is result of cheap developing the generic sensors. These sensors are designed with open standards, which can lead to different security risks.

5 Conclusions

Implementation of AAL system with portable independent sensors enables easy reconfiguration depending on different use case scenarios. Benefits by using this approach are high confidence in tracking movement, smoke detections and temperature tracking. We provide active monitoring of patient health status with relatively low data traffic in our AAL prototype system. Presented AAL system can be easily implemented and used in different cases. Designing AAL system architecture based on open standards and technologies allows the AAL system to be affordable for anyone who wants systems for notifications and recommendation but not for medical usage.

Acknowledgement. The authors would also like to acknowledge the contribution of the COST Action IC1303 - AAPELE, Architectures, Algorithms and Platforms for Enhanced Living Environments.

References

1. Trajkovik, V., Vlahu-Gjorgievska, E., Koceski, S., Kulev, I.: General Assisted Living System Architecture Model,Mobile Networks and Management Lecture Notes of the Institute for Computer Sciences, Social Informatics and Telecommunications Engineering Volume 141, (2015), pp 329-343
2. Frejlichowski, D., Gos ́ciewska, K., Forczman ́ski, P.,Hofman, R.: "SmartMonitor" — An Intelligent Security System for the Protection of Individuals and Small Properties with the Possibility of Home Automation, *Sensors* (2014), *14*, 9922-9948; doi:10.3390/s140609922
3. Blasco, R., Marco, A., Casas, R., Cirujano, D.,Picking , R.: "A Smart Kitchen for Ambient Assited Living", *Sensors* (2014), *14*, 1629-1653; doi:10.3390/s140101629
4. Oh, Y.: "Context Awerness of Smart Space for Life Safety", (2015)
5. Vlahu-Gjorgievska, E., Trajkovik, V.: "Personal Healthcare system model using collaborative filtering techniques". Advances, Inform. Sci. Ser. Sci., 3: 64-74. DOI: 10.4156/aiss.vol3.issue3.9.
6. Chakravorty, R. MobiCare: "A Programmable Service Architecture for Mobile Medical Care". Proc. 4th IEEE Conference on Pervasive Computing and Communications Workshops (PerCom 2006 Workshops). IEEE Computer Society; pp. 532-536, (2006).
7. Blount M, Batra VM, Capella AN, Ebling MR, Jerome WF, Martin SM, Nidd M, Niemi MR, Wright SP.: "Remote health-care monitoring using Personal Care Connect". IBM Systems Journal. 46 (1): 95-113, (2007).

8. Ricci, F., Rokach, L.,Shapira, B., B. Kantor., P.: "Recommender Systems Handbook", Springer, (2011).
9. Melville, P., Sindhwani, V.: "Recommender Systems, Encyclopedia of Machine Learning", (2010).
10. Balabanovic, M., Shoham, Y. Fab: "Content-based, collaborative recommendation". Communications of the ACM, 40(3):66-72, (1997).
11. Schafer, J. B., Frankowski, D., Herlocker, J.,Sen, S.: "Collaborative filtering recommender systems". The adaptive web: methods and strategies of web personalization, Springer-Verlag, Berlin, Heidelberg, (2007).
12. Alpaydin, E. 2004. "Introduction to Machine Learning". MIT Press, Cambridge, MA.
13. Han, J., Kamber, M.: "Data Mining: Concepts and Techniques". Second edition. Morgan Kaufmann, San Mateo, CA. (2006).
14. Haskellw. L., I. M. Lee, R. R. Pate, K. E. Powell, S. N. Blair, B. A. Franklin, C. A.Macera, G. W. Heath, P. D. Thompson, and A. Bauman.: "Physical activity and public health: updated recommendation for adults from the American College of Sports Medicine and the American Heart Association". *Med. Sci. Sports Exerc.* 39:1423–1434, (2007).
15. Warburton, D. E., Nicol, C. W. Bredin, S. S.: "Health benefits of physical activity: the evidence". *Can. Med. Assoc. J.* 174:801–809; (2006).
16. Choi, Y.S., Moon, B.R.: "Feature Selection in Genetic Fuzzy Discretization for the Pattern Classification Problems", in: IEICE - Trans. Inf. Syst. E90-D, 7 1047-1054, (2007).
17. Choi, Y.S., Moon, B.R., Seo, S.Y.:"Genetic fuzzy discretization with adaptive intervals for classification problems, in Proceedings of the 2005 conference on Genetic and evolutionary computation", Beyer, H.G. (Ed.). ACM, New York, USA, 2037-2043, (2005).
18. Amo, A., Montero, J., Biging, G.S., Cutello, V.:"Fuzzy classification systems", European Journal of Operational Research 156(2), pp. 495-507, (2004).
19. Roubos, J. A., Setnes, M., Abonyi, J. "Learning fuzzy classification rules from labeled data", Inf. Sci. Inf. Comput. Sci. 150(1-2), pp. 77-93, (2003).
20. Beddit, Sleep Monitor,http://www.beddit.com/
21. Arduino, Arduino Nano, http://www.arduino.cc/en/Main/ArduinoBoardNano
22. Sain Smart, Arduino I2C RTC DS1307, http://www.sainsmart.com/arduino-i2c-rtc-ds1307-at24c32-real-time-clock-module-board-for-avr-arm-pic.html
23. Adafruit, DHT11 basic temperature humiditysensor,http://www.adafruit.com/products/386
24. Espruino, ESP8266 WiFi Module, http://www.esp8266-projects.com/

Novel Connected Health Interoperable Layered (CHIL) Model

Monika Simjanoska[1], Jugoslav Achkoski[2], Ana Madevska Bogdanova[1], and Vladimir Trajkovik[1]

[1] Ss. Cyril and Methodious University, Faculty of Computer Science and Engineering, Rugjer Boshkovikj 16, 1000 Skopje, Macedonia
{monika.simjanoska,ana.madevska.bogdanova,trvlado}finki.ukim.mk
[2] Military Academy "General Mihailo Apostolski", Vasko Karangelevski bb, 1000, Skopje, Macedonia
jugoslav.ackoski@ugd.edu.mk

Abstract. In this paper we present a novel design of a Connected Health paradigm that solves the interoperability and transitivity issues by introducing layers - the Connected Health Interoperable Layered Model (CHIL Model). The goal of our cylindric CHIL model is achieving a new quality of telemedical systems whether they are already developed or in progress. Building new systems can be performed by direct implementation of its components in each disk (layer) of the cylinder. The existing systems can be mapped in two moduses, a Distributed mapping along multiple disks or Focused mapping on a single disk. Considering both cases, using the CHIL Model, result in a complete stand-alone system that can be successfully included in the Connected health Environment. As a prove of its leverage, we discuss two case studies that comprise both distributed and focused mapping of existing systems.

Keywords: Connected Health, Telemedicine, Layered Model.

1 Introduction

The World Health Organization (WHO) definition of health that has not been amended since 1948, states that health is a state of complete physical, mental and social well-being and not merely the absent of disease or infirmity [1]. This definition has been attacked as inadequate because of its social dimension making it to correspond much more closely to happiness than to health. To avoid the conflict, Sarraci proposes another definition stating that health is a condition of well being free of disease or infirmity and a basic and universal human right [2]. This description provides an intermediate concept linking the WHO's ideal to the real world of health and disease as measurable by means of appropriate indicators of mortality, morbidity, and quality of life [2]. Those measurable indicators are something according to which we can make models nowadays and use them to predict the health condition.

Shukla et al. [3] made an effort to appropriately define Health Informatics (HI). They present the Americal Medical Informatics Association definition

S. Loshkovska and S. Koceski (eds.), *ICT Innovations 2015*,
Advances in Intelligent Systems and Computing 399,
DOI: 10.1007/978-3-319-25733-4_13

which says that HI is "all aspects of understanding and promoting the effective organization, analysis, management and use of information in health care". Canada's HI defined it as "intersection of clinical, IT and management practices to achieve better health". They all agree that HI can be divided into four main subfields:

1. Clinical care;
2. Administration of health services;
3. Medical research, and
4. Training.

Once the HI is defined, the health-care becomes a combination of alarms, sensors and other equipment to monitor people's physiological data as blood sugar, pressure, heart rate, stress level, lung function, etc., and help them live independently. That combination is referred to as telehealth/care, whereas the delivery of health-care services when the distance is a critical factor is called telemedicine [4]. All these terms are covered by an approach to health-care delivery that uses a wide range of information and collaboration technologies to facilitate the accessing and sharing of information, as well as to allow subsequent analysis of health data derived from electronic medical records and connected biomedical devices across healthcare systems called Connected Health (CH) [5]. CH does not encourages only management of patient's clinical data, but also the communication and collaboration among all the entities involved in a patients health. This feature introduces some barriers that can stand in the way to CH:

- Systems and policies;
- Organization and management;
- Clinicians and end users, and
- Patients and the public.

The ambition of CH is to ensure the confidentiality of personal health data and to connect all parts of a healthcare delivery system through interoperable health information system, so that critical health data is available anytime and anywhere [5]. Kovac in his study [6] elaborate interoperability in a way that it shouldn't be understood as simply technical connectivity, but true interoperability enables health information and medical expertise to be exchanged among clinicians, patients and others to further understanding. In order to achieve interoperability, the CH system need to sustain security and privacy, transparency, preservation of information, re-usability, technological neutrality and patient centricity.

In this paper we present the scope of CH by introducing a novel designed layered model, each layer presenting different entities, their sub-layers and the interconnections between them.

The goal is to create CH model that will overcome the before mentioned barriers and achieve interoperability as described by Kovac [6]. Even more, the proposed design allows transitivity between the different layers without preserving the hierarchy of the model. Because of its properties, the proposed model is referred to as Connected Health Interoperable Layered model - CHIL model.

In order to show both the horizontal (on one layer) and vertical (among the layers) transitivity and the interoperability, we analyze two case studies. The first case study presents a distributed integration of a new state-of-the-art telemedical IS - SIARS (Smart I (eye) Advisory Rescue System) [7] and the second one describes a case of a focused implementation of an existing classification system [8] in a single layer of the CH model.

The rest of the paper is organized as follows. In Sect. 2 we present the existing connected health models as well as telemedicine systems and prediction approaches. The novel CHIL model and its components are presented in Sect. 3. In Sect. 4 we present two case studies to show the flexibility of CHIL when integrating existing systems. The conclusion and the possibilities for further development of CHIL are presented in Sect. 5.

2 Related Work

In this section we present some recent health models that leverage the information technology and have contributed to better understanding of the necessity of this kind of models integration in people's lives.

Collaborative Health-Care System (COHESY) is a novel model for monitoring health parameters and physical activities developed by Trajkovik et al. [9]. The approach itself is original by the fact that it is not only predictive, but also introduces collaborative filtering module which means that the entities can communicate and exchange experience via social networks. Designed to increase self-care regarding the health, the model combines data from patient's current state and environmental sensors. Given the measure, the model automatically adjust threshold parameters and sends notifications or emergency call if necessary. The collaborative algorithm behind COHESY is explained via two types of scenarios. A simple one that considers sensors connected to a smartphone and data presented to patients and doctors, and a smarter scenario where the model makes conclusions about the current health condition of the patient based on both the sensors and integrated expert knowledge [10].

When talking about an integrated advice giving system based on prior knowledge, Jovanovik et al. [11] did a remarkable work by stressing out the importance of knowing how the negative food - drug interactions are spread in various cuisines. In the focus of their research is the patient who is under a treatment with particular drug and an advisory mechanism of the possible negative food - drug interactions and also the cuisines that should be avoided while the patient is under treatment.

Besides the challenge of making a system that is capable of giving advices, the authors [10] aim to automate the process. An accurate predictive algorithm is still a great challenge in the healthcare environment. As the authors state in [12], healthcare environment is information rich, but knowledge poor. There is still lack of analysis tool to discover hidden relationships and trends in data.

The current state of chronic disease data and analysis is being investigated by Sullivan et al. [13]. They define a simple model for chronic disease to be disease free at the beginning, then preclinical, afterwards comes the clinical manifestation and the final stage is everything that follows-up. By using statistics and Machine Learning (ML) techniques to model the chronic disease, they inspect the probability of manifestation of the each clinical stage based on the risk factors as age, gender, smoking, blood pressure, cholesterol etc.

Another very challenging field of research is the creation of prediction models in terms of risk and costs of diabetes. Farran et al. [14] performed a research on data from Kuwait to assess the risk of type 2 diabetes. They present a fact that one in ten suffers from diabetes and one in three from hypertension. Thus, their aim was to model the increased proneness in diabetic patients to develop hypertension and vice versa. In achieving the goal they use four ML techniques including logistic regression, KNN, multifactor dimensionality reduction and SVM. Their work ascertain the importance of ethnicity.

Health-care cost projections for diabetes and other chronic diseases are discussed in [15]. The work examines the relationship between chronic disease on current and projected health care costs. Diabetes is found to be most suitable for epidemiological modeling to analyze long-term nature of the development. The models are combined with clinical trial data and are used to easily extrapolate clinical trials over the lifetime of patients. There are variety of models proposed. Most notable is the Archimedes model that differs from the others in that it sets out to account for basic cellular and organ functioning to predict the risk of complications.

3 The Novel CHIL Model

The model proposed in Fig. 1 is based on the challenges that are mostly elaborated in [5]. The idea behind the cylindric representation of CH is to depict all sub-layers and the relations amongst them.

3.1 Policy Layer

A clear strategies and protocols that are aligned with the wider health reforms are necessary for achieving desired health outcomes. The policies should be flexible and thus the regions or local institutions will be encouraged to be innovative and take their own approaches to the development of health-care IT systems that will meet the local demands. Policies comprise education and training as essential in improving the expertise for achieving appropriate integration of the clinicians and administration in the CH system. This layer serves to the economy layer by providing it with the suitable costs plans for actions.

3.2 Economy Layer

The economy layer consists of protocols for management, funding mechanisms and business cases. It connects the policies and the users by providing the appropriate and cost-efficient strategies to maximize the potential of limited funding

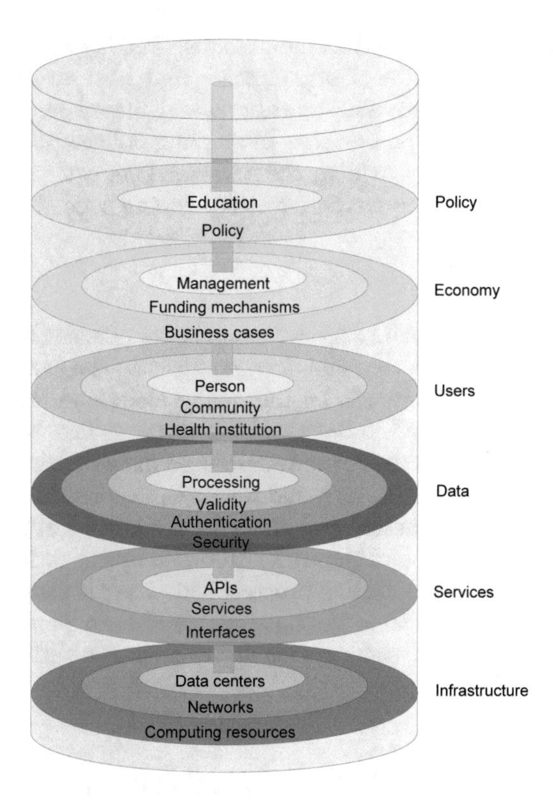

Fig. 1. The CHIL model

to ensure quality and efficiency of the health institutions. All the activities in this layer aim towards linking the investments directly to the achievement of health outcomes. The management should rely on innovation schemes that encourage integration and coordination across organizations by creating coordinating bodies to manage required cultural and organizational changes.

3.3 Users Layer

Developing plan for communication and collaboration by which the benefits of the healthcare IT will be demonstrated to the persons and to the physicians involved in the processes is necessary. Training is a significant component that suppose to encourage the patients to play a role in managing their own healthcare. The main idea is to create multi disciplinary teams that will be regularly consulted during the process of designing the systems, whether they are portals, mobile technologies or remote monitoring devices, in order to encourage patients to take an active role in managing their own health. The strategies should also take into consideration the local opinion on privacy, protection, security and confidentiality of the patients data to assure appropriate sharing and use of information for the benefit of the patients. The layer uses the advantages of the economy layer in terms of organization, but also serves to improve the economy policies taking into consideration a real life cases. Those experience and involvement of the different entities in creating policies also affects the layer that concerns the data by providing the policymakers with information of the properties that need to be preserved when developing such.

3.4 Data Layer

The use of structured data is necessary to enable an efficient leverage of it. The term structured data means that the data is stored in the same format and according to the specified standards by the policy makers in the first layer. The Data layer stands in between the users and the services. The users, whether they are patients or health institutions, need structured data to achieve interoperability from an information point of view and the services need standardized formats to enable a meaningful exchange and analysis of clinical data. The Data layer enables using Data Mining techniques for resource optimization in the prevention process as well as in the healing process. This layer is also beneficial for the patients self care since once they will have access to their personal information, the people will take greater responsibility for their own health.

3.5 Services Layer

All the healthcare applications for monitoring, decision or diagnostics, may have been developed in various technologies or by different standards. The Service layer should provide technical standards to which IT healthcare systems need to conform and use standardized data formats and common medical terminology in order to exchange the information between similar and dissimilar healthcare applications. A standardized communication platform would guarantee the accessibility and usability advantages to both patients and physicians [16]. CH can be enabled only by working with the service providers and ensuring technological capabilities.

3.6 Infrastructure Layer

Building the databases, getting the hardware right and understanding how to link the systems are all necessary actions for appropriate interaction of all layers, achieving interoperability and transitivity of services among each others.

4 Case Studies

By introducing our novel model we achieve new quality in two existing multifunctional systems. The first one, SIARS [7], is designed in multiple layers and consists of multiple services. When mapping SIARS, the system gains new quality by using the interoperability of our CHIL model. The second is a Bayesian

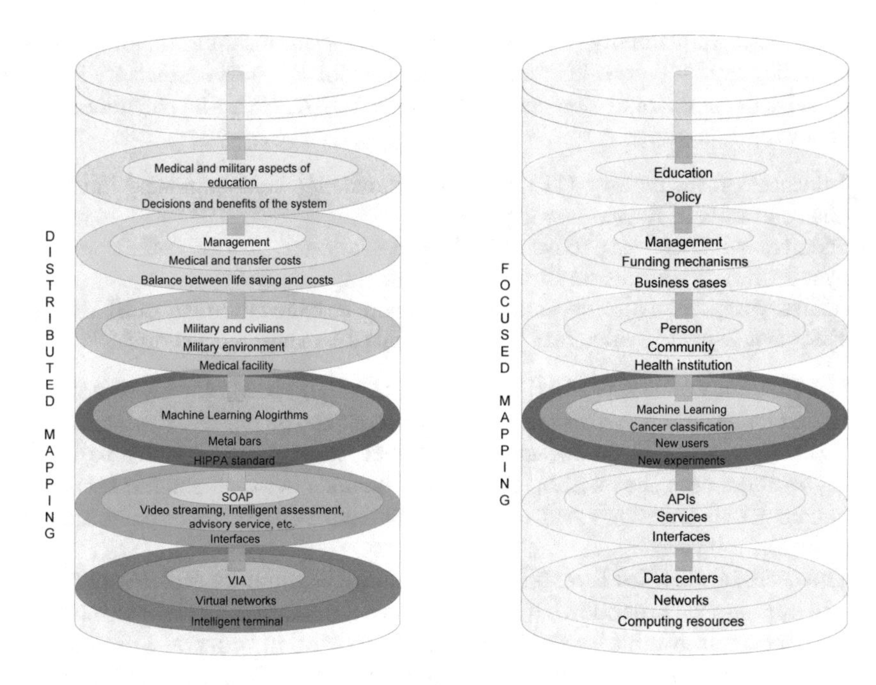

Fig. 2. Distributed vs Focused mapping

classification system that relies on gene expression data to perform diagnostics on the patients' health condition [8]. This classification system cannot be distributed on all the CHIL layers, instead it is focused only in the Data layer. However, the CHIL model still adds new quality by allowing the system to use its transitivity and easily face the issues as policies, education, its usage by the health institutions, costs for maintenance, services and infrastructure, and thus helping its upgrading to a stand-alone platform in a HC environment.

4.1 Distributed Mapping

This section presents an implementation concept of the cylindrical representation of CHIL as prove of concept for telemedical system (Smart I(eye) Advisory Rescue System SIARS) [7] development in military environment. SIARS works in two modes: on line, in the cloud infrastructure, consult the e-medical file of the injured and support on-line video stream with the medical specialist, or locally, off-line, with the Advisory Intelligent Module (AIM). The SIARS platform will be incorporated in the existing NATO projects and platforms, and will also be easily extensible with additional services. The distributed mapping of SIARS is presented in Fig. 2.

First layer for policy wraps up decisions for telemedical system development with its goals, objectives and outcomes that society will have after the system implementation. Additionally, the benefits from the system should be implemented in the policy layer. These elements are the basis for the initial phase and can be upgraded with additional elements. Education in SIARS can be considered from the following four aspects:

- Information technology (IT) aspects - knowledge that engineers have when the system is in development phase;
- Medical aspects - expertise that is necessary for proper definition of the system functionalities and user requirements;
- Military aspects - also necessary for appropriate development of SIARS, and
- End-user aspects - adequate training related to the final product.

The Economy aspect of SIARS is evident through the Management optimization when the wounded person is to be transfered to the nearest medical facility, since it requires usage of substantial technical items. CHIL enables to build a strong business case that will be the perfect balance between the life saving and the medical and transfer costs. Additionally, this layer directly communicates with the Data layer.

The users in the third layer encompass person, community and health institution. In our case we are following CHIL, but we are focused on the establishment of this concept in a military environment. Therefore, in order to have more robust and comprehensive SIARS development we can enhance the third layer of the CHIL model by adding another horizontal level - Environment. Users that are selected in SIARS development are military persons (soldiers, doctors and patients) and wounded civilians. In this case study soldiers (life savers) are users that can give first aid by using the SIARS and transfer the data to the medical facility. Doctors in medical facility can receive transferred data and prepare for further actions. Previously mentioned roles are users in the system.

Data layer in our cylindrical presentation encompass validity, authentication and security. In our case study we consider all three elements for SIARS development. Gathering and processing the data is done by ML tools, especially the knowledge engineering approach. We find authentication to be very sensitive element in a military surrounding. The identification of the military personnel is plain and simple because their data is already available in the system. The problem with authentication arises with civilians since their personal data is not previously available. For SIARS, we propose the authentication to follow HIPAA standard as a security rule. Considering the security of data, DDL triggers and DML triggers should be customized on appropriate level, thus each operation for changing certain data in the database will be caught and digital forensic can be applied.

In [7] authors show a model for telemedical information system for blood type analysis in a military environment based on Service Oriented Architecture (SOA). SOA based multi-tier approach provides legacy systems to be hooked up in a new infrastructure where new systems and legacy systems can communicate without complexity in communication protocol [17]. We followed the authors' [7] idea and based our case study on the SOA concept. The horizontal communication in CHIL indicates interoperability which means that the SIARS services written in different programming languages are capable to communicate. In this connotation, processing elements (web servers, sensors for collecting data, etc.) can create distributed environment for sharing information. SIARS will provide prototype platform with basic functionalities as services in the cloud (on-line) and local (off-line) modules that will be tested in real situations. The main novelties are the video-streaming of the injured person, advisory/learning service and intelligent assessment of the level of injury. The platform will be extensible providing basis for implementing additional telemedical services.

The proposed distributed application requires fast and reliable exchange of information across a network to synchronize operations and to share data. Traditional communication architectures do not provide the safe and secure communication links required by the proposed application. Therefore, we plan to realize a system as a stand-alone intelligent interactive terminal. In the first phase, we intend to develop a system for the local operation only and later we intend to connect the intelligent terminal to NATO telemedicine information system. Given the fact that we do not have access to the NATO standards, we plan to upgrade the system communications in the form of a popular Virtual Interface Architecture (VIA). The VIA specification describes a network architecture that provides userlevel data transfer. Hardware implementations of VIA support context management, queue management, and memory mapping functions in addition to standard network communication functions.

4.2 Focused Mapping

The methodology presented in [8] is based on statistical analysis of colorectal gene expression data and is applied in the Bayes' theorem for classification of colorectal tissues. The methodology can be implemented in the Data layer of the CHIL model, particularly in the processing and validity sub-layers of the model as presented in Fig. 2. The processing layer is fulfilled by the methodology itself that provides methods for statistical preprocessing and ML techniques for building the classification model. The model has been tested with new patients and the obtained results prove that the used techniques are valid, thus it also fits the validity sub-layer.

Given the horizontal opportunity for transitivity, the Bayesian classifier can be easily integrated in a tool for colorectal cancer analysis that will demand both the authentication and security sub-layers for authenticating new users and secure the patients confidentiality when performing new experiments.

The vertical possibility for transitivity and interoperability provides the advantage to develop the tool as a service and deploy it in a cloud infrastructure. This opportunity demands the activation of both the Service and the Infrastructure layers. Furthermore, the development of a user-friendly tool's interface demand an involvement of the physicians and clinicians that are part of the User layer. The maintenance of the tool in terms of cost-efficiency is a target of the Economy layer. The Policy layer is needed to provide appropriate training for its usage and teach on its benefits.

This implementation clearly presents the leverage of the structure of the CHIL model in terms of building blocks that can be a good indicator for upgrading and integrating a healthcare tool in the CH paradigm.

5 Conclusion and Future Work

This research examines the problem of incomplete CH models and proposes a novel model that solves the interoperability and transitivity issues reported in the literature. The introduced CHIL model, follows cylindric design that comprises different layers organized in sub-layers. Even though the design is elaborated in a hierarchical manner, it is flexible and allows communication between the layers without the necessity of consulting the neighboring layers first. Therefore, the actions triggered by any layer can affect the rest of the layers.

To test the ability of the CHIL design to implement any kind of medicine system in the CH paradigm, we analyze two types of existing systems developed in a different manner. The first case study analyze a distributed implementation of a telemedical advisory rescue system and the second case study exhibit a focused implementation of a methodology for classification of patients in a single layer of the model and its potential to use its leverage in terms of interoperability and transitivity to spread over the rest of the layers and obtain its distributed form.

In our future work we will upgrade the COHESY model, described in Sect. 2, by following our CHIL design. The goal is to achieve maximum utilization of its potential to collect, analyze and derive decisions on behalf of improving the health in the way defined [2] and thus confirm the validity of our approach.

Acknowledgment

This work was supported by TD COST Action TD1405 - "European Network for the Joint Evaluation of Connected Health Technologies" (ENJECT) and the NATO Science for Peace and Security Program Project ISEG.IAP.SFPP 984753.

References

1. Organization, W.H., et al.: www. who. int/bulletin/archives/80 (12) 981. pdf who definition of health. In: Preamble to the Constitution of the World Health Organization as adopted by the International Health Conference, New York. (1946) 19–22
2. Saracci, R.: The world health organisation needs to reconsider its definition of health. Bmj **314** (1997) 1409
3. Shukla, D., Patel, S.B., Sen, A.K.: A literature review in health informatics using data mining techniques. Int. J. Softw. Hardware Res. Eng. IJOURNALS (2014)
4. Wragge, Co: Connected health white paper (2014)

5. Accenture: Connected health: The drive to integrated healthcare delivery (2012)
6. Kovac, M.: E-health demystified: An e-government showcase. Computer **47** (2014) 34–42
7. Tasič, J.F., Bogdanova, A.M., Achkoski, J., Glavinov, A.: Service–oriented architecture model for blood type analysis (smart i (eye) advisory rescue system) in military environment. In: ICT Innovations 2013. Springer (2014) 149–156
8. Simjanoska, M., Madevska Bogdanova, A., Popeska, Z.: Bayesian posterior probability classification of colorectal cancer probed with affymetrix microarray technology. In: Information & Communication Technology Electronics & Microelectronics (MIPRO), 2013 36th International Convention on, IEEE (2013) 959–964
9. Trajkovik, V., Koceski, S., Vlahu-Gjorgievska, E., Kulev, I.: Collaborative health-care system (cohesy) model. mHealth Multidisciplinary Verticals (2014) 101
10. Trajkovik, V., Koceski, S., Vlahu-Gjorgievska, E., Kulev, I.: Evaluation of health care system model based on collaborative algorithms. In: Mobile Health. Springer (2015) 429–451
11. Jovanovik, M., Bogojeska, A., Trajanov, D., Kocarev, L.: Inferring cuisine-drug interactions using the linked data approach. Scientific reports **5** (2015)
12. Soni, J., Ansari, U., Sharma, D., Soni, S.: Predictive data mining for medical diagnosis: An overview of heart disease prediction. International Journal of Computer Applications **17** (2011) 43–48
13. Sullivan, L.M., et al.: Chronic disease data and analysis: Current state of the field. Journal of Modern Applied Statistical Methods **1** (2002) 32
14. Farran, B., Channanath, A.M., Behbehani, K., Thanaraj, T.A.: Predictive models to assess risk of type 2 diabetes, hypertension and comorbidity: machine-learning algorithms and validation using national health data from kuwaita cohort study. BMJ open **3** (2013) e002457
15. O'Grady, M.J., Capretta, J.C.: Health-care cost projections for diabetes and other chronic diseases: the current context and potential enhancements. Partnership to Fight Chronic Disease (2009)
16. Bellazzi, R., Montani, S., Riva, A., Stefanelli, M.: Web-based telemedicine systems for home-care: technical issues and experiences. Computer Methods and Programs in Biomedicine **64** (2001) 175–187
17. Achkoski, J., Trajkovik, V.: Service design and distributed system reliability in intelligence information system based on service-oriented architecture. In M. Ganzha, L. Maciaszek, M.P., ed.: Position Papers of the 2014 Federated Conference on Computer Science and Information Systems. Volume 3 of Annals of Computer Science and Information Systems., PTI (2014) 211–217

A Survey on User Interaction Mechanisms for Enhanced Living Environments

Andrej Grguric[1], Alejandro M. Medrano Gil[2] Darko Huljenic[1], Zeljka Car[3], and Vedran Podobnik[3]

[1] Ericsson Nikola Tesla d.d., Krapinska 45, 10002 Zagreb, Croatia
E-mail: {andrej.grguric, darko.huljenic}@ericsson.com
[2] Universidad Politécnica de Madrid, Avenida Complutense 30, 28040 Madrid, Spain
E-mail: amedrano@lst.tfo.upm.es
[3] Faculty of Electrical Engineering and Computing, Unska 3, 10000 Zagreb, Croatia
E-mail: {zeljka.car, vedran.podobnik}@fer.hr

Abstract. Designing user interfaces (UIs) for elderly and/or disabled is a big challenge since they have various special needs that have to be considered in every aspect of a design process. Intelligent UIs that tailor interaction and presentation capabilities based on user needs, preferences and expectations can significantly increase the usability of information and communication technology (ICT) for Enhanced Living Environment (ELE). This paper deals with evolution of interaction in ELE systems together with a survey of the state-of-the-art user interaction frameworks. Using the proposed analysis framework, the survey shows how a challenge of user interaction is approached in different Ambient Assisted Living (AAL) platforms. Finally, based on an elaborated analysis the paper discusses ongoing challenges faced by developers of user interaction frameworks in ELE.

Keywords: Ambient Assisted Living, Enhanced Living Environment, human-computer interaction, elderly persons, disabled persons

1 Introduction

User interface (UI) design is a complex multidisciplinary activity which deals with many different details and aspects that have to be very well thought out, including interaction, navigation, notification as well as information and communication technology (ICT) behaviour in general. Otherwise, utilization of computing equipment can lead to stress and consequently reduce user quality-of-experience (QoE). This is specifically true for elderly and/or disabled people.

Elderly and/or disabled people have the same need to access information as everybody else (e-Inclusion) but require different approaches during equipment, applications and UI design. A potential approach are *Ambient Assisted Living* (AAL) technologies, which show a great potential in coping with challenges of greying societies. AAL technologies strive towards ICT-enabled independence of elderly and/or disabled through building *Enhanced Living Environments* (ELEs)

S. Loshkovska and S. Koceski (eds.), *ICT Innovations 2015*,
Advances in Intelligent Systems and Computing 399,
DOI: 10.1007/978-3-319-25733-4_14

and thus extending living in their homes by increasing their autonomy and preventing social isolation.

The vast majority of existing interfaces does not adequately fulfill accessibility requirements of (elderly) users suffering from visual, hearing or any other impairment. It is rather common for an individual to suffer from more than just one impairment at the same time [1], what additionally increases complexity of ELE design. Although many AAL platforms have been presented over the years, a trust of target users (i.e., elderly and/or disabled individuals) is still missing. A user acceptance of ELE systems is additionally hampered by a fear of technology combined with a lack of experience as well as possible user physical or cognitive impairments. In order to mitigate these issues, a lot of effort has still to be invested towards making the usage of ELE systems easier and more enjoyable. One of solutions is to use multimodal interfaces that can address many of the target group requirements. According to Oviatt [2], "multimodal interfaces process two or more combined user input modes (such as speech, pen, touch, manual gesture, gaze, and head and body movements) in a coordinated manner with multimedia system output". Different means of interaction and different communication channels between a user and an ELE system greatly improve system's usability.

This paper makes a contribution to the multidisciplinary research field through a survey on user interaction mechanisms for ELE systems based on AAL technologies. In the next section, an evolution of interaction in using technology is addressed. Section 3 presents the methodology used in a survey of user interaction frameworks for ELEs, obtained results and discusses findings which are used in Section 4 for identification of ongoing challenges in the field of UIs in ELE systems. Finally, Section 5 concludes the paper.

2 Survey

Over the last few decades we are witnessing an evolution of *one-to-one* interaction relationship towards *many-to-many* interaction relationship among multiple users and multitude of devices in a dynamically changing environments as illustrated on Fig 1. *Multimodality* is becoming more and more important and with further technological advancements interaction possibilities enabled by multimodal interfaces will be really exciting. Multimodal UIs [3] comprising different complementing types of input methods (e.g. natural language and gestures) support the trend of ubiquitous interaction and e-Accessibility very well.

UIs are one of key factors that enable elderly users and/or users with disabilities to accept technological aid offered to them. Therefore, UIs must support simple, intuitive and natural interaction what puts the following requirements on the UI design: i)*Effectiveness* - accuracy and completeness of interaction; ii) *Efficiency* - user can finish a task with a minimum effort; iii) *Acceptance* - UI provides information in a format easy understandable for users; iv) *Learnability* - well-designed UI can speed up a learning process and reduce training effort time; v) *Human errors* - user friendly design helps with avoiding so-called

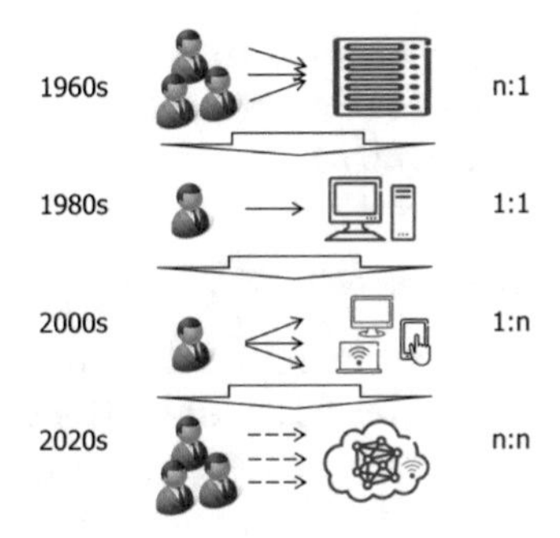

Fig. 1. Evolution of the interaction over the years

human errors because users do not have to deal with inconsistencies or ambiguities; vi) *Satisfaction* - user attitude towards a used ICT system has a high influence on, previously mentioned, acceptance; vii) *Sales* - well-designed UI can bring a competitive edge to an ICT system.

2.1 Analysis Parameters

Based on technology achievements a number of different ELE systems have been suggested for elderly and/or disabled people. The performed analysis was focused towards understanding: i) how the input was gathered from users and how the ICT system presented output, as well as which devices and/or modalities were used during the interaction process, ii) whether pluggable UIs were used, iii) whether modality independent UI model was used and, finally, special focus was put on iv) support for adaptation.

Pluggable UIs promote an idea which becomes more and more popular [4] that UIs should be easily changeable (pluggable) without (or with minimal) changes in the ICT system application layer. When coming to a separation of application logic and presentation layers, modality neutral UI descriptions (*Abstract User Interfaces*, AUIs) become very useful since they can be transformed to Final User Interfaces (FUIs) and applied to a wide range of devices. There is an obvious trend towards easily adaptable and customizable UIs [5].

After literature review, the following criteria for comparison of UI frameworks were defined: i) *Support for multimodal fusion* - UI is supported by at least two different input modalities of interaction; ii) *Support for multimodal fission* - UI is supported by at least two different output modalities of interaction; iii) *Support for pluggable UIs* - UI should be easily replaceable by another without changing application functionality; *Modality independent UI description* - UI description language capable for an abstract description of an UI is used within the framework; iv) *Support for adaptivity and adaptability* - terms *adaptivity* and *adaptability*, often mentioned when discussing adaptation of UIs, are often mixed. An ICT system is *adaptive* [7] if it is able to automatically change its own characteristics according to end user needs and *adaptable* if it provides the end user with tools that make it possible to change system characteristics; v) *Object of adaptation* - three ways of adaptation, as proposed by Kobsa at. al. [6] are: *structure*, *presentation* and *content*.

2.2 Platform Selection Methodology

For the selection of AAL platforms mainly the two extensive overviews were used: i)*Next Generation Services for Older and Disabled people* [8] containing summaries of over 100 research and development projects and initiatives; and ii) *AAL JP Catalogue of Projects 2013* [9] containing 120 projects funded by the AAL JP. From both documents firstly projects with the aim of producing platform with at least simple UI were selected. After initial selection, for all remaining projects *corresponding web sites* were tried to be reached, at which point it became obvious that a lot of the projects (especially those finished more than a few years ago) were not maintained anymore. An *extent and availability of public documentation* was used as the next filtering parameter. During this particular step yet another big set of projects was lost since a lot of them did not have any useful information about results and prototypes publicly available. The remaining projects were taken under consideration for further analysis together with the additional five projects (namely: *universAAL*, *HOMEdotOLD*, *PeerAssit*, *I2HOME* and *MonAMI*) that were not mentioned in the two initial overview documents but which authors considered were relevant for the survey. The platforms selected for the analysis do not, by any means, include the comprehensive list of the all available AAL platforms with some kind of UI for the elderly but they can, in many ways, be considered a representative set.

The **universAAL** [15] developed a reference open source platform for building distributed assistive services. **HOMEdotOLD** [16] developed a television (TV) platform that enhanced the interaction of older people and thus prevented isolation and loneliness. **PeerAssist** [17] designed, implemented and demonstrated a flexible peer-to-peer platform that enables older people to dynamically construct virtual communities based on shared interests and communication needs, while **I2HOME** [18] focused on the intuitive interaction with home appliances based on industry standards for people with mild cognitive impairment and the elderly. **MonAMI** [19] used personalized UIs for applications related to home control, activity planning, monitoring of users, activities detection, visitors recognition and communication and **Soprano** [20] designed innovative and context-aware smart services with natural UI for older people. **Amigo** [21] developed a platform for integration of different devices and services that enable people to share activities and experiences in a personalized way, regardless of location and communication device and **OASIS** [22] developed an architecture and platform to facilitate interoperability, seamless connectivity and sharing of content between more than 12 different types of services for the elderly users. **MPOWER** [23] developed an open platform for easy development and delivery of services for elderly persons and persons with cognitive impairments while **Persona** [24] developed an open platform for AAL service development. **Diadem** [25] provided people with reduced cognitive ability an effective interaction through a customizable interface of an expert system that monitors users at work and at home, while **Domeo** [26] developed an open platform for integration and personalization of care services at home and cognitive and physical help. **IS ACTIVE** [27] developed a solution for chronic patients, especially elderly persons, on the basis of miniaturized wireless inertial sensors for motion capture and intelligent recognition of activities and situations in real time and **ALICE**[28] explored communication practices of older people and tried to simplify electronic communication and interaction video conference via TV. **FoSIBLE** [29] developed a TV platform based on the technology of games and smart furniture that provides social support through a virtual community and entertainment applications. **Join-In** [30] developed a social platform that allows online socializing, multiplayer gaming, group exercises, games involving physical exercise, correspondence and the like through the interaction with a computer or a TV with a set-top box with custom controllers. **CogKnow** [31] created a cognitive auxiliaries for people in early stages of dementia and **Easy Line +** [32] developed a prototype of home appliances to support elderly people. **HERMES** [33] resulted in home and mobile devices preventing cognitive decline, while **Connected Vitality** [34] resulted in a personalizable videoconference device for elderly persons. **EasyReach** [35] developed a TV-based system for social interaction and **AALuis** [36] developed a framework for user interaction allowing terminal UIs on different platforms with AAL services. **Care@Home** [37] developed a communication system capable of personalization based on the smart TV for elderly users to be able to communicate with family, friends and health professionals. **GoldUI** [38] developed a secure user profile in the cloud for personalization of representation of multimedia content and enabling natural interaction through existing devices. **ChefMySelf** [39] developed a expandable system built around automatic cookers as solutions to support elderly users in preparing meals. **EDLAH** [40] developed a tablet-based system for elderly users for monitoring user health and proposing the most appropriate recipes or delivering reminders for taking medications. **CapMouse** [41] resulted in a UI for elderly and/or disabled persons with tongue controlled keyboard/mouse functionality, while **AEGIS** [42] developed an accessibility support using generally accepted ICT devices and applications. **Long Lasting Memories** [43] resulted in a platform with 3 interoperable components for physical training, cognitive training and independent living. **PAMAP** [44] developed a system for precise monitoring of physical activity of older people allowing health workers more objective diagnosis as well as end users feedback applications with web and TV UIs. Finally, **Sociable** [45] used a new approach to personalized cognitive training and social activation via assisted ICT technologies for older people.

2.3 Overview of Selected User Interface Frameworks

Table (Table 1) summarizes findings of the performed analysis. Places without special marks do not necessarily always mean that the observed characteristic was not supported by the platform, but reflect the fact that there was no hard evidence of observed characteristic which authors could find.

Table 1. Selected UI frameworks

	Input										Output										Adap.		
System	**Voice**	**TV remote**	**Gesture**	**Touch screen**	**Mobile phone**	**Keyboard**	**Camera**	**Mouse**	**Trackball**	**Tablet**	**GUI**	**TV**	**Loudspeakers**	**Mobile phone**	**Tablet**	**Avatar**	**Pluggable user interfaces**	**Modality independent UI model**	**Adaptability**	**Adaptivity**	**Structure**	**Presentation**	**Content**
SOPRANO[20]	✓	✓		✓							✓	✓	✓							✓		✓	✓
AMIGO[21]	✓		✓								✓	✓		✓				✓		✓		✓	
OASIS[22]	✓			✓	✓	✓		✓			✓		✓	✓			✓			✓		✓	✓
MPOWER[23]				✓	✓	✓		✓			✓			✓									✓
PERSONA[24]	✓		✓	✓	✓	✓					✓			✓			✓	✓	✓	✓		✓	
universAAL[15]			✓	✓	✓	✓		✓		✓	✓		✓	✓	✓		✓	✓	✓	✓	✓	✓	
DIADEM[25]	✓			✓		✓		✓			✓		✓					✓		✓			✓
DOMEO[26]	✓			✓							✓												
IS-ACTIVE[27]			✓		✓					✓		✓			✓								
ALICE[28]	✓	✓					✓				✓	✓	✓										
FoSIBLE[29]		✓	✓	✓					✓		✓	✓											
HOMEdotOLD[16]	✓	✓										✓											
PeerAssist[17]	✓	✓		✓	✓		✓	✓			✓	✓	✓	✓						✓			✓
JOIN IN[30]		✓			✓					✓	✓	✓		✓	✓								
CogKnow[31]				✓	✓							✓	✓	✓					✓			✓	✓
EasyLinePlus[32]		✓				✓							✓							✓		✓	
HERMES[33]				✓	✓		✓				✓												
I2HOME[18]		✓			✓		✓					✓					✓	✓					✓
MonAMI[19]		✓		✓	✓						✓	✓		✓									
Connected Vitality[34]				✓			✓			✓	✓		✓		✓								
easyReach[35]	✓	✓					✓				✓	✓							✓				
AALuis[36]	✓			✓	✓						✓		✓			✓	✓	✓	✓	✓	✓	✓	✓
Care@Home[37]	✓	✓			✓	✓	✓	✓			✓	✓	✓	✓					✓			✓	
GoldUI[38]				✓	✓	✓		✓		✓	✓		✓	✓	✓				✓			✓	
ChefMyself[39]		✓		✓						✓	✓	✓	✓		✓								
EDLAH[40]				✓	✓					✓	✓			✓	✓								
Capmouse[41]	✓							✓		✓													
AEGIS[42]			✓	✓	✓	✓	✓	✓			✓		✓	✓					✓			✓	✓
Long Last. Memories[43]	✓		✓	✓		✓	✓	✓			✓		✓			✓			✓			✓	
PAMAP[44]	✓		✓			✓	✓	✓			✓	✓		✓		✓							
SOCIABLE[45]				✓						✓	✓												

2.4 Survey Summary and Discussion

Since considered projects producing different platforms had different scopes and goals, their architectures and user interaction frameworks were different but always focused towards responding to end user needs as much as possible. From the survey results it is obvious that more traditional ways of output (such as graphical UIs, keyboards and computer mouses) are used in most cases but also that new ways of interaction are starting to be more and more popular. Touch screens as well as voice and gesture interaction are especially being recognized to be suitable for input.

Possibility to change one UI with another (pluggability) is not supported by most of the platforms. Separation of presentation and application logic as a precondition for pluggable UIs was addressed only in a small number of platforms. Possible reason for that is the fact that such an approach makes development of UIs more complex and consequently was only tackled in projects that devoted more effort towards user interaction framework. A lot of work and research effort is still needed in order to make possible development of easy to use and intuitive UIs that are completely separated from underlying functionality. It can be observed that - considering adaptation - most emphasis was put on adaptation of the presentation while a little less on adapting structure and content of information being delivered to end users.

In order to be capable of smart responses, user interfaces will have to know their users and their environment and be able to adapt. In order to achieve good UI adaptation underlying user models and contextual awareness are essential. A comprehensive user profile, as a representation of user model, should include different categories of user data. Context of use includes: i) *user*, with all needs as well as specific abilities and preferences; ii) *environment*, in which interaction occurs; and iii) *ICT system*, composing of hardware and supporting software. Expansion of user modelling field in recent years has also in part been triggered due to the growing amount of data that can be collected from users and their environments. Availability of this new data has, in turn, created new opportunities in the field of user interaction.

Having in mind that targeted users are often very heterogeneous (e.g., have different demographic, cultural, knowledge backgrounds, as well as desires, preferences, impairments or needs) it is clear that *one-user-interface-fits-all* is very difficult, if not impossible, to achieve. A degree of adaptation depends on a number of adaptation stages performed by a user. In most cases, four stages of adaptation process are discussed: i) *initiation* (of the adaptation), ii) *proposal* (of possible changes), iii) *decisions* (on actions to be taken) and iv) *execution* (of the adaptation). The system is *self-adaptive* if it performs all mentioned stages autonomously. For adaptivity to be effective, system has to track user intents, otherwise it will be almost impossible to determine which actions will help rather than hinder the user. On one hand, some generic applications have no idea what a user is attempting to do and hence offer almost no support for adaptation. On the other hand, some ICT systems that try to anticipate user intent are not doing this very well, what can be more annoying than helpful.

3 Ongoing Challenges

Designing for elderly users is a big challenge. Base factors that have to be considered are age, gender and education, but also social and cultural background which are in great deal underestimated.

Support for multimodality is one of the key requirements for ubiquitous home systems. To relieve application developers from dealing with different end user devices and different interfaces general frameworks for multimodal user interaction are becoming of a pivotal importance.

Intelligent UIs have to incorporate some sort of intelligence in order to deal with the increasing amount of information quantity that is arriving from increasing number of different applications offering more and more functionalities. The principles of the *user always in control* over predictable and transparent systems should be considered at all times. Abilities of a system to learn through user interaction and to self-adapt are needed to best meet users expectations. For this aspect, modelling user(s) and their behaviour becomes increasingly important.

User modelling for UI adaptation represent another important challenge. Growing demand for making ICT systems more accessible to users means that user requirements have to be captured by more powerful user models [12]. Depending on a purpose of a specific ELE system user model is adapted and, since ELE is addressing target group whose requirements change in time, this adaptation usually happens continuously and dynamically.

Standardization and guidelines are very important as well. There are quite a few studies that deal with older population and UIs [13],[14]. Although a standard approach for developing UIs for elderly is still lacking, some may argue that a more important problem is acceptance and use of provided standards and not so much their availability.

Trust and acceptability is the final ongoing challenge we identified. The uptake of solutions and platforms developed in AAL projects is still not good enough. There are ethical and technical concerns, as well as trust challenges, that need to be addressed properly. Fear of technology can also lead to reluctance to accept new solutions. Therefore, trust is of crucial importance for acceptability of such solutions. Benefits should be obvious for people what means that apart from interfaces being very appealing they also have to be very usable.

4 Conclusion

With an advancement of technology user interaction has also evolved and will continue to do so. Our activities are increasingly becoming mediated by different devices such as computers or mobile phones. Many solutions that are being developed are rather complex what results in the fact that more attention has to be directed towards designing usable UI. Ability for some sort of self-refinement and dynamic adaptation are very important for UIs in Enhanced Living Environments (ELEs) and this is the reason why good user models are needed.

Many advances in science and technology allow developing services that were formerly not possible, what also adds pressure in designing usable and intuitive

UIs. Interface adaptation is a challenge to accomplish and with the advances in Ambient Intelligence difficulties are constantly growing. Together with these advancements user interaction is also evolving.

Many approaches are used in building different kinds of ELE systems. Results of the performed analysis of user interaction in existing Ambient Assisted Living (AAL) platforms, that are presented in this paper, lead to the conclusion that there are many projects but not yet common platform(s) that can really be considered the standard. In most user interaction frameworks for ELEs certain common characteristics can be identified, but still a lot of work is needed in order to identify the best approach.

Acknowledgements. This study was supported by the "COST Action IC1303 - AAPELE, Architectures, Algorithms and Platforms for Enhanced Living Environments", funded by the European Union and the research project "Managing Trust and Coordinating Interactions in Smart Networks of People, Machines and Organizations", funded by the Croatian Science Foundation.

References

1. Chappell, N.L., Cooke, H.A.: Age Related Disabilities - Aging and Quality of Life. International Encyclopedia of Rehabilitation (2012)
2. Oviatt, S.L.: Advances in Robust Multimodal Interface Design. IEEE Computer Graphics and Applications, vol. 23 (2003)
3. Dumas, B., Denis, L., Oviatt, S.: Multimodal Interfaces: A Survey of Principles, Models and Frameworks. Lecture Notes in Computer Science, Human Machine Interaction, Vol. 5440, pp. 3–26 (2009)
4. Medrano Gil, A., Salvi, D., Abril Jimenez, P., Grguric, A., Arredondo Waldmeyer, M.T.: Separating the content from the presentation: the universAAL UI framework and the Swing UI Handler. 4th International Symposium on Ambient Intelligence, ISAmI 2013, Salamanca, Spain (2103)
5. Armentano, M. G., Amandi, A.A.: Personalized detection of user intentions. Knowledge-Based Systems, Volume 24, Issue 8, pp. 1169-1180 (December 2011)
6. Kobsa, A., Koenemann, J., Pohl, W.: Personalized hypermedia presentation techniques for improving online customer relationships. The Knowledge Engineering Review, 16, pp. 111-155 (2001)
7. Jameson, A.: Adaptive Interfaces and Agents. The HumanComputer Interaction Handbook: Fundamentals, Evolving Technologies and Emerging Applications (2nd ed.), pp. 433-458. (2008)
8. i2 media research limited: Ofcoms Advisory Committee on Older and Disabled People Next Generation Services for Older and Disabled people, ANNEX B: R&D activities http://www.ofcom.org.uk/files/2010/09/ACOD_NGS_ANNEX-B.pdf (13th September 2010)
9. AAL Joint Programme: Catalogue of Projects 2013. http://www.aal-europe.eu/wp-content/uploads/2013/09/AALCatalogue2013_Final.pdf

10. Obrist, M., Bernhaupt, R., Tscheligi, M.: Interactive TV for the home: An ethnographic study on users requirements and experiences. International Journal on Human-Computer Interaction 24 (2), pp. 174–196(2008)
11. Moritz, N., Goetze, S., Appell, J-E.: Ambient Voice Control for a Personal Activity and Household Assistant. Ambient Assisted Living, pp. 63–74 (2011)
12. Arnold, M., Hopewell, P., Parry, P., Sustache, N., Paddison, C.: User Centred Design How to design accessible products. European Usability Professionals Association Conference, Vol. 3. British Computer Society, London, pp. 22–31 (2002)
13. Hawthorn, D.: Possible implications of aging for interface designers. Interacting with Computers, Volume 12, Issue 5, pp. 507–528 (2000)
14. Morgan Morris, J.: User interface design for older adults. Interacting with Computers, Volume 6, Issue 4, pp. 373–393 (December 1994)
15. universAAL project, UNIVERsal open platform and reference Specification for Ambient Assisted Living: 7h Framework Programme of the European Union, Grant Agreement No. 247950, 2010-2014, http://www.universaal.org/
16. Perakis, K., Tsakou, G., Kavvadias, C., Giannakoulias A.: HOMEdotOLD, HOME Services aDvancing the sOcial inTeractiOn of eLDerly People. Lecture Notes in Computer Science Vol. 6693, pp. 180–186, (2011)
17. PeerAssist project: http://www.cnl.di.uoa.gr/peerassist/ [Accessed February 2014]
18. I2HOME project: http://www.i2home.org/ [Accessed February 2014]
19. MonAMI project: http://www.monami.info/ [Accessed February 2014]
20. Wolf, P., Schmidt, A., Klein, M.: SOPRANO - An extensible, open AAL platform for elderly people based on semantical contracts. 3rd Workshop on Artificial Intelligence Techniques for Ambient Intelligence (AITAmI08), 18th European Conference on Artificial Intelligence (ECAI 08), Patras, Greece (2008)
21. AMIGO project, Ambient intelligence for the networked home environment: EU FP6, Grant Agreement No. 004182, 2004-2008, http://www.amigo-project.org/
22. OASIS, Open architecture for Accessible Services Integration and Standardization: EU FP7, Grant Agreement No. 215754, 2008-2011, http://www.oasis-project.eu/
23. MPOWER project, Middleware platform for empowering cognitive disabled and elderly: EU FP6, Grant Agreement No. 034707, 2006-2009, www.mpower-project.eu
24. PERSONA Project, Perceptive spaces promoting independent aging: EU FP6, Grant Agreement No. 045459, 2007-2010, http://www.aal-persona.org/
25. DIADEM project: http://www.project-diadem.eu/ [Accessed January 2014]
26. DOMEO project: http://www.aal-domeo.eu/ [Accessed January 2014]
27. IS-ACTIVE project: http://www.is-active.eu [Accessed February 2014]
28. ALICE project: http://aal-alice.eu/ [Accessed January 2014]
29. FoSIBLE project: http://fosible.eu/ [Accessed January 2014]
30. JOIN IN project: http://www.aal-europe.eu/projects/join-in/ [Accessed Feb 2014]
31. CogKnow project: http://www.cogknow.eu/ [Accessed January 2014]
32. EasyLinePlus project: http://www.easylineplus.com/ [Accessed January 2014]
33. HERMES project: http://www.fp7-hermes.eu/ [Accessed January 2014]
34. Connected Vitality: http://www.connectedvitality.eu/ [Accessed February 2014]
35. easyReach project: http://www.easyreach-project.eu/ [Accessed February 2014]

36. AALuis project: http://www.aaluis.eu/ [Accessed February 2014]
37. Care@Home project: http://www.careathome-project.eu/ [Accessed Feb 2014]
38. GoldUI project: http://www.goldui.eu/ [Accessed February 2014]
39. ChefMyself project: http://www.chefmyself.eu/ [Accessed February 2014]
40. EDLAH project: http://www.edlah.eu/ [Accessed February 2014]
41. Capmouse project: http://www.brusell-dental.com/aal/ [Accessed February 2014]
42. AEGIS project: http://www.aegis-project.eu/ [Accessed April 2014]
43. Long Lasting Memories: http://www.longlastingmemories.eu/ [Accessed Apr 2014]
44. PAMAP project: http://www.pamap.org/ [Accessed April 2014]
45. SOCIABLE project: http://www.cognitivetraining.eu/ [Accessed April 2014]

Processing and Analysis of Macedonian Cuisine and its Flavours by Using Online Recipes

Aleksandra Bogojeska, Slobodan Kalajdziski, and Ljupco Kocarev

Faculty of Computer Science and Engineering, Ss. Cyril and Methodius University, Skopje, Republic of Macedonia
{aleksandra.bogojeska,slobodan.kalajdziski,ljupcho.kocarev}@finki.ukim.mk

Abstract. Culinary data that are available online can be analysed from many different aspects. In this paper we provide methods for portraying the Macedonian cuisine, as a representative of the South-European cuisine, but highly influenced from the Middle-Eastern and Eastern-European cuisine. By performing different analyses on the Macedonian recipe dataset, we look into the food dietary habits in our country, identify its specific ingredients, their combinations, characterise the flavour trait and depict the Macedonian flavour network. We propose a metric that will address the contribution of a specific ingredient in a recipe by combining the extracted recipes data with existing flavour data. This metric reflects the difference of shared flavours in a recipe when a specific ingredient is present or not, and thus allows us to identify the positive and negative contributing ingredients in a cuisine, providing essential information while comparing two different cuisines. The methods provided in this work can be easily applied for analysis of any cuisine of interest.

Keywords: flavour analysis, food science, Macedonian cuisine, ingredients

1 Introduction

Selection of food is one of the basic instincts of all living organisms, including humans. This ability was mainly developed to avoid food poisoning, to provide energy for the body, to fight illness and to contribute to the overall well being. Nowadays, human diet is not only influenced by the food's nutritional value or the body's energy needs, but also by other factors, such as climate and culture. The climate factors include the terrain, the soil quality, the robustness and the availability of the crops and livestock. For example, the usage of spices in the hot climate regions comes from the need of keeping the food resistant to bacteria for longer time periods [1]. The culture factors include beliefs, religion, and socio-economic status. Many spices, ingredients and therefore meals are unique for specific regions and cultures. The Chinese cuisine includes rice as a main ingredient in every dish, the Muslims don't eat pork meat or don't drink alcohol regardless where they live, the Thanksgiving holiday is always associated with turkey meat as a main dish, etc [2].

S. Loshkovska and S. Koceski (eds.), *ICT Innovations 2015*,
Advances in Intelligent Systems and Computing 399,
DOI: 10.1007/978-3-319-25733-4_15

Flavour and flavour combinations can be treated as another factor that has great impact in the selection of food and meals. The flavour can be represented by group of sensations like taste, odour and freshness [3]. It is believed that these characteristics are one of the crucial in our choice of ingredient combinations. This is the so called *flavour compounds profile* of an ingredient and is consisted of list of chemical compounds, which can range from few to few hundred [4] [5]. What is interesting about the flavour compounds, is that they are associated with the hypothesis, which states that ingredients sharing more flavour compounds are likely to taste well if combined together. When the chef Heston Blumenthal discovered that the salty-sweet combination of caviar and white chocolate is a perfect match, he eagerly investigated their chemical structure to find that this is due to their sharing compound *trimethylamine* [6]. Therefore, the food pairing hypothesis can be used to find many novel combinations of ingredients that taste well together. Furthermore, it triggered a study which examines the specific attributes, flavours and combinations of the world cuisines determining whether a cuisine incorporates ingredients that share similar flavours or not [7].

In this paper we analyze the characteristics of the Macedonian cuisine based on different statistical measurements, we assess the food pairing hypothesis for this specific cuisine and extract its flavour network. Starting from gathering of online recipes data, detailed processing and various aspects of data analysis, we provide general framework for assessing the main characteristics of a local cuisine. As a result here we present the most used ingredients, ingredient pairs and flavour contributing ingredients of the local Macedonian cuisine. All this information together with the flavour network provides a insights about the culinary and flavour preferences of the local region, and enables comparing and contrasting geographically or culturally diverse cuisines.

This paper is organized as follows. Section 2 presents the related work in the field of food data mining and analysis, in Section 3 the methods used in this work are given and Section 4 is devoted to the obtained results and their discussion. The paper is concluded in Section 5.

2 Related Work

An exciting work published by Ahn et al.[7] analyses the different world cuisines by testing the hypothesis saying that humans tend to combine ingredients that share same flavour compounds. They also provide the food ingredient network of the world cuisines based on the acquired ingredients and flavour data.

Furthermore, the findings of this work triggered a research in IMB where an algorithm and application were developed enabling more creative combinations of ingredients while innovating new recipes. By analysing big data and connecting the ingredients by their co-existence in recipes and flavour sharing, the algorithm proposes a recipe based on the human preferences for surprise, pleasantness of odour and flavour pairing [1]. The model and analysis results from this research are present in [8], [9].

[1] IBM-A new kind of food science

The flavour pairing of the Medieval European cuisine and the problems arising when dealing with "dirty" data are addressed in [8]. The authors here look trough cookbooks from the medieval period and add existing flavour information. Using the data to test the food pairing hypothesis they extract the main reasons and effects of dirty data and how they alter the analysis results.

In the work of Y.X. Zhu, et al.[10], a specific analysis for the Chinese cuisine is provided. The authors are focused on the many different regional cuisines and investigate their similarity in geographical and climate terms. The findings in this work state that geographical distance increases the usage of similar ingredients, while the climate conditions don't play any significant role.

The work presented here gives a complete framework for data-driven analysis of cuisine specific online recipes, starting from processing raw recipe data, to gaining valuable results using different analysis aspects and statistical measurements, thus providing a detailed characterisation of the flavours of a local cuisine, with the Macedonian cuisine as an example.

3 Methods

Today with the advance of the IT, a lot of online recipe repositories emerged on the world-wide web scene, providing the opportunity for data-driven analysis of food and recipes preferences for specific world regions and cultures.

3.1 Dataset and Preprocessing

Existing Data. The information for the chemical compounds of the ingredients (flavours), was used from the published data available in [7]. The dataset includes three documents: a document with list of ingredients, second document with list of chemical compounds, and the third, a list of ingredient-flavour identifier pairs, where a pair is present if the ingredient is characterized with a flavour.

Recipe Crawler. The quest of available online Macedonian recipe dataset resulted in looking into 10-15 websites containing from one hundred to few thousands recipes per web site. All of the sites have different structure, and some of them using even 'iframes', making them unfeasible to extract data from.

The final decision resulted in crawling eight web sites (`www.kulinar.mk`, `www.moirecepti.mk`, `www.migusto.mk`, `www.tikves.com.mk`,`www.mojatakujna.mk`, `www.chasovipogotvenje.mk`,`www.somelikeitraw.mk`, `www.surovoivkusno.com`). We have defined special crawling function for extracting recipe data for every web site. For the purposes of this work we only stored the ingredient name, the measurement units and some descriptive words. At the end, a total of 3925 recipes records were added to the initial dataset.

Parsing. Once the recipe dataset was collected the process of extracting valid information was divided in three parsing phases:

Stemming. Since the official web sites language is Macedonian and the alphabet is Cyrillic and currently there is no available corpus or any NLP technique that helps defining the word structures and the meaning, the recipes were subject to extensive formatting. First the measurement quantities were removed and after that a stemmer was designed to group the similar words. Each group was assigned with a root word. The stemmer is very simple, words shorter than 5 letters are merged if are same. For longer words it looks in the first 2/3 of the letters in the word, and groups them if they are the same. Additionally, a Levenshtein distance is calculated and taken into account to avoid typing, or different charsets errors. The groupings were manually checked for synonymous ingredients. From the resulting ingredients grouping, two dictionaries were formed. The first dictionary includes the measurement units (kilograms, cups, spoons), preparation and cooking process (chopped, washed, peeled etc.), adjectives (sweet, big, fresh etc.), which is named the *blacklist of words.* The other group of words includes only valid ingredients and presents the *ingredients dictionary.* Using this two dictionaries any new set of recipes can be easily purified and delivered to a "only root word ingredients" form.

Phrases. The second phase look-ups for word phrases as olive oil, lemon juice, black pepper, white wine, vanilla seed etc. and distinguishes them from one word ingredients and ingredients that tend to appear together, as: salt and pepper, onion and garlic, or ingredients records that were not well structured or include multiple ingredients. This process also includes searching for the reverse combination of words in a phrase, like lemon juice or juice from a lemon, and substituting it with a single form. From 5308 found phrases the set was brought to number of 500 valid phrases of ingredients.

Translation. In order to include the flavour information available, it was necessary to perform Macedonian-English translation or vice versa. We decided to translate the Macedonian ingredients dictionary and match them with the existing English dataset. The Macedonian ingredients data set includes 744 ingredients (including phrases) which were then translated mostly many-to-one to English or leaved out, ending with 492 translated and matched English ingredients. Some ingredients were not present in the English dataset, and some were referring to a compact product (ketchup, curry, poppy etc.).

After the deletion of the duplicate ingredients in recipe, and duplicate recipes in the dataset, the final dataset consists of 3628 recipes for analysis, and 3557 translated recipes.

3.2 Analysis Parameters

The resulting recipes dataset was analysed using standard statistics including: number of ingredients per recipe, average ingredients distribution, number of shared compounds per recipe, prevalence (percentage) of each ingredient, pairs and triplets of ingredients in the cuisine, positive and negative contributing ingredients, nodes neighbours and weights distribution of the flavour network.

Ingredient Contributions. To assess the food pairing hypothesis first we calculate the mean number of shared compounds per recipe ($N_s(R)$) using the following formula as in [7]:

$$N_s(R) = \frac{2}{n_R(n_R - 1)} \sum_{i,j \in R, i \neq j} |C_i \cup C_j| \quad (1)$$

where n_R is the number of ingredients per recipe R, and C_i is the set of compounds for ingredient i, and C_j is the set of compounds for ingredient j.

To calculate the specific flavour contributing ingredients for the cuisine we first used the provided equation for ingredients contributions from same paper. This analysis resulted in poor outcomes for the negative contributors, giving ingredients which are rarely used (ex. radish, sauerkraut, sage, pumpkin, all with less than 0.7% prevalence). Therefore, a new formula for calculating the ingredient contributions was created. The individual contribution for each ingredient can be calculated as:

$$\chi_i = \frac{1}{N_c} \sum_{R \ni i} \frac{N_s(R) - N_s(R_{-i})}{n_R} \quad (2)$$

where $N_s(R_{-i})$ is the mean number of shared compounds of the recipe R if ingredient i is excluded and N_c is the total number of recipes in the cuisine. The equation emphasise the influence of each ingredient in the recipe by looking into the recipe flavour structure with or without the specific ingredient. Using the proposed formula we can see whether one ingredient contributes to higher or lower values of the mean number of shared flavour compounds in the overall recipes, i.e. cuisine.

Food Pairing Hypothesis. The testing of the null hypothesis was performed as suggested in [7] by generating 1000 random recipe datasets with same number of recipes and ingredients per recipe, using four different types of null models:

- *Frequency conserving*: For each recipe in the real dataset consisting of N_i ingredients, a random recipe was generated with the same number of ingredients, giving each ingredient probability proportional to its frequency (f_i) in the cuisine.
- *Frequency and category conserving*: This model takes into account the category of the ingredient in the real recipe (meat, vegetable, herb, flower, nut seed, animal, alcohol, fish, cereal, crop, dairy, plant, plant-derivative, fruit, spice) and again chooses the ingredient from the frequency distribution. In this way we preserve not only the prevalence of each ingredient in the cuisine but also the combination of categories.
- *Uniform*: The ingredients in the random recipe are chosen uniformly giving same probability to frequently used and rarely used ingredients.
- *Uniform and category conserving*: An ingredient is selected uniformly from the set of ingredients in the matching category.

Using the above defined measure $N_s(R)$, we calculate the mean number of shared compounds, $N_s = \sum_R \frac{N_s(R)}{N_c}$ for the real dataset and for each of the random datasets obtaining the values of N_s^{real} and N_s^{random}, respectively. The N_s^{random} value for each null model is calculated by averaging the values from the 1000 datasets per null model. Since each of the null models gives great changes in the frequency and type of ingredients in the random recipes this analysis will provide information about the statistical significance of the real dataset, and whether the ingredients in the studied cuisine share more flavour compounds or not than expected. To estimate the significance of the difference between these values, $\Delta N_s = N_s^{real} - N_s^{random}$, a Z-test is used.

3.3 The Food-Interaction Network

The weighted matrix of ingredients and their number of shared compounds was extracted from the available data in [7] including only the ingredients that are present in the Macedonian recipes. At first, a bipartite network was formed using ingredients and flavours as different nodes and each ingredient was connected to the flavour compounds it contains. The projection of this network in ingredient space results in network where two ingredients are connected if they share one or more flavour compounds [11][12][13]. This presents the so called *flavour network.*

The weight of each link in the flavour network ω_{ij} is the merely number of compounds shared between the two nodes (ingredients) i and j. The resulting backbone flavour network [14] of the Macedonian cuisine clustered using two different algorithms [15][16] can be accessed at `http://bogojeska.finki.ukim.mk/projects/flav_net.html`.

4 Results and Discussion

In this section we show the statistical results calculated on the recipes dataset and the flavour ingredient network, Fig. 1.

The distribution of ingredients per recipes presented on Sub-fig. 1a depicts similar results with the distributions of the European cuisine. The average value of the number of ingredients per recipe is 6.76, similar to the values for the North American and European cuisines (ranging from 6.82 - 8.83) [7]. The translated recipe dataset has average ingredients value of 6.17 and this number shows that the translation of the Macedonian ingredients is good, although includes many-to-one matching into English and number of Macedonian ingredients which are not present in the ingredient-flavour compounds database are excluded.

Another extracted distribution was the flavour compound sharing between the ingredients in the recipes, calculated using Eq. 1. The plot represents the percent of ingredients in a recipe that share same number of flavours, Sub-fig. 1b. In this figure we can see that most of the ingredients used together in recipe share at least few chemical compounds, and a small number of ingredients have 40 or more same chemical compounds which define their flavour. Following this figure one can extract conclusion that the cuisine is characterised with recipes that include ingredients that don't share many flavours, but this is a wrong conclusion, explained in the following paragraph.

The flavour network characteristics were assessed using the analysis of the number of links of each ingredient i.e. neighbours and the weights of the links in the network. Sub-fig. 1c shows that there exist many nodes with 100-200 neighbours confirming that the network is dense. Sub-fig. 1d shows the distribution of the link weights in the flavour network, where we can see that many of the links share small number of compounds, and there are only few ingredients that share more than hundred flavours. Therefore, it is also natural characteristic of the flavour sharing distribution in the recipes to show that there are more ingredient combinations that share small number of flavour compounds.

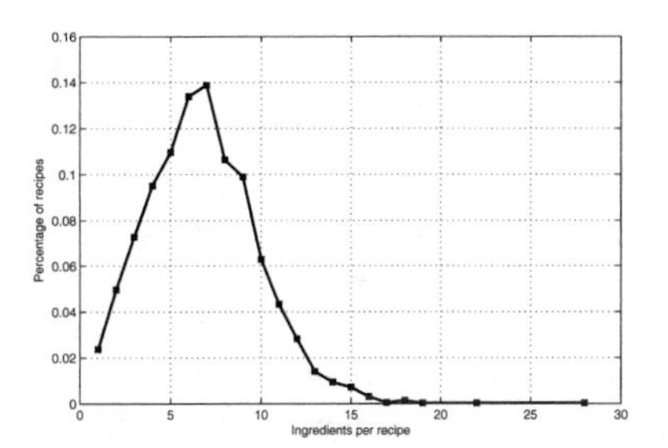

(a) The distribution of the number of ingredients per recipe

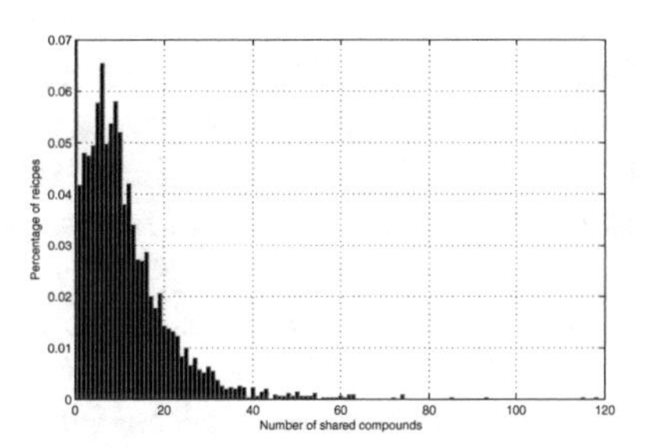

(b) Flavour compound sharing distribution of the ingredients per recipe

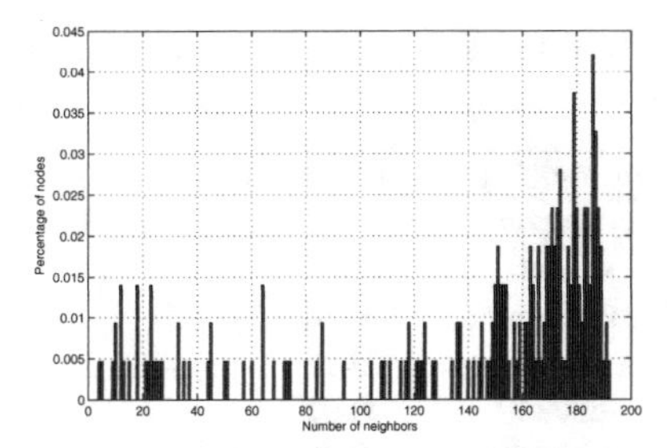

(c) Distribution of the number of neighbours of the flavour network

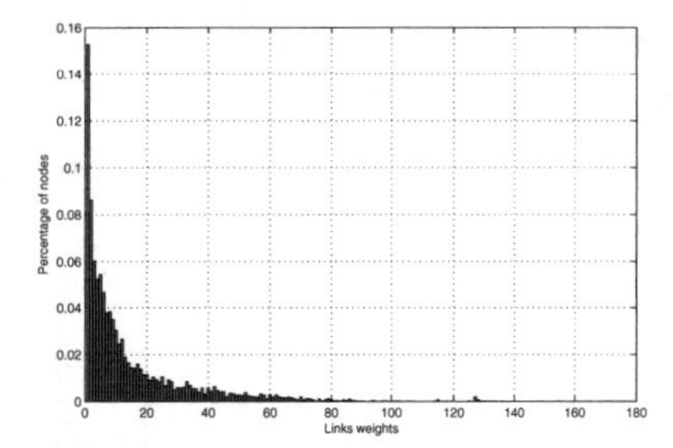

(d) Distribution of the link weights of the flavour network

Fig. 1. Statistical plots

Prevalence. Table 1 shows the 10 most used ingredients in the Macedonian cuisine, which include: black pepper, egg, flour, sunflower oil, onion, milk, garlic and etc. The most common ingredients in the cuisine include the classic European ingredients as milk, eggs, flour, butter, but also specific spices and vegetables as black pepper, parsley, onion and garlic. The top 5 ingredient pairs in the recipes are egg and flour, black pepper and onion, egg and butter, butter and flour and black pepper and garlic, with only butter and egg sharing flavour compounds. The first 5 triplets of most used ingredients are all some combination of flour, butter, egg and milk. The sixth triplet consists of black pepper, garlic and onion.

Table 1. The top 10 most used ingredients in the recipes

Ingredient	Percentage (prevalence)
black pepper	33%
egg	31%
flour	29%
sunflower oil	27%
onion	23%
milk	20%
garlic	20%
butter	19%
parsley	15%
olive oil	14%

Contributions. The main fifteen positive and negative contributing ingredients are given in Table 2, calculated according Eq. 2. The positive contributors include the classical West European and American ingredients (milk, egg, butter), South-European (tomato, cheese, apple, parmesan), Middle-Eastern (lemon, orange), Eastern-European (rum, potato), the negative contributors from the South-European ingredients (garlic, parsley, olive oil), Middle-Eastern (walnut, dates, onion), Eastern-European (onion, yeast). In the negative ingredients we have more unique representers as the walnut, almond, hazelnut, dates which are mostly used in cakes and cookies. These ingredients also have higher frequency of usage in the cuisine.

Table 2. The fifteen most positive and the fifteen most negative contributors in the Macedonian cuisine.

Ingredient	contribution	Ingredient	contribution
Positive		Negative	
butter	0.35	wheat	-0.23
milk	0.30	sunflower oil	-0.16
cocoa	0.18	black pepper	-0.12
cheese	0.16	cacao	-0.07
strawberry	0.11	walnut	-0.06
parmesan cheese	0.09	garlic	-0.06
apple	0.08	olive oil	-0.05
coffee	0.08	lemon juice	-0.05
rum	0.07	date	-0.04
tomato	0.07	parsley	-0.03
lemon	0.07	almond	-0.03
orange	0.06	yeast	-0.03
bacon	0.06	hazelnut	-0.03
potato	0.06	fruit juice	-0.03
banana	0.06	cream	-0.02

Food-Pairing Hypothesis. The Macedonian recipe dataset was analysed against the food-pairing hypothesis. The real dataset has mean number of shared compounds $N_s = 12.2$, which was higher compared to the mean number of all compound sharing distributions of the tested null models (which range from 9-11). According this information it can be concluded that the Macedonian cuisine supports the positive income of the food-pairing hypothesis, although as it can be seen in the prevalence section the most used ingredients and the pairs and triplets do not share many flavours. The tested Z-scores were positive with high values giving the statistical significance of ΔN_s and therefore showing that the selection of the ingredients in the recipes is not random and the number of compound sharing ingredients is greater than expected in random recipe dataset. The results from this analysis can be assessed at `http://bogojeska.finki.ukim.mk/projects/null_models.html`.

5 Conclusion

In this paper we give an overlook to the complete process of collecting, parsing and thorough analysis of online recipes, which provides information for accessing the characteristic flavour trait of a specific cuisine. Here, we depict the Macedonian cuisine, composed as a mixture of ingredients and meals, mostly common for the European and Middle-Eastern cuisines.

The statistical results on the collected recipes show analogue results as the analysis conducted on the European cuisines, with similar number of ingredients per recipe. The top most used ingredients in the cuisine confirmed the statement that the meals in the cuisine are comprised from the main ingredients of the European cuisine (milk, butter, egg), South-European (tomato, garlic, cheese, parsley, olive oil) and Middle-Eastern cuisine (lemon, onion, walnut). The results presented in the pairs and triplets analysis affirm that the ingredients inherited from the European cuisine (milk, butter, egg) share more similar flavours compared to the others. These ingredients are also main factor for contributing to the positive income of the food pairing hypothesis.

Additionally we were interested in finding the characteristical ingredients that add to the positive or negative flavour sharing interactions in a recipe. To achieve this goal we provided a new measure which calculates the contribution for each ingredient by comparing the number of shared compounds in a recipe with and without this ingredient, provided that the original recipe includes this ingredient. Using this formula we identified the top 15 positive and negative flavour contributing ingredients.

In overall, the work presented here gives a full methodology for detailed data-driven analysis and characterisation of a local cuisine, represented by the Macedonian cuisine. The process of analysis includes the initial steps of collection, parsing and translation of online recipes to a form suitable for many aspects of analysis including: basic statistical information, ingredients contribution and flavour network analysis. The results extracted using these methods give a detailed information about the specific cuisine characteristics, its flavour trait information and enable visualisation of the cuisine flavour network. The information extracted using this approach can be used for many diverse experiments as comparing the flavours of different cuisines and their mutual influence, up to proposing recipes with specific flavours.

Our research will continue with analysis of the health benefits from the recipes present in our cuisine by looking into their nutritional values. This analysis will contribute to everyday selection of highly nutritive meals and even providing more healthier substitute ingredients in one recipe.

Acknowledgments

The work in this paper was partially financed by the Faculty of Computer Science and Engineering, Ss. Cyril and Methodius University in Skopje, as part of the project "Knowledge Discovery in Annotated Graphs".

References

1. Billing, J., Sherman, P.W.: Antimicrobial Functions of Spices: Why Some Like it Hot. Q. Rev. Biol. 73, 3–49 (1998)
2. Edelstein, S.: Food, Cuisine, and Cultural Competency for Culinary, Hospitality, and Nutrition Professionals. Jones and Bartlett, Sudbury MA (2011)
3. This, H.: Molecular Gastronomy: Exploring the Science of Flavor. Columbia University Press (2005)
4. Breslin, P.A.S., Beauchamp, G.K.: Suppression of Bitterness by Sodium: Variation Among Bitter Taste Stimuli. Chem. Senses. 20, 609–623 (1995)
5. Burdock, G.A.: Fenaroliś Handbook of Flavor Ingredients, 5th edn. CRC Press (2004)
6. Blumenthal, H.: The Big Fat Duck Cookbook. Bloomsbury, London (2008)
7. Ahn, Y.Y., Ahnert, S.E., Bagrow, J.P., Barabási, A.-L.: Flavor Network and the Principles of Food Pairing. Sci. Rep. 1, 196 (2011)
8. Varshney, K.R., Varshney, L.R., Wang, J., Meyers, D.: Flavor Pairing in Medieval European Cuisine: A Study in Cooking with Dirty Data. In: International Joint Conference on Artificial Intelligence Workshops, pp. 3–12 (2013)
9. Pinel, F., Varshney, L.R.: Computational Creativity for Culinary Recipes. In: CHI'14 Extended Abstracts on Human Factors in Computing Systems, pp. 439–442 (2014)
10. Zhu, Y.X., Huang, J., Zhang, Z.K., Zhang, Q.M., Zhou, T., Ahn, Y.Y.: Geography and Similarity of Regional Cuisines in China. PLOS ONE. 8, e79161 (2013)
11. Newman, M.E.J., Barabási, A.-L., Watts, D.J.: The Structure and Dynamics of Networks. Princeton University Press (2006)
12. Caldarelli, G.: Scale-Free Networks: Complex Webs in Nature and Technology. Oxford University Press, USA (2007)
13. Barrat, A., Barthélemy, M., Pastor-Satorras, R., Vespignani, A.: The Architecture of Complex Weighted Networks. Proc. Nat. Acad. Sci. 101, 3747 (2004)
14. Serrano, M.A., Boguñá, M., Vespignani, A.: Extracting the Multiscale Backbone of Complex Weighted Networks. Proc. Natl. Acad. Sci. 106, 6483–6488 (2009)
15. Rosvall, M., Bergstrom, C.T.: Maps of Random Walks on Complex Networks Reveal Community Structure. Proc. Natl. Acad. Sci. 105, 1118–1123 (2008)
16. Ahn, Y.Y., Bagrow, J.P., Lehmann, S.: Link Communities Reveal Multiscale Complexity in Networks. Nature, 466.7307, 761–764 (2010)

Balancing Performances in Online VM Placement

Sonja Filiposka[1,2], Anastas Mishev[1], Carlos Juiz[2]

[1] Faculty of Computer Science and Engineering, Ss. Cyril and Methodius University – Skopje, Republic of Macedonia
{sonja.filiposka,anastas.mishev}@finki.ukim.mk
[2] Architecture and Performance of Computer and Communication Systems Group, University of the Balearic Islands, Palma de Mallorca, Spain
cjuiz@uib.es

Abstract. Optimal usage of data center resources has become one of the most important topics in the cloud computing research community. Increased efficiency alongside decreased power consumption becomes a desired goal. Crucial point in achieving this goal is the process of virtual machine placement. In this paper, we analyze and compare several heuristics aiming to evaluate their capabilities with special attention to balanced resource usage versus total number of used physical machines. The presented results identify the preferred placement heuristic that achieve maximum balancing performances based on the data center characteristics, size of the cloud services and their diversity.

Keywords: Cloud data center, heuristics, performances, VM placement

1 Introduction

Cloud computing becomes today's prevalent computing trend. The centralized resources that reside inside the data centers are flexibly answering to the elastic online demand from cloud users [1]. The key technology that enables cloud computing is virtualization, facilitating the separation of the physical servers from the operating systems and user applications, thus making the usage of computing resources more economically consolidated. While seeking to maximize the utilization of the available hardware resources, datacenters are simultaneously striving for two potentially diverging targets: maximum performance and minimum energy usage.

The cloud users' demand in Infrastructure as a Service (IaaS) environment is represented as a set of tightly coupled virtual machines (VMs) that are governed by the same user. This set of user controlled VMs represents a cloud service [2] that can be comprised of one or multiple VMs with possibly different resource demands (CPU, memory, etc.).

Key component of the cloud datacenter physical machines (PMs) resource manager is the VM placement algorithm that maps the demanded virtual machines resources onto carefully selected target PMs. The mapping abilities of these algorithms are crucial for achieving the best physical resources consolidation and maximizing the profit. Opposed to traditional grid computing and the problem of job scheduling, in the cloud

S. Loshkovska and S. Koceski (eds.), *ICT Innovations 2015*,
Advances in Intelligent Systems and Computing 399,
DOI: 10.1007/978-3-319-25733-4_16

environment the arrival of cloud service demands can not be controlled by the broker. This situation makes the employment of a batch offline method for deciding the best placement of all cloud services at once an unacceptable solution. Thus, the VM placement algorithm in the case of cloud computing VM placement needs to work *online*: dynamically deciding on the placement of the VMs belonging to a given cloud service independently as the services arrive in the requests queue.

The VM placement problem represents an instance of the bin-packing problem, which is known to be strongly NP-hard [3]. In our case the PMs represent the bins that are to be packed with items, i.e. VMs. Finding optimal solutions to this problem has been a big challenge for the research community which is intensified in the recent period by considering the most general case of bin-packing where both bins and items are described as a vector in the$n$-dimensional space, thus allowing the VMs and PMs to be defined with their resources, each dimension representing one type of resource (e.g., CPU, RAM, etc.)[4]. Opposed to the one-sized bins problem, where once a placement is made on a given PM, it becomes not-usable even though there are still available resources on it; the variable bin size across resources [5] enables the representation of non-homogenous data centers that have potentially different resources available after each cloud service placement. The usual methodology for solving such problems is to build a mathematical representation or a metric of the resource utilizations by different VMs and PMs [6]. This metric is typically a function of the normalized resource utilization of individual resource types, sometimes called resource utilization vector. Some approaches use metric that is a weighted sum of the resources [7], while others use a more complex mathematical function of resources [8].

The final goal of all VM placement algorithms is to map the cloud service into a minimum number of bins available, which is commonly implemented as a heuristic approach that aims to minimize or maximize a given objective function based on the metrics used to describe the problem. Thus, the most popular approaches fall into the greedy types of First Fit or Best Fit heuristics, wherein the ordering is defined using a size function which returns a scalar for each bin and item. Note that all more complex approaches using multi-objective functions are still based on the combination of the heuristic approaches that are examined in this paper. While striving for most efficient packing, the difference in the implementations can also be in whether they take into account balanced resource utilization [9]. Although, load balancing seems indifferent on the small scale of one cloud service placement, it has major repercussions on the overall resource utilization and performances of the entire datacenter. The main objective in this case is to minimize the number of used PMs but in such a way that the used resources are optimally utilized, i.e. the PMs have a small amount of wasted unutilized resources along any dimension of the resource vector.

Thus, in this paper we aim to analyze the performances of the most popular online VM placement heuristic algorithms from the balancing efficiency point of view and how it is influenced by the different characteristics of the datacenter PMs and the demanded cloud services from the user side. The rest of the paper is organized as follows: In the next section we describe the variable size n-dimensional implementation of four different heuristics. In section 3 the results from the extensive performance analysis are presented. The final section concludes the paper.

2 VM Placement Heuristics

The most commonly implemented VM placement algorithms are based on the following heuristics: BinCentric(BC), DotProduct (DP), Vector Based Load Balancing (VBLB) and Vector Based Consolidation (VBC). Since we are mostly interested in the balancing performance of these heuristics, in the first part of this section we give just a brief overview of their packing strategy. For more information please refer to the corresponding references.

The Bin centric [10] heuristic belongs to the Best Fit Decreasing group. The packing starts from the smallest bin and iterates while the list of bins is not empty. It continuously places the biggest item that can fit into the selected bin until such items no longer exist, after which the selected bin is removed from the list. The scalar sizes of the items and bins used for ordering inside this heuristic are calculated as weighted sums of the respective vector components (requirements for items, and remaining capacities for bins). Among the different proposed scaling coefficients used for the weighted sums, we used the best performing BC with capacity normalized fitting implementation where the normalization is based on bins capacities.

The DotProduct [11] approach is an example of First Fit Decreasing heuristic. Its goal is to maximize the weighted similarity of the bin and the item, i.e. the scalar projection of the item requirements onto the bin remaining capacities. In our DP implementation, we normalize both requirements and capacities, thus minimizing the angle between the bin and item vectors. Note that, in order to determine the maximum similarity, dot products for all pairs of bins and items must be computed, which lowers the performance of this heuristic in terms of computational time.

Opposed to the previous approaches, the Vector Based Load Balancing [6] heuristic aims at balancing the load of the bins. Inside this heuristic, the current load of the bins falls into one of three categories: low, medium and high, with respect to the normalized resource utilization vector. When placing a new item, the heuristic tries to discover the least loaded bin that has complementary resource usage with respect to that item.

In the cases when the main goal is to minimize the number of used bins, instead of load balancing, Vector Based server Consolidation [6] heuristic can be used. In this situation, when placing a new item, the heuristic tries to find the bin with the highest load that has complementary resource usage with respect to the item.

2.1 Dynamic Online VM Placement Illustration

As a first step towards discovering the underlying VM placement mechanisms of the discussed heuristics we present an overview of the online placement efficiency in terms of balanced physical machines for 3 sample cloud services (see Fig.1).

Each quadrant represents a single PM described with two normalized physical resources (CPU – x axes, and RAM memory – y axes). Each cloud service is comprised of different VMs and their placement across the PMs is color coded (yellow, green, blue). The light blue rectangles represent the free capacity still available for further use on the PM.

If during VM placement, one of the PM's resources becomes depleted (the VM rectangle reaches the borders of the PM's quadrant, as marked on Fig.1-a), the rest of the PM's resources are being wasted. For achieving efficient use of the data center resources this type of placement is not desirable and eventually leads to using a larger number of PMs that increases the costs and power usage of the datacenter. Thus, one of the important characteristics of the chosen VM placement heuristic must be uniform, i.e. balanced, usage of the PMs that should (in ideal cases) follow the main resource utilization vector represented as the main diagonal of the PM quadrant.

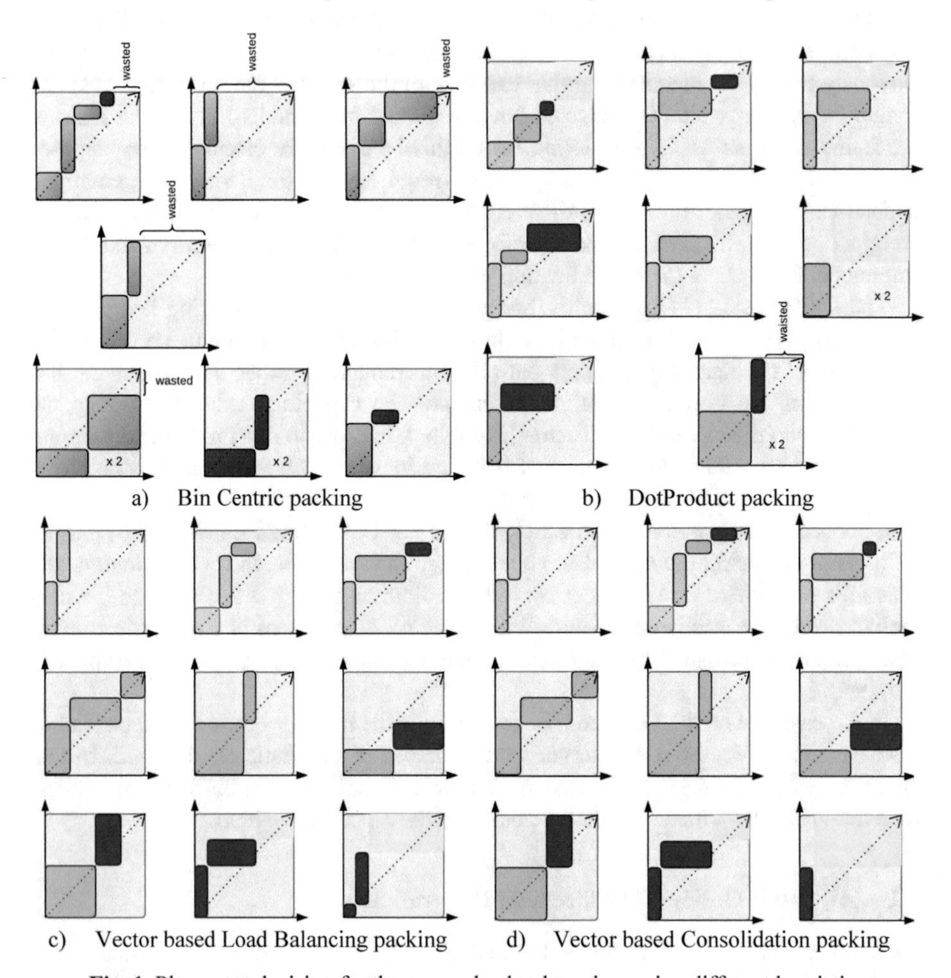

Fig. 1. Placement decision for three sample cloud services using different heuristics

When comparing the different heuristics, we can conclude that the Bin Centric (BC) packing heuristics exhibits the worst performances on balanced packing of the presented sets of VMs, while on the other hand the DotProduct (DP) heuristic achieves the maximum possible balance. However, in order to achieve the maximum balancing DotProduct uses 10 PMs compared to the other 3 heuristics that need only 9 PMs to accommodate the same VMs. The two variations of the Vector based packing differ in the placement of two very small VMs due to the consolidation effort of the second heuristic, which results with slightly better packing. Another remark that should be noted for the presented placement is the small number of variations in the placement decisions across all heuristics, which leads to the conclusion that, when compared on a larger scale, the heuristics should have similar performances, with DotProduct using a slightly larger number of PMs in order to achieve better balancing. However, as presented below, this is not the case.

3 Performance Analysis

In order to analyze the performances of the four heuristics in the case of online VM placement of a large number of cloud services thus recreating a typical cloud datacenter setting, we defined a number of different simulation scenarios by varying the main cloud service description parameters, as well as, the datacenter PMs resources. The results presented in the rest of the paper are obtained different cases of online placement of 1000 to 8000 cloud services, each defined with minimum 5 and maximum 20 VMs. Each VM is randomly generated with 1, 2 or 4 cores and 2, 4, or 8 GB RAM. The VMs are to be placed inside a 5000 or 10000 PMs homogenous cloud datacenter wherein each PM has 8 cores and 16 GB RAM, or 16 cores and 32 GB RAM. Note that the heuristics are deciding on the placement on each cloud service separately, one by one, i.e. online VM cloud service placement, as opposed to the batch mode where all cloud services are placed at once as a complete set of VMs.

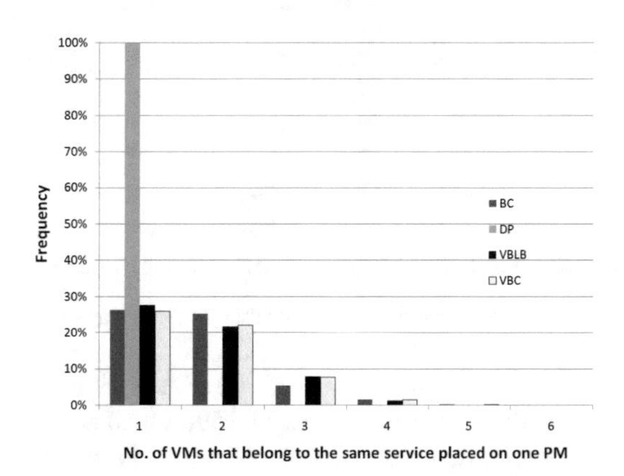

Fig. 2. VM placement diversity across PMs

We first tested the persistence of the DP's typical behavior that was already noticed in Fig. 1. Namely, while DP places one set of VMs that belong to the same cloud service, it aims at placing each VM on a different PM. This however is not a regular case for the other heuristics, where there are also large number of cases when 2, 3 or more VMs that belong to the same cloud service are placed on the same PM as it is represented in Fig. 2. This behavior exhibited by DP is one of the main reasons for achieving the best balancing compared to the other heuristics and is due to the DP's aim towards a global minimum when observing the total placement of all cloud services. However, this strategy's pitfall can be manifested in the case of having a fraction of cloud services consisting of an extremely high number of VMs. In this case DP, following the motto of 1 VM on 1 PM per cloud service, will have to allocate new PMs, while the other heuristics will consolidate the placement better and yield to better resource usage.

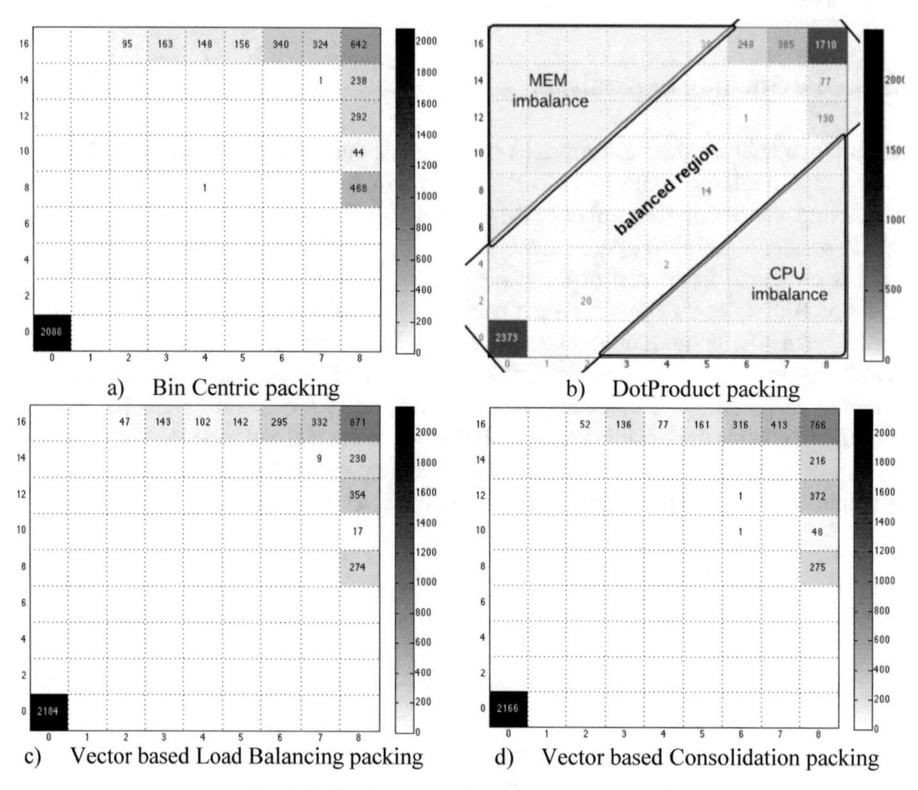

a) Bin Centric packing b) DotProduct packing

c) Vector based Load Balancing packing d) Vector based Consolidation packing

Fig. 3. Balancing VM placement "heat" maps

The resource usage of a cloud datacenter with 5000 PMs is given in the corresponding heat maps in Fig.3after the online placement of 1500 cloud services, each with max 10VMs. The cell annotation represents the number of PMs that have the corresponding used resources (CPU x-axis, RAM y-axis).Note that lowest leftmost cell (0,0) represents the empty, not-used PMs, and the top right cell (8,16) contains the number of fully occupied PMs that have no wasted resources.

All four heat maps depict the dense packing ability of the chosen heuristics, where there is a very small number of PMs that are not close to fully packed according to at least one resource dimension. When considering the performances of the different heuristics via the number of used PMs only, the absolute winner is DP, followed by VBLB and VBC that show slight differences, and BC as the worst performer. We use the represented heat maps in order to gain a deeper insight on the way these performances are achieved, especially from the point of view of balanced or wasted resources, and future usage of the not fully used PMs.

Following the examples from Fig. 1, we define the usage of the resources to be *balanced* if the majority of the PMs are within the region around the main diagonal (consider the annotation on Fig. 3-b). Outside this region the PMs can be unbalanced due to the higher number of cores used while a larger portion of the memory remains unused, the so-called *CPU imbalance* region, or due to the higher quantity of memory used while there is a large number of cores still available, the *MEM imbalance* region.

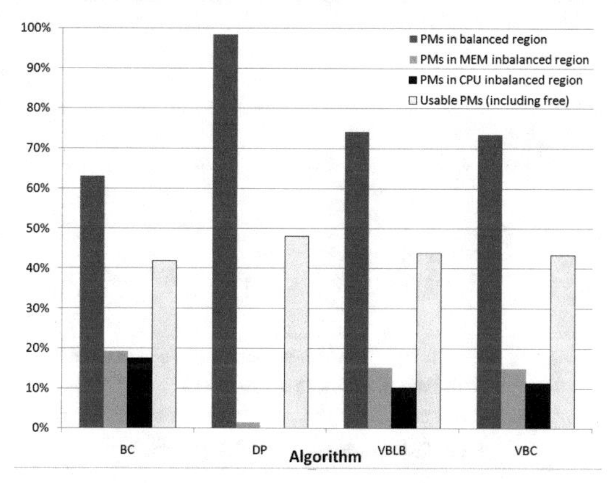

Fig. 4. Comparison of the balanced PM resource usage

The performances of the different heuristics in terms of number of PMs in the balanced versus imbalanced region, are presented in Fig. 4. As it is expected, DP has almost all of the PMs (98%) placed in the balanced region, with only a few in the memory unbalanced region. Also, DP has the highest number of still usable PMs (used CPU<8, and used Memory<16) upon the placement of the full set of cloud services. Next in performances are VBLB and VBC, while BC is last, having lowest number for both balanced and usable PMs.

In order to further understand the influence of the PM resource capacity on the heuristics behavior, we compared the balancing performances of the placement decisions for the cases when the cloud data center is built using PMs with 8 cores and 16 GB RAM vs. 16 cores and 32 GB RAM. As shown in Fig. 5, the PM capacity does not strongly affect DP's balancing performances, although it performs slightly better when the PMs have smaller capacity. Aside from DP, when increasing the PM resource capacity, the heuristic performances are decreasing because of the accentuated non-balanced packing when compared to DP.

An interesting observation is that VBC down-performs relative to the rest of the heuristics, with its performances falling in the case when larger PMs are used. Thus, when working with PMs with higher resource capacity, the difference between the two vector-based approaches is more pronounced.

The differences in performances per heuristic that can be observed in Fig. 5 are due to the different nature of the cloud services that are to be placed, or more precisely, the size in terms of maximum VMs per cloud service. While this parameter has no influence on DP's performances, the changes in balancing performances for the other three heuristics are especially pronounced for smaller number of total VMs. These variations are largest for the BC's performances, and this is the reason why this heuristic has been chosen as a representative for the results presented in Fig. 6. VBLB and VBC exhibit high to moderate performances with similar, less obvious, behavior.

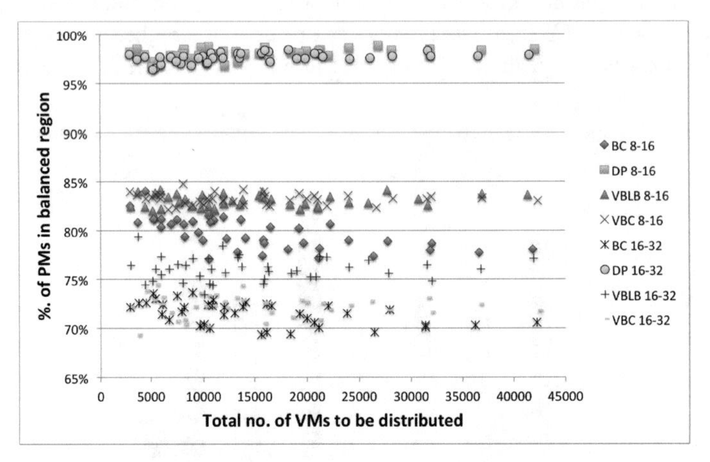

Fig. 5. Balanced PMs distribution

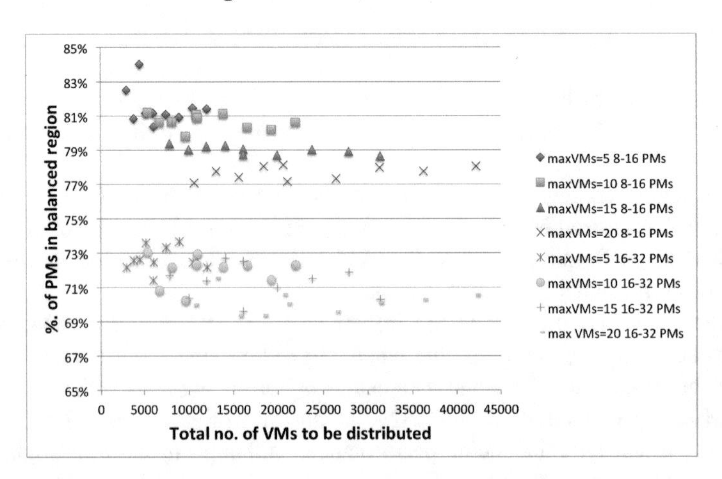

Fig. 6. Cloud service size influence on BC's balancing performances

Fig. 6 clearly demonstrates that in both cases for different initial resource capacities of the underlying PMs, BC's balancing performances are best in the cases when the online cloud service placement is done for small cloud service sizes (max 5 VMs). As the cloud service size grows the heuristic balancing performances are dropping due to the fact that in each service placement the heuristic tends to fill the PMs up their maximum capacity thus making more imbalanced packing decisions.

However, one must bear in mind that the total number of PMs used to accommodate the same set of cloud services changes drastically with the change of initial PM resource capacity. Fig.7 represents the overall packing performances of the four heuristics, when larger and smaller PMs are used. It is expected that when using larger PMs the number of used PMs is lower. Yet, the figure shows that the overall performances of the DP heuristic are deeply influenced by the PM's initial resource capacity. When using smaller PMs (8 cores and 16 GB RAM), the DP performances are the highest, using the fewest PMs of all compared heuristics. But, in the case of larger PMs, DP underperforms even the so far worst heuristic BC, using the largest number of PMs compared to the other heuristics. This is due to the same fact that makes DP the best balancer: choosing to place all VMs from one cloud service on different PMs makes DP use more PMs per cloud service. In the case of high capacity PMs, reaching the full used capacity requires a huge number of cloud services to be distributed.

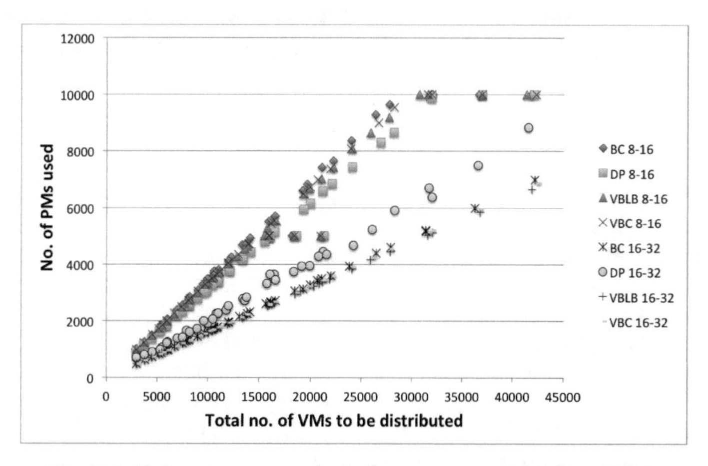

Fig. 7. Initial resource capacity influence on number of used PMs

For the rest of the heuristics, their performances are tending to equalize for higher capacity PMs, showing slight variations in the number of large capacity PMs used. All of the observed differences in performances are due to the fact of tasking the heuristics to pack proportionally much smaller items into large bins that becomes an exceptionally difficult problem for DP, while BC is most resilient to changes compared to the other three, although still exhibiting worst performances.

4 Conclusion

In this paper we analyzed the balancing performances of the most popular online VM placement heuristics used today by cross comparing them to the overall global performances. Our goal was to determine the influence different factors like: characteristics of the cloud services that are to be placed, and features of the physical resources capacity, have on the overall and balancing performances.

Balancing performances of online VM placement heuristics in cloud data centers is crucial for determining the long-term behavior and efficiency of the data center as a whole. While using heuristics that provide best balancing (DP) ensures the best possible usage of the PMs resources, there are cases when due to dimensioning mismatches the price that will be paid for an efficient resource usage is the engagement of a larger number of PMs leading to higher power consumption inside the datacenter.

Thus, a careful highly tailored choosing of the VM placement heuristic that is going to be employed needs to be made in order to align the datacenter physical characteristics with the users demand in the form of cloud services. The overall results show that the BC heuristics is the worst choice for all analyzed cases no matter the type of cloud services or underlying resource capacities. On the other side, DP holds best performances for well-matched cloud service – physical capacities. Hence, if there is no prior knowledge on the compatibility of cloud service demands with the available physical resources, the conservative approach would be to use vector load balancing, while vector consolidation exhibits slightly lower performances.

References

1. Mell, P., Grance, T.: The NIST definition of cloud computing. National Institute of Standards and Technology, 53(6), 50 (2009)
2. Tordsson, J., Montero, R. S., Moreno-Vozmediano, R., Llorente, I. M.: Cloud brokering mechanisms for optimized placement of virtual machines across multiple providers. Future Generation Computer Systems 28:2, 358-367 (2012)
3. Magazine, M. J., Chern, M.-S.: A note on approximation schemes for multidimensional knapsack problems. Mathematics of Operations Research, 9:2, 244-247 (1984)
4. Xi Li; Ventresque, A.; Murphy, J.; Thorburn, J.: A Fair Comparison of VM Placement Heuristics and a More Effective Solution.2014 IEEE 13th International Symposium on Parallel and Distributed Computing (ISPDC), 35:42, 24-27 (2014)
5. Gabay, M., Zaourar, S.: Variable size vector bin packing heuristics - Application to the machine reassignment problem. OSP (2013)
6. Mishra, M., Sahoo, A.: On Theory of VM Placement: Anomalies in Existing Methodologies and Their Mitigation Using a Novel Vector Based Approach. IEEE 4th International Conference on Cloud Computing, 275-282 (2011)
7. Arzuaga E. et al: Quantifying load imbalance on virtualized enterprise servers. In Proc. of the 1st joint WOSP/SIPEW int. conf. on Performance engineering, 235–242. ACM(2010)
8. Wood,T. et al: Black-box and gray-box strategies for virtual machine migration. In Proc. Networked Systems Design and Implementation, 17-17 (2007)
9. Chen, L., Zhang, J., Cai, L., Li, R., He, T., Meng, T.: MTAD: A Multitarget Heuristic Algorithm for Virtual Machine Placement. International Journal of Distributed Sensor Networks, Article ID 679170, (2015)
10. Panigrahy, R., Talwar, K., Uyeda, L., Wieder, U.: Heuristics for vector bin packing. research.microsoft.com (2011)
11. Singh, A., Korupolu, M., Mohapatra, D.: Server-storage virtualization: integration and load balancing in data centers. In Proc. of the ACM/IEEE conf. on Supercomputing (2008)

Synergistic User $\leftrightarrow$ Context Analytics

Andreea Hossmann-Picu[1], Zan Li[1], Zhongliang Zhao[1], Torsten Braun[1], Constantinos Marios Angelopoulos[2], Orestis Evangelatos[2], José Rolim[2], Michela Papandrea[3], Kamini Garg[3], Silvia Giordano[3], Aristide C. Y. Tossou[4], Christos Dimitrakakis[4], and Aikaterini Mitrokotsa[4]

[1] University of Bern, Switzerland
[2] University of Geneva, Switzerland
[3] SUPSI, Switzerland
[4] Chalmers University, Sweden

Abstract. Various flavours of a new research field on *(socio-)physical* or *personal analytics* have emerged, with the goal of deriving semantically-rich insights from people's low-level physical sensing combined with their (online) social interactions. In this paper, we argue for more comprehensive data sources, including environmental and application-specific data, to better capture the interactions between users and their context, in addition to those among users. We provide some example use cases and present our ongoing work towards a synergistic analytics platform: a testbed based on mobile crowdsensing and IoT, a data model for representing the different sources of data and their connections, and a prediction engine for analyzing the data and producing insights.

Keywords: crowd-sensing; information fusion; crowd-sensing analytics

1 Introduction

The goal of *(socio-)physical* or *personal analytics* [4,15,14,13] is to derive semantically rich insights about people (high-level activity, preferences, intentions) from low level measurements (e.g., location, type of activity), from their (online) social interactions, or from a combination of these. The results of such analytics could be used to improve customer engagement for businesses, provide space and event planning that accounts for the self-organising phenomena in crowds, and create higher value location-based services for users. People's behavior is influenced by their environment e.g., weather, infrastructure, air quality. For example: on a *rainy day*, one may take the bus rather than cycle. In some application scenarios, specialized information may also be useful. For example, when analyzing shopper behavior, if a *big sale* is announced, someone may reschedule her regular shopping to attend the sale. We argue here for building a much more comprehensive user context. We propose the concept of *synergistic user* $\leftrightarrow$ *context analytics*, illustrated in Fig. 1, as a way to promote the generalizability of an analytic initiative. Synergistic Analytics (SA) is a modular construction, consisting of the above-cited personal analytics core (based on smartphone and online

S. Loshkovska and S. Koceski (eds.), *ICT Innovations 2015*,
Advances in Intelligent Systems and Computing 399,
DOI: 10.1007/978-3-319-25733-4_17

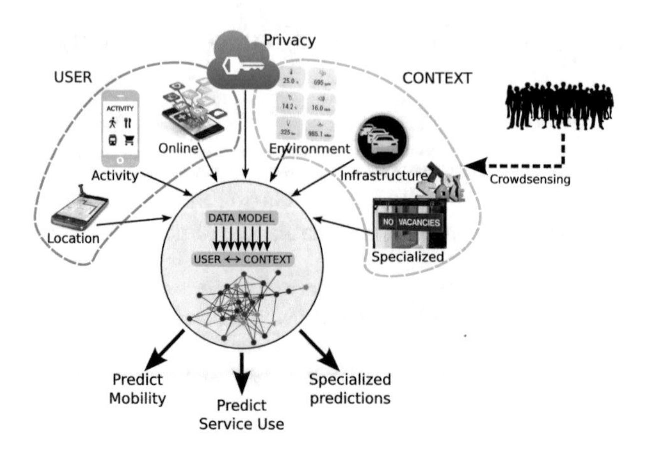

Fig. 1. Synergistic User ↔ Context Analytics

media data), enriched with extra layers of additional information, such as environmental, infrastructure-related or specialized data (e.g., retail). The results of Synergsitic Analytics will be much more than the sum of its parts: instead of isolated predictions of limited scope, deeper, semantically richer inferences are possible. Privacy-protection must be developed alongside and in full synergy with other system's components. Our platform for providing privacy-preserving, location- and context-based services to users aims to support a variety of applications, as discussed in Section 3. The scenarios for synergistic analytics underscore several scientific challenges, to be addressed by relying on the following research pillars, cf. Fig. 1: (i) location and activity prediction; (ii) context (environment, infrastructure etc) awareness via crowdsensing analytics; (iii) social profile and behavioral analytics; and (iv) privacy-preservation methods for each of the above. We present our ongoing work on three main aspects of the platform: a testbed with two units that we aim at integrating (a crowdsensing unit with smartphones and an Internet of Things (IoT) unit with sensors/actuators) in Section 4.1; a data model and storage solution, for efficiently representing and processing the highly heterogeneous information collected from the smartphones and from the sensors in Section 4.2; and a predictive analytics engine in Section 4.3.

2 Related Work

(Socio-)physical analytics. [14] presents a system that can integrate mobile sensing data with data from online social networks, to provide insights into user mobility and their interactions (both online and physical). *SocialFusion* [4] focuses on the immediate context of individuals, rather than on their interactions. In [13] a personal analytics engine generates high-level user states (e.g., emotions, preferences), which can be used to intervene in user actions. In [15] authors recognize the importance of a more comprehensive user context.

Mobile crowdsensing. A user's context consists of many variables: immediate (e.g., location), personal (e.g., activity, heart rate) and of a broader nature (e.g., weather, pollution). Traditionally, these variables are measured via standalone specialized sensors. Through *mobile crowdsensing (MCS)* [8], smart mobile devices can be used to infer or measure the above variables. This is achieved via the devices' sensing capabilities. While this solves the sensing problem, it creates new challenges: resource limitations (energy, bandwidth, computation), privacy and security issues, and the lack of a unifying architecture. The latter is important not only for making the best use of sensor data across applications and devices, but also for guaranteeing privacy and security. A common framework will allow seamless integration of both sensory and security information. We already defined the fundamental components of a mobile crowdsensing system – *crowd*, *server*, *task*) – and their interactions [2], and we have identified incentives for engaging the *crowd*. Our proposed synergistic analytics platform will tackle the additionally raised issues by integrating and jointly analyzing data from different MCS sensors to extract comprehensive patterns and predictions about user behavior and/or their context.

Privacy and security. Mobile crowdsensing (including location and activity sensing) raises many privacy and security concerns. The *crowd* provides sensed data to a *server*, which may or may not be trusted. If the server is not trusted, computation must be performed on encrypted data, which can be achieved via homomorphic encryption [10] or through secure multi-party computation [11]. Even if the server is trusted, private information may still leak, e.g., when a third party constructs clever queries that, if answered truthfully, cause the server to divulge private information. A characterisation of resistance to this is given by the concept of differential privacy [6]. These issues have not yet been addressed in the context of mobile crowdsensing, and it is our goal to design efficient algorithms, fitted for these cases. Our platform integrates privacy and security seamlessly, by embedding privacy and security guarantees within the graph that describes the relations between measured and inferred variables. For privacy, a simple solution is to utilise Bayesian posterior sampling for message passing [5], which allows us to trade off communication costs with privacy and accuracy.

3 Synergistic Analytics Use Cases

Highly Personalized Navigation. Current navigation applications are typically limited to a few transportation modes and miss complex context and user related data. Exploiting data on user preferences, transportation modes, and the environment, allows a more effective user-oriented navigation and recommender system. The data may include real-time traffic data, public transportation, rental vehicles, air quality, weather conditions, safety ratings and user habits. The system suggests places to visit, transportation modes, as well as important traffic and environmental data to city officials. Users benefit by improved social interactions, handling mobility more sustainably and efficiently. Security and privacy issues may arise, such as untruthful users and non-trusted local infrastructure.

User-Optimized Coupon Dispensing. An empirical study [19] found that *proximity* drives coupon redemption. It considered the behavior of people, while moving into the proximity of Subway restaurants: the authors showed that the distance to a restaurant is inversely proportional to the amount of monetary incentive needed to prompt people to redeem the restaurant coupons. However, the physical distance to a shop is not the only driving factor for an optimized coupon distribution. In fact, a better insight into potential customers' profile would allow a more effective dispensing. Along with proximity, other user-related information may be important driving factors, for example: *personal preferences* (i.e., a user who likes Italian food is most likely to visit nearby Italian restaurants) and *social network* (i.e., a user tends to go where their friends have already been). Consequently, a coupon distribution service could optimize the process of customer selection and coupon distribution, by exploiting the our synergistic platform, for retrieving location- and context-related user information.

Recommendation systems. Synergistic analytics could be exploited to make recommendations to users, according to their location and social profile similarity. For example, in a crowded touristic city, the dissemination of localized recommendations (i.e., interesting events and places in the city) among users would be more effective than static provider-based data distribution, in terms of both resource usage (downlink) and time for the recommendations to reach the target users [9]. Such an environment is usually populated by people with various social profiles and interests. The availability of rich information about users may improve the dissemination of localized recommendations by identifying the people and/or communities with similar profiles and interests.

4 Synergistic Analytics: Early Experiences

We present our efforts on three main aspects of the proposed privacy-preserving location- and context-based platform: a testbed with two units that we aim at integrating; a data model and storage solution, for efficiently representing and processing the highly heterogeneous information collected from the smartphones and from the sensors; and a predictive analytics engine.

4.1 Data Collection

For our generic platform for location- and context-based services, we need access to real(istic) data and to be able to easily develop, deploy and debug software on real(istic) end devices. We are building VIVO, a novel human- and sensor-based testbed with volunteers.

The VIVO volunteer testbed The VIVO testbed is based on the concept of *enrolled crowdsourcing*, which allows the deployment of several experiments, as opposed to the traditional usage of crowd-sourcing for a single experiment. VIVO provides a secure and privacy-respecting platform for *testbed users* to

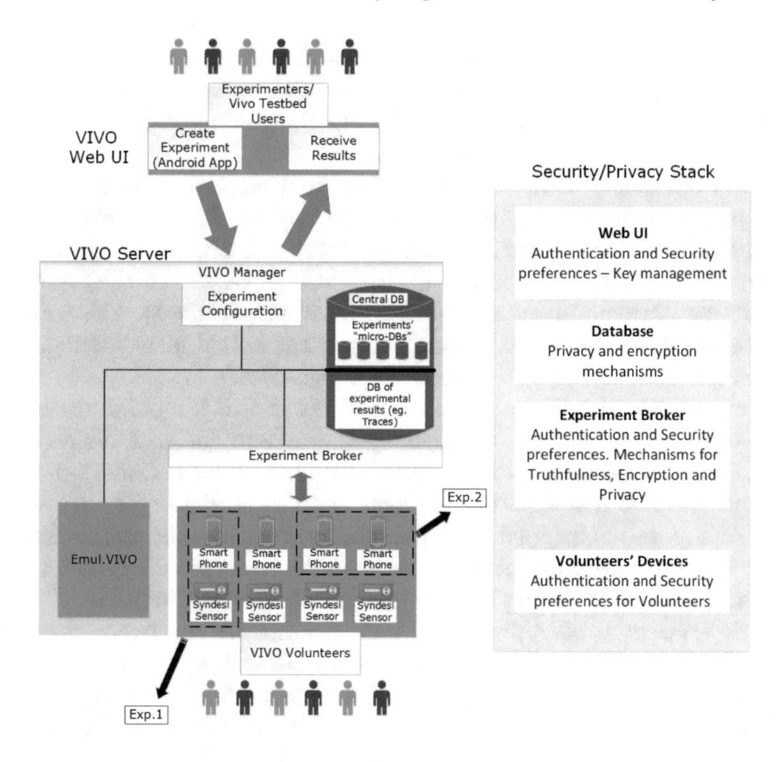

Fig. 2. VIVO Testbed Architecture

collect social, physical and environmental information. The information can be accessed remotely, as in traditional testbeds. However, VIVO differs from traditional testbeds in that it allows testing algorithms and solutions by scheduling and running them *in real time* on real mobile phones of people participating in the testbed (also called *volunteers*, not to be confused with *users*). Further, VIVO also provides an emulation environment for *users* to run and test experiments on already existing data, stored in the VIVO database. Unlike LiveLab [3] and SmartLab [12] (where a single specific and static application is installed on each smartphone to constantly save the data collected from the sensors), VIVO aims to offer more flexibility. More precisely, VIVO *testbed users* can dynamically deploy their own application on each *volunteer*'s device, as in PhoneLab[16]. However, while PhoneLab requires *volunteers* to run a modified version of the Android OS on their mobile (thus limiting the set of potential participants), VIVO applications run on normal Android versions without extra hardware requirements. In addition, VIVO promotes reproducibility of the experiments via its emulation environment. Fig. 2 depicts the VIVO architecture. At the top level, experimenters and researchers are provided with a **Web User Interface** for access. They can define new experiments, upload the corresponding source code and parameterise them; e.g. define the number of volunteers to be engaged or the environment in which the experiment will be conducted (indoor, outdoor, in a smart building, etc). At this layer the front-end management of users' authentication takes place and corresponding security preferences are defined. The main back-end platform noted as the **VIVO Server** lies below the Web user interface. It consists of the following elements:

1. The *VIVOManager* handles requests from the testbed users and, based on their preferences, forwards experiments to be run either on real devices or on an emulation environment provided by the *EmulVIVO* component. Once an experiment has been completed, it sends the results to the testbed user conducting the experiment in a secured and anonymised way. This component also performs the back-end management of the actual identification keys as well as the authentication and security preferences.
2. The central database of the system constitutes the anchor point via which the other components are able to exchange data. Here, for each defined experiment the corresponding data structures are maintained. Collected data are then provided to the experimenter and are also available for "a posteriori" analysis; e.g. to be stripped from potentially sensitive information and to be stored in a repository for future reference. The database will also be equipped with mechanisms enforcing privacy and handling encrypted data.
3. The *Experiment Broker* provisions and orchestrates the experiments to be conducted by using devices provided by *VIVO Volunteers*. This component also takes care of aspects such as the time scheduling of the experiments as well as load balancing issues among the available volunteers. While the experiments are running, data collected from *VIVO Volunteers* is stored in the corresponding micro-DB of each experiment. At this layer authentication and security issues related to the *VIVO Volunteers* are addressed. Also, mechanisms regarding truthfulness, encryption and privacy are implemented, thus mitigating such issues from the side of the volunteers.
4. *EmulVIVO* offers an environment to run experiments on existing data, available in the *EmulVIVO-DB*. The reasoning component of this module is the *EmulVIVO Manager*, which is in charge of receiving requests from *VIVO-Manager*, retrieving the corresponding data from the *EmulVIVO-DB* and allocating the emulation-running environment for the requested experiments.

The final layer includes the **VIVO Volunteers**, who are people equipped with smartphones or other personal devices able to run experiments and who accept to run VIVO experiments. Volunteers provide their characteristics (e.g. socioeconomic profile) and also define their availability. The experiments proposed by the VIVO platform must first be checked and validated (during an alpha testing phase) in terms of respecting privacy and trust issues. Also, authentication and security mechanisms are incorporated in the experiments' source code.

The Crowd-augmented Experimenting Facility. Syndesi 2.0 [1] based on [7] is an IoT testbed architecture for smart buildings, which enables the seamless and scalable integration of crowdsourced resources, provided by the end-users of the facility. The end-users of the smart building are equivalent to VIVO *volunteers*, and different from *testbed users*. End-users are not necessarily VIVO volunteers and vice versa. This integration increases the awareness of the facility both in terms of sensory capabilities as well as in terms of end-user preferences and experienced comfort. The purpose of integrating crowdsourced resources, such as smartphones and tablets, is two-fold. First, the sensory capabilities of the resources provided by the crowd are combined with those of the building for smart actuations. Second, the system is able to interact directly with end-users, both to incentivize them to provide sensory data from their devices and to receive feedback. functionalities will be abstracted to the experimenters as services via RESTful APIs, thus enabling their usage in the context of webservices. Given the testbed APIs, an experimenter canto use them while being agnostic of the technical details. Such architectures, in which testbed functionalities are exposed as services, have led to the notion of Testbed as a Service (TBaaS). Thanks to its modular architecture, Syndesi 2.0 can be integrated into the VIVO testbed presented above. All testbed resources of Syndesi 2.0, along with the accompanying mechanisms (e.g. defining the incentivizing strategies towards the end-users) are exposed as services via RESTful APIs. These services can be consumed by the VIVO testbed, thus leading to the integration of VIVO and Syndesi.

VIVO Privacy and Security. The security issues facing the VIVO testbed (including the IoT unit) can be defined by specifying different trust models. First, we can assume that the user trusts the application, but may not trust the central VIVO Server. The user definitely does not trust the intervening network. The server, on the other hand, cannot be sure that the application (or users) are providing truthful information. Many security components are available to make sure that the system is functioning properly. Mechanism design can be used to give incentives to users to provide truthful information. Differentially private statistical models can be used to optimally trade off user privacy requirements with utility of the barometric service, in a task-dependent manner. Finally, cryptographic methods can be used for secure communication between the server and the users. The particular provisioning of the VIVO testbed for trust and privacy preserving issues along with the capability of supporting a heterogeneous set of information will enable the facility to be used in more diverse experiments, by a higher number of end-users, e.g., for monitoring and collecting data on environmental conditions in out-door settings (via sensors for ambient noise and luminance levels, pressure, etc.) and their correlation to user preferences. The extracted data can then be utilized in order to emulate and study more populous crowds in the EmulVIVO running environment.

4.2 Tackling the Heterogeneous Data Challenge

In addition to the challenges of collecting and unifying the data, our proposed platform also needs an appropriate data model that allows easy and efficient querying, processing and analytics. Efficiently storing, processing and analyzing continuous streams of heterogeneous and dynamic data is a complex task [4,13,14]. The goal of analytics is to identify and exploit relationships in data. A graph-based model is the natural data model choice, as widely recognized (e.g.

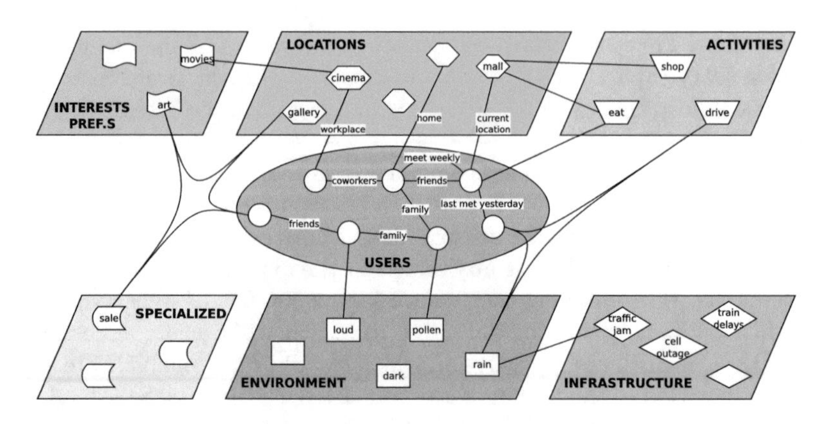

Fig. 3. Graph model for synergistic user $\leftrightarrow$ context analytics

Google's knowledge graph, Facebook's social graph and Twitter's interest graph) Other growing commercial uses include cloud management, bioinformatics, content management, and security and access control.

In the case of Synrgistic Analytics, we are dealing with multiple node types (users, locations, activities etc.) and multiple link types ("knows", "is interested in", "is currently at" etc.). In addition, both nodes and links may have attributes, such as demographic information for users, usage for locations or statistical information for links. Finally, while graphs normally only support edges between two nodes, it would be clearly beneficial to be able to represent links among several nodes, forming hypergraphs. For example, as shown in Fig. 3, an *interest* in art is connected both to the interested *user* and to a *gallery*. Storing this type of information in an efficient, but easy to handle manner is challenging. The two main options are: (i) the new generation (hyper)graph databases and the RDF (Resource Description Framework) databases. Choosing between the two (or additional options) will highly depend on the type of processing to be done on the graph, which we discuss in the next section.

4.3 Prediction Tasks

The prediction engine of our synergistic analytics platform enables different types of predictions, such a user mobility, behavior or service use predictions, as shown in Fig. 1. This engine uses social and physical data, environmental and infrastructure information, and application-specific data to predict the users' next place and behavior, the users' service usage, as well as any required application-specific predictions. Social data mainly consists of the user's social profile (e.g., habits, interests, etc.) and social contacts/activity information. Physical data contains the user's mobility history, activity, sensed data from different embedded sensors in mobile phones and physical contacts with other people. Environmental and infrastructure information may include ambient noise levels, ambient luminance, barometric pressure, public transportation schedules, road traffic data etc.

Finally, the application-specific data (e.g., shopping behavior) should be provided by the contracting entity (e.g. retailer, hotel owner).

The heterogeneity of the collected data gives high potential to the prediction engine, which is then able to perform a deeper analysis of the user and context related data. In terms of mobility, it predicts the user's next-visited physical location together with its semantic meaning (i.e., where the user is willing to go), and it predicts the user's next physical contact. The behavioral prediction includes user activity (i.e., what the user is willing to do), mood (i.e., how the user feels), social contacts, and activity (i.e., who the user is willing to meet).

The prediction methodology is based on both historical and current data. The historical data is analyzed to create a user-dynamic mobility and behavioral model. This allows a user characterization in terms of mobility aspects (i.e., more active or sedentary persons) and the identification of the locations that are relevant for both the user itself and the social community he belongs to (according to users' social profile similarities). The model dynamically adequate to changes in the user mobility and behavior. The current data allows adaptivity to the current user's context, providing so more accurate predictions.

The potentials of including social aspects to location prediction was confirmed in some preliminary study: we showed that with the analysis of the user's mobility history we can classify the visited locations according to their relevance to the user. This classification is then used to retrieve the user's mobility and behavioral *characteristics*. Even this simple information about the user profile already improves the next-visited location prediction [18,17]. The synergistic platform will further combine our initial results with personality and social behavior information to improve the location prediction.

5 Conclusions

We introduced synergistic user $\leftrightarrow$ context analytics, a concept extending recent proposals for (socio-)physical or personal analytics by including more comprehensive data sources. We argued that, in addition to smartphone sensors and (online) social interactions, the environment and application-specific information is valuable for gaining insights into interactions between users and their context. We presented a testbed, based on mobile crowdsensing and the IoT, a data model for representing the different sources of data and their connections, and a prediction engine for analyzing the data and producing the insights.

Acknowledgments. We thank Steven Mudda and Alan Ferrari for their contributions. This work is supported by the Swiss National Science Foundation via the SwissSenseSynergy project and by the COST Action IC1303.

References

1. Angelopoulos, C.M., Evangelatos, O., Nikoletseas, S.E., Raptis, T.P., Rolim, J.D.P., Veroutis, K.: A user-enabled testbed architecture with mobile crowdsensing support for smart, green buildings. In: IEEE ICC 2015, London, U.K., 2015.

2. Angelopoulos, C.M., Nikoletseas, S.E., Raptis, T.P., Rolim, J.D.P.: Characteristic utilities, join policies and efficient incentives in mobile crowdsensing systems. In: IFIP Wireless Days, Rio de Janeiro, Brazil, 2014. pp. 1–6
3. Balan, R.K., Misra, A., Lee, Y.: Livelabs: Building an in-situ real-time mobile experimentation testbed. In: Proceedings of the 15th Workshop on Mobile Computing Systems and Applications. p. 14. ACM (2014)
4. Beach, A., Gartrell, M., Xing, X., Han, R., Lv, Q., Mishra, S., Seada, K.: Fusing mobile, sensor, and social data to fully enable context-aware computing. In: Proceedings of the Eleventh Workshop on Mobile Computing Systems & Applications. pp. 60–65. ACM (2010)
5. Dimitrakakis, C., Nelson, B., Mitrokotsa, A., Rubinstein, B.: Robust and private Bayesian inference. In: Algorithmic Learning Theory (2014)
6. Dwork, C.: Differential privacy. In: Encyclopedia of Cryptography and Security, pp. 338–340. Springer (2011)
7. Evangelatos, O., Samarasinghe, K., Rolim, J.: Syndesi: A framework for creating personalized smart environments using wireless sensor networks. In: DCOSS. pp. 325–330 (2013)
8. Ganti, R.K., Ye, F., Lei, H.: Mobile crowdsensing: current state and future challenges. IEEE Communications Magazine 49, 32–39 (2011)
9. Garg, K., Giordano, S.: Towards developing a generalized modeling framework for data dissemination. EWSN 2015 p. 9 (2015)
10. Gentry, C.: A fully homomorphic encryption scheme. Ph.D. thesis, Stanford University (2009)
11. Goldreich, O.: Secure multi-party computation (1998)
12. Larkou, G., Mintzis, M., Taranto, S., Konstantinidis, A., Andreou, P.G., Zeinalipour-Yazti, D.: Demonstration abstract: Sensor mockup experiments with smartlab. In: IPSN-14. pp. 339–340. IEEE (2014)
13. Lee, Y., Balan, R.K.: The case for human-centric personal analytics. In: ACM MobiSys WPA 2014
14. Misra, A., Jayarajah, K., Nayak, S., Prasetyo, P.K., Lim, E.p.: Socio-physical analytics: Challenges & opportunities. In: Proceedings of the 2014 workshop on physical analytics. pp. 19–24. ACM (2014)
15. Nandakumar, R., Rallapalli, S., Chintalapudi, K., Padmanabhan, V.N., Qiu, L., Ganesan, A., Guha, S., Aggarwal, D., Goenka, A.: Physical analytics: A new frontier for (indoor) location research. Tech. Rep. MSR-TR-2013-107 (2013)
16. Nandugudi, A., Maiti, A., Ki, T., Bulut, F., Demirbas, M., Kosar, T., Qiao, C., Ko, S.Y., Challen, G.: Phonelab: A large programmable smartphone testbed. In: First International Workshop on Sensing and Big Data Mining. pp. 1–6. ACM (2013)
17. Papandrea, M., Giordano, S.: Location prediction and mobility modelling for enhanced localization solution. Journal of Ambient Intelligence and Humanized Computing 5(3), 279–295 (2014)
18. Papandrea, M., Zignani, M., Gaito, S., Giordano, S., Rossi, G.P.: How many places do you visit a day? In: PERCOM Workshops. pp. 218–223. IEEE (2013)
19. Spiekermann, S., Rothensee, M., Klafft, M.: Street marketing: how proximity and context drive coupon redemption. Journal of Consumer Marketing 28(4), 280–289 (2011)

Machine Learning Approach to Blocking Effect Reduction in Low Bitrate Video

Ana Stojkovikj[1], Dejan Gjorgjevikj[2], Zoran Ivanovski[1]

[1] FEEIT Skopje Macedonia
anazstojkovic@gmail.com, zoran.ivanovski@feit.ukim.edu.mk
[2] FINKI Skopje Macedonia
dejan.gjorgjevikj@finki.ukim.mk

Abstract. This work presents an approach for blocking artifacts removal in highly compressed video sequences using an algorithm based on dictionary learning methods. In this approach only the information from the frame content is used, without any additional information from the coded bit-stream. The proposed algorithm adapts the dictionary to the spatial activity in the image, by that avoiding unnecessary blurring of regions of the image containing high spatial frequencies. The algorithms effectiveness is demonstrated using compressed video with fixed block size of 8x8 pixels.

Keywords: Image compression, Video compression, Coding schemes, Blocking artifacts, Super-resolution, Dictionary learning methods, Machine learning methods.

1 Introduction

Digital video is essential part of human interaction today. Its widespread was made possible by introduction of fast and efficient standards for video compression. The most popular and most widely used today is H.264/MPEG-4 AVC, while the new HEVC standard is still in the phase of slow acceptance by the industry due to its complexity. The compression algorithms used in most standards are prone to introduction of artifacts in the final compressed video sequence that can be especially noticeable at low bitrates. The nature of different types of artifacts, as well as the reasons for their introduction, is described in details in [1,2]. Among the different types of artifacts probably the most perceptually annoying are the blocking artifacts. To cope with this problem, compression standards for digital video of the H.264 series have embedded deblocking filter. Another widely used approach is post-processing, performed on the decompressed video sequence. In that direction many algorithms for reduction of blocking artifacts were proposed [3,4,5,6,7]. They use spatial filtering techniques [4] in the area where blocking effect appears or techniques in which the discontinuity in the luminance level is modeled with 2D linear function [5,6]. In [4], three filtering modes depending on the spatial activity and the characteristics of the human visual system (HVS model) are proposed. The algorithm depends on the coding information extracted from the bit-stream. These algorithms treat only the fixed blocking effect

S. Loshkovska and S. Koceski (eds.), *ICT Innovations 2015*,
Advances in Intelligent Systems and Computing 399,
DOI: 10.1007/978-3-319-25733-4_18

introduced at the boundaries of the block, and not the blocking effect inside the block. In [7,8], fixed blocking artifacts, as well as displaced ones, that are result from motion compensation between frames, are effectively treated and reduced. In [7] a technique that utilizes 1D spatial filtering is proposed. It is implemented in two phases, detection of presence of the blocking artifacts and adaptive directional filtering. In [8], a fast algorithm for detection and reduction of displaced and fixed blocking-artifacts that considers only the luminance samples of the frame was proposed. Compared to [8], [7] is more computationally expensive, due to the fact that spatial filtering is applied on all 64 pixels in the block. Although many algorithms for adaptive filtering were proposed, still one of the major problems in these algorithms is introduction of blurring in the areas with high spatial activity.

Another very pronounced artifact of video compression is blurring due to the high frequencies suppression in the quantization phase of the compression algorithm. This artifact is usually coped with using image restoration and super-resolution techniques. Many algorithms for single image super-resolution are based on the concept of joint dictionary learning and sparse representation [9,10,11,12]. These techniques are effective in boosting of high frequencies and, thus, sharpening the image. However, when applied to images containing blocking artifacts they often increase the visibility of the artifacts.

In our approach an algorithm similar to those utilized for super-resolution is used. The algorithm aims to restore the compressed frame, with an intention of reducing the blocking artifacts and increasing the high-frequency content at the same time. Its novelty is in combining the adaptive filtering approach [8] and the dictionary learning methods via sparse representation of an image patch [9,10,11].

In the Section 2 of the paper, a short overview of the nature of different compression artifacts is presented, after which the proposed algorithm is described. Experimental results are presented in Section 3 and Section 4 contains conclusions and directions for future research.

2 Proposed Algorithm

In order to better present the proposed algorithm, a short description of the blocking artifacts nature is presented in the following text.

The utilization of blocks, as base units in processes of transformation, quantization and motion estimation generates unreal discontinuities in the block boundaries in the reproduced frame of the video sequence. These discontinuities can be classified into three sub-categories, usually designated as mosaic effect, staircase effect and false edge [1]. Mosaic effect appears in regions with low spatial activity, i.e. smooth regions. On a block level, in the process of quantization, very often almost all alternate components (AC) from the DCT transform are quantized to zero, therefore, in the reconstruction stage blocks are reconstructed from the DC components. The fusion of these reconstructed blocks produces mosaic effect, and it is characterized with abrupt changes of the luminance level at the block edges. Staircase effect appears along a diagonal line or curve, in the form of fake vertical and horizontal edges at the block boundaries. False edge appears in the vicinity of real edge, and it is due to the motion estimation and compensation between frames in the video.

Restoration of compressed images is a real challenge due to the existence of compression artifacts. Since there is no available information about the uncompressed image, there is a need of a priori knowledge that can be obtained using machine learning approaches. Most intuitive approach in knowledge based image restoration is the dictionary learning approach that is widely used in single-frame SR approaches.

The proposed algorithm was implemented to work with image blocks of size 8x8 pixels; nevertheless, the same approach is applicable for different block sizes. In this paper only grayscale frames (Y component) are considered. The algorithm can be easily extended to consider color frames.

The approach consists of three steps, shown in Fig.1 a). In the first step, image patch of size 8x8 pixels is extracted from the area around each pixel of the frame from the compressed video. For each extracted patch the procedure in the second step, shown in Fig.1 b), is applied separately for horizontal and vertical direction.

In order to make a better distinction between the different types of compressed image patches, we trained three separate dictionaries depending on the spatial activity in the region around the pixel of interest. In the first step, the spatial activity is calculated and then depending on the activity one of the following cases applies. In case when spatial activity is very high, the extracted image patch remains unchanged and there is no need for reconstruction. If the activity is not very high, recovery patch is estimated using one out of three dictionaries, depending on the level of measured activity as described in subsection 2.1. After selecting one of the three dictionaries, a sparse representation of the recovery patch is estimated, as a linear combination of the available dictionary pairs. Iterative estimation of the sparse representation is performed by minimizing the error between the extracted compressed image patch and the estimate of the patch. As a minimizing function, L_2 norm with regularization term is used. In the third step back projection is performed by averaging the luminance of the overlapping areas of neighboring pixels. At the end the frames restored carrying out the procedure in horizontal and in vertical direction are averaged.

2.1 Measuring the Local Spatial Activity

The proposed algorithm uses three types of dictionaries. The selection of the dictionary to be used is determined by the values of the parameters calculated from the luminance values of the neighboring pixels, following the approach of the filtering algorithm described in [8]. For the vertical direction these parameters ($L_{i,j}$, $R_{i,j}$ and $D_{i,j}$) are calculated as shown by the equations (1), (2) and (3). Similar equations are used for the horizontal direction.

$$D_{i,j} = f_{i,j} - f_{i,j+1} \tag{1}$$

$$L_{i,j} = \sum_{m=1}^{3} \left| f_{i,j-m} - f_{i,j-m+1} \right| \tag{2}$$

$$R_{i,j} = \sum_{m=1}^{3} \left| f_{i,j+m} - f_{i,j+m+1} \right| \tag{3}$$

Here $f_{i,j}$ is the luminance value of the pixel at the coordinates i and j from the compressed image. The value of $D_{i,j}$ reflects luminance difference at the border between columns j and $j+1$, and the values of $L_{i,j}$ and $R_{i,j}$ reflect the activity in the region of size 3 pixels left and right of the border, respectively.

The same thresholds as in [8] were used, in order to distinguish which dictionary to use. As shown in Fig.1 b), the first dictionary is used in image regions with low spatial activity, where blocking artifacts are most noticeable. The second dictionary is used in regions with medium spatial activity, weak edges and textures. The third dictionary is used for regions with high spatial activity, sharp edges and clear textures.

Very high values of these measurements imply occurrence of natural edge, in which case the image pattern should be left unchanged.

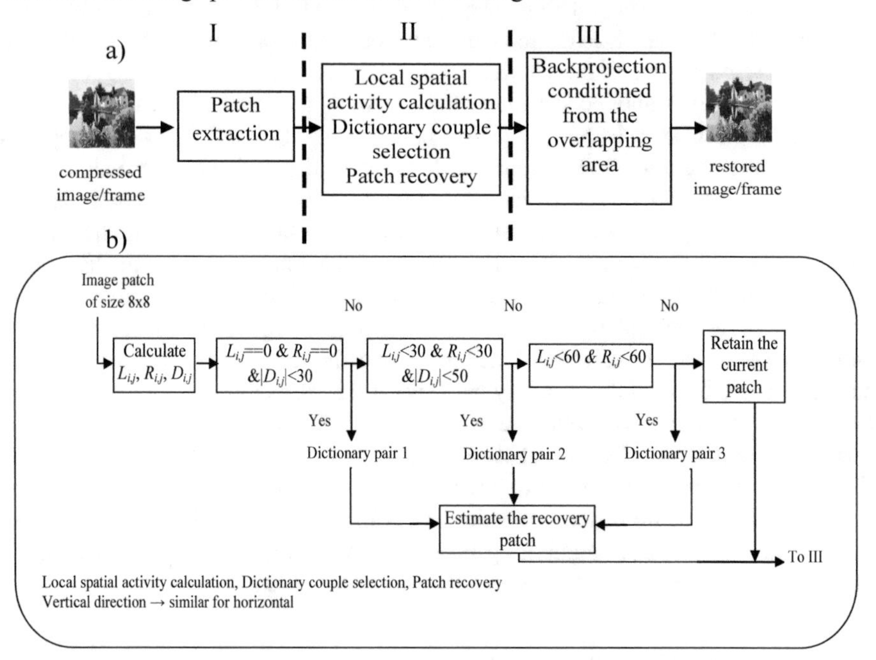

Fig. 1. Block diagram of the proposed algorithm

2.2 Training Process

The aforementioned dictionaries are constructed during the training process. 51 cropped images of size 256x256, taken from frames of 10 different low bitrate videos were used for training. Different types of dynamic and static scenes, with big content variety, were considered.

In order to employ the idea for joint dictionary learning that is usually applied in single-image super-resolution, we used the same concept as in [9] and [10]. Every dictionary is a set of pairs of patches - dictionary pairs. Each pair consists of a patch extracted from the uncompressed image and a corresponding patch from the compressed image. All patches in a dictionary extracted from uncompressed frames are forming a subset denoted as $\mathbf{D}_u$, and the corresponding parts of the dictionary pairs, extracted from the compressed frames are forming a subset denoted as $\mathbf{D}_c$. Training set of dictionary pairs will be denoted with $\mathbf{P} = \{\mathbf{x}_i, \mathbf{y}_i\}_{i=1}^{W}$, where $\mathbf{X}^u = \{\mathbf{x}_i\}_{i=1}^{W}$ represents the subset of uncompressed image patches, and $\mathbf{Y}^c = \{\mathbf{y}_i\}_{i=1}^{W}$, is the subset consisted of compressed image patches. W is the number of patterns in the set. The sparse representation is denoted with $\mathbf{Z}$.

Joint Dictionary Learning. Joint dictionary learning in the training stage is usually performed with utilization of (a) K-SVD algorithm, or (b) k-means algorithm, or simply by (c) alternate minimization of particular cost function of two variables, the estimated set $\{\mathbf{D}_u, \mathbf{D}_c\}$ and estimated sparse representation $\mathbf{Z}$.

Joint Dictionary Learning Using L2 norm Minimization. The estimation of the dictionary is achieved by minimizing the cost functions of the form:

$$\mathbf{D}_u = \arg\min_{\{\mathbf{D}_u, \mathbf{Z}\}} \|\mathbf{X}^u - \mathbf{D}_u\mathbf{Z}\|_2^2 + \lambda\|\mathbf{Z}\|_1 \quad (4)$$

$$\mathbf{D}_c = \arg\min_{\{\mathbf{D}_c, \mathbf{Z}\}} \|\mathbf{Y}^c - \mathbf{D}_c\mathbf{Z}\|_2^2 + \lambda\|\mathbf{Z}\|_1 \quad (5)$$

by forcing the uncompressed and compressed representations to share same minimization code, as proposed in [9]. Minimization procedure, for both dictionary pair, and sparse representation is performed iteratively with appropriate alternation of the variables (method (c) from above). This type of dictionary learning was performed with the Matlab package developed in [13] that utilizes Quadratically Constrained Quadratic Programming Package.

Joint Dictionary Learning Using Joint k-means Algorithm. Another approach for dictionary learning known as joint k-means clustering (JKC) is presented in [10]. In this approach the main idea is to jointly cluster both types of image patches, i.e. image patches from the compressed frame and the appropriate image patches from the uncompressed frame.

The procedure is similar to the classical k-means clustering. For k clusters, we can define a set of cluster centers $\{\mathbf{c}_j\}_{j=1}^{k}$, where each center $\mathbf{c}_j$ consists of uncompressed and compressed parts, $\mathbf{c}_j^x$ and $\mathbf{c}_j^y$, respectively. According to the algorithm joint patch vector $\mathbf{xy}_i = (\mathbf{x}_i, \mathbf{y}_i)$ belongs to certain cluster if both $\mathbf{x}_i$ and $\mathbf{y}_i$ share the same center. The algorithm is consisted of four steps with the two alternating steps (cluster assignment and cluster re-centering), as follows:

1. Arbitrarily initialize the k centers.
2. (Cluster assignment) For each $i \in \{1, \dots N\}$, $L(i) = j'$ if both $\mathbf{c}_{j'}^x$ and $\mathbf{c}_{j'}^y$, are the closest centers to $\mathbf{x}_i$ and $\mathbf{y}_i$, respectively; otherwise $L(i) = 0$.
3. (Cluster re-centering) For each $j \in \{1, \dots k\}$, a related cluster is defined as $C_j = \{\mathbf{z}_i \text{ s.t. } \mathbf{L}(i) = j\}$ and the joint center $(\mathbf{c}_j^x, \mathbf{c}_j^y)$ is recomputed.
4. Repeat steps 2 and 3 until $\boldsymbol{L}$ no longer changes.

In this procedure, $\boldsymbol{L}$ is a vector of labels that contains, element by element, the index of the assigned cluster. We set $\boldsymbol{L}$=0 for those vectors that do not find any placement, i.e. do not belong to the same neighborhood (cluster) of compressed and uncompressed patches.

Additionally, for each obtained dictionary pair, in order to counter-balance the negative effect of the pruning, simple geometrical transformations of the patches should be considered. These are: rotation of 90°, 180° and 270°, horizontal and vertical reflection, as well as the two types of diagonal reflection.

2.3 Patch Recovery and Image Restoration Process

In this step of the proposed approach, the aim is to estimate the recovery patch by using the sparse representation as a linear combination from the patches in the $\mathbf{D}_\mathrm{u}$ subset of the dictionary. The coefficients of the sparse representation $\boldsymbol{\alpha}$ are estimated by solving the optimization problem, as shown below. After that, estimation of the recovery patch $\boldsymbol{x}$ is performed using estimated coefficients. The procedure is as follows:

Input: The appropriate trained dictionary consisted of $\mathbf{D}_\mathrm{u}$ and $\mathbf{D}_\mathrm{c}$ and the extracted patch $\boldsymbol{y}$ for each pixel of the compressed frame.

1. Subtract the DC component from the particular image patch.
2. Solve the optimization problem defined with: $\min_{\boldsymbol{\alpha}} \|\mathbf{D}_\mathbf{c}\boldsymbol{\alpha} - \widehat{\boldsymbol{y}}\|_2^2 + \lambda\|\boldsymbol{\alpha}\|_1$.
5. Estimate the restoration patch $\boldsymbol{x}=\mathbf{D}_\mathrm{u} \cdot \boldsymbol{\alpha}$.
6. Backprojection: put the estimated patch back into the restored image $\widehat{\mathbf{X}}$ by averaging all estimated values for each pixel. Multiple values are estimated for each pixel due to overlapping blocks.

Output: Restored image $\widehat{\mathbf{X}}$.

3 Results

For the performance testing of the proposed approach, nine different video sequences were taken from the Consumer Video Library database site [14]. They were compressed to constant bitrates in the range of 512 to 1200 kbps, and from each sequence one frame was extracted and converted to grayscale. The original uncompressed sequences labeled with 3, 4, 7 and 9 are VGA sequences (640x480p), and the sequences labeled with 1, 2, 5, 6 and 8 are HD videos (1920x1080p). Most of the testing frames were taken from parts of the videos (sequences labeled with 1, 2, 4, 6 and 8) where the scene was static and the camera wasn't moving. In sequence 3, the scene is static and there is a considerable zooming present, while in the sequences 7 and 9, the scenes are very dynamic and the camera is not moving. The sequence labeled with 5 has a very dynamic scene and moving camera (football terrain). Content from natural scene is considered in sequences 1, 2, 3, 7 and 8, and the sequences labeled with 1, 4,

5, 7 and 8 are abundant with details. Faces, as most searched content in an image, are considered in sequences labeled with 4, 6 and 9.

As a measure of quality we have used Peak Signal to Noise Ratio (PSNR) and Mean structural similarity index (MSSIM). These measures are frequently used when the objective and subjective quality are discussed, despite the fact that they do not correspond to the amount of blockiness in a particular image. In order to measure the amount of blockiness in the restored frame we have used Blockiness Measure (BM), as proposed in [15].

In all tests the regularization factor λ was estimated using extensive search in the range [0, 1]. Visually most pleasing results were achieved using λ=0.1.

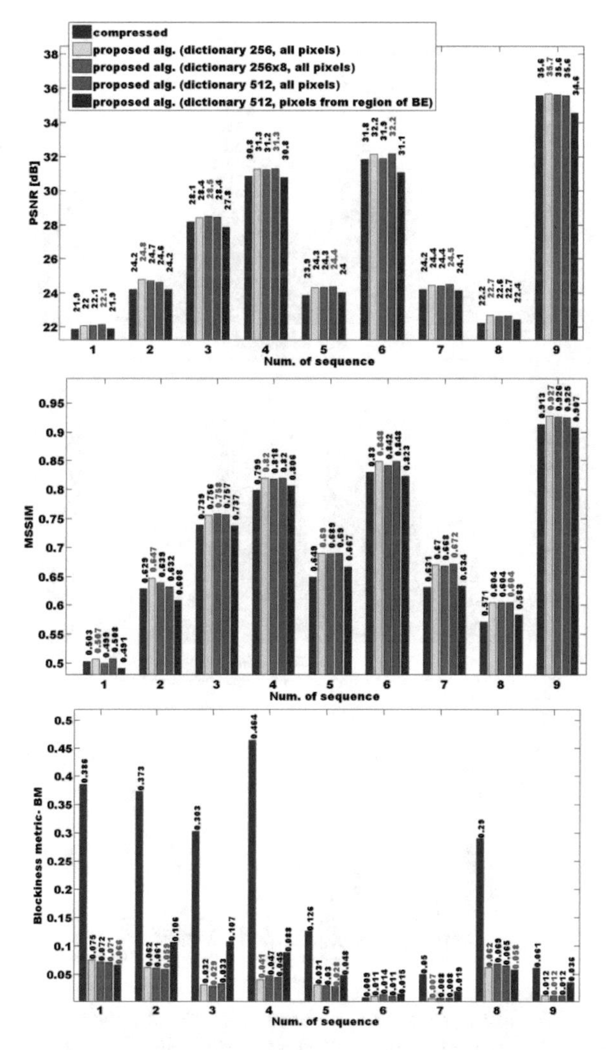

Fig. 2. PSNR, MSSIM and BM values for the compressed frames, and the restored ones with four versions of the proposed algorithm

Two variants of the proposed approach were considered in the performance testing. In the first variant the restoration is performed for each pixel in the frame, and in the second the restoration is performed only for the pixels where the blocking effect (region of BE) was detected, with the detection procedure described in [8]. Results from this comparison are shown in Fig.2. It can be noticed that when algorithm is applied to each pixel, the performance is better in terms of measured quality as well as visual quality.

Two different algorithms for dictionary learning were considered. The results of using dictionaries constructed by algorithm labeled with (c), and the algorithm labeled with (b), (both described in Section 2) were compared.

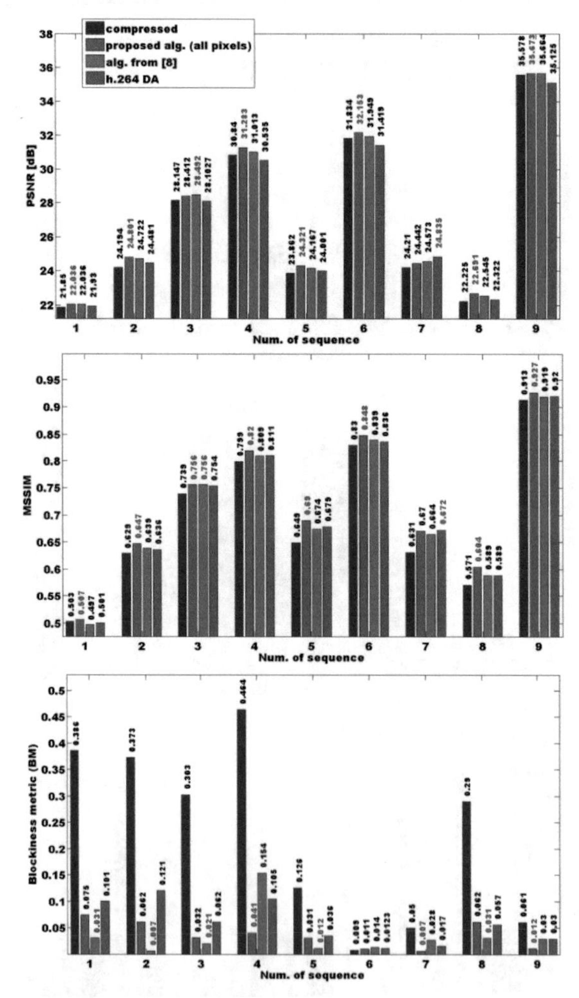

Fig. 3. PSNR, MSSIM and BM values for the compressed frames, the restored frames with the proposed algorithm, the restored frames with the algorithm from [8] and the restored frames with h.264 DA

Because the differences in the obtained results are almost unnoticeable, only numerical values of the quality improvement for the algorithm (b) are presented here.

The size of the dictionary and the variability of image patterns may affect the quality of the restored image. For that purpose dictionaries of size 256, 256x8 (the acquired 256 patterns were geometrically transformed in 8 versions) and 512 were trained. The results are presented in Fig.2. It can be noticed that the improvement in quality compared to compressed frames is achieved in all cases, except for sequence 6 when the quality is measured using BM. This is due to the fact that compressed sequence 6 does not contain significant amount of blocking artifacts. The restoration procedure for this sequence introduced smoothing of some textured regions (ex. grass, leaves etc.), however, the level of smoothing is very low and the restored frame is as pleasant as the compressed one. For all other sequences the blockiness is reduced, but it is not completely eliminated, as can be seen from the values of the BM presented in the graphics from Fig.2 and Fig.3, as well from the results presented in Fig.4.

Fig. 4. Frames from sequence 1, 4, 6, 7, 8, 9 (column-wise); Frames: compressed, restored with dictionary of size 256, restored with dictionary 256x8; restored using [8]; restored using the h.264 deblocking alg. (row-wise)

If we compare the numerical results of different size dictionaries of different sizes, presented in graphics on Fig.2, we can notice that in most cases they have higher values when the images are restored using dictionaries of size 512. It can also be noticed that results achieved with dictionaries of size 256, don't differ too much from those achieved with dictionaries of size 512. This fact brings us to a conclusion that in cases where a particular dictionary is descriptive enough, increasing its size does not affect significantly the video quality. On the other hand, the usage of smaller dictionaries is more efficient in terms of computations and time consumption. Also, from Fig.2, it can be noticed that in the cases when all versions of geometrical appearance of the patch prototypes are considered in the dictionary, PSNR and MSSIM have smaller values compared to those when using dictionaries of sizes 256 and 512, suggesting that dictionaries without geometrical variations are more effective. At the same time, adding the geometrical variations to the dictionaries increases the variability of patterns that are used in the restoration procedure, and due to this fact, the final estimate of the image has more details and distinguishable edges, thus better visual quality. In what follows only the results obtained using the dictionary 256 will be presented.

Comparison results of different algorithms can be seen in Fig. 4. The fourth and the fifth row of Fig.4 show results obtained using the algorithm proposed in [8] and the in-loop adaptive deblocking algorithm implemented in h.264 (h.264 DA), [16], applied as a post-processing algorithm, respectively.

The numerical results for the approach proposed in [8], and h.264 DA (mode 4 - strongest filtering), in comparison with the proposed algorithm are presented in Fig.3. As can be seen in Fig.3, our approach has achieved better results than [8] in 7 out of 9 sequences in PSNR terms and in all 9 sequences in MSSIM terms. In terms of BM the proposed algorithm outperforms [8] only in four cases (sequences labeled with 4, 6, 7 and 9). For the rest 5 sequences, the reduction of the blocking effect is obvious, but the numerical values show that the performance of the proposed and the algorithm from [8] are comparable. Considering h.264 DA, the proposed algorithm shows better performance in terms of PSNR and MSSIM for all sequences except for sequence 7. In this sequence, considering that the camera is moving and also the movement of the bees is rapid, applying stronger filtering with h.264 DA produced smoother outcome in which the blocking artifacts were reduced, while some details were lost. This caused higher PSNR and MSSIM values, compared to the results achieved with the proposed algorithm.

4 Conclusion and Future Work

In this paper an algorithm for adaptive restoration using dictionary learning methods, targeting blockiness reduction in highly compressed videos, was presented. From the presented results it can be concluded that higher values of PSNR and MSSIM for the proposed algorithm are result of the performed restoration, which cannot be obtained using only adaptive low-pass filtering. The presented results also demonstrate significant blocking-effect reduction. The overall performance of the proposed algorithm is comparable and, in some cases, superior to the algorithm proposed in [8] and h.264 DA. Considering the computational cost of the algorithms, the proposed algorithm is computationally more expensive compared to other two algorithms. However, it is a choice of tradeoff between the achieved higher quality and performance speed.

The future research will focus on expansion of the algorithm to work with color videos and different sizes of compression blocks. The research will also address the problem of computational complexity through optimization of the descriptive power of the dictionary.

References

1. Zeng, K., Zhao, T., Rehman, A., Wang, Z.: Characterizing Perceptual Artifacts in Compressed Video Streams. In: Proc. of SPIE, Human Vision and Electronic Imaging XIX, vol.9014 (2014)
2. Randhawal, K.S., Kumar, P.: A Novel Approach for Blocking Artifacts in Compressed Video Streams. In: International Journal of Emerging Technology and Advanced Engineering, ISSN 2250-2459, vol.2. (2012)
3. Kong, H.S., Vetro, A., Sun, H.: Edge map guided adaptive post-filter for blocking and ringing artifacts removal. In: Proc. of International Symposium on Circuits and Systems (ISCAS), vol.3, pp. III-929-932. (2004)
4. Tai, S.C., Chen, Y.Y., Sheu, S.F.: Deblocking Filter for Low Bit Rate MPEG-4 Video. In: IEEE Trans. Circuits Syst. Video Technol., vol.15, no.6, pp.733-741. (2005)
5. Liu, S., Bovik, A.C.: Efficient DCT-Domain Blind Measurement and reduction of Blocking Artifacts. In: IEEE Trans. Circuits Syst. Video Technol., vol.12, no.12, pp.1139-1149. (2002)
6. Petrovski, A., Kartalov, T., Ivanovski, Z., Panovski, Lj.: Blind Measurement and Reduction of Blocking Artifacts. In: 48th International Symposium ELMAR on Multimedia Signal Processing and Communications, pp.73-76. (2006)
7. Kochovski, B., Kartalov, T.,Ivanovski, Z., Panovski, Lj.: An Adaptive Deblocking Algorithm for Low Bitrate Video. In: Proc. of IEEE 3rd International Symposium on Communications, Control and Signal Processing (ISCCSP), pp.888-893. (2008)
8. Petrov, A., Kartalov, T., Ivanovski, Z.: Blocking Effect Reduction in Low Bitrate Video on a Mobile Platform. In: Proc. of IEEE 16th International Conf. on Image Processing (ICIP), pp.3937-3940. (2009)
9. Yang, J., Wright, J., Huang, T., Ma, Y.: Image Super-resolution via Sparse Representation. In: IEEE Trans. on Image Processing, vol.19, no.11, pp.2861-2873. (2010)
10. Bevilacqua, M.: Algorithms for Super-resolution of Images and Videos Based on Learning Methods. In: Image Processing, University of Rennes 1. (2014)
11. Bevilacqua, M., Roumy, A., Guillemot, C., Alberi Morel, M.-L.: Compact and Coherent Dictionary Construction for Example-based Super-resolution. In: Proc. of IEEE International Conf. on Acoustics, Speech, and Signal Processing (ICASSP), pp.2222-2226. (2013)
12. Thiagrajan, J., Ramamurty, K., Spanias, A.: Multilevel Dictionary Learning for Sparse Representation of Images. In: Proc. of IEEE DSP/SPE Workshop, pp.271-276. (2011)
13. Lee, H., Battle, A., Raina, R.., Ng, A.: Efficient Sparse Coding Algorithms. In: Advances in Neural Information Processing Systems, pp.801-808. (2007)
14. Consumer Digital Video Library, http://www.cdvl.org/
15. Wang, Z., Bovik, A.C., Evans, B.L.: Blind Measurement of Blocking Artifacts in Images. In: Proc. of International Conf. on Image Processing, vol.3, pp.981-984. (2000)
16. List, P., Joch, A., Lainema, J., Bjontegaard, G., Karczewicz, M.: Adaptive Deblocking Filter. IEEE Trans. Circuits Syst. Video Technol, vol.13, no.7, pp.614-619. (2003)

Secure and Efficient Automatic Ruling in Three-Level Network Architecture

Pawani Porambage[1], An Braeken[2], Matthias Carlier[2], Pardeep Kumar[1], Andrei Gurtov[3], and Mika Ylianttila[1]

[1] Centre for Wireless Communications, University of Oulu, Finland
[2] Department of Industrial Engineering INDI, Vrije Universiteit Brussel, Belgium
[3] Department of Computer Science and Engineering, Aalto University, Finland

Abstract. Wireless sensor and actuator networks (WSANs) allow intelligent healthcare for homecare environments in ambient assisted living (AAL) systems. Typically, WSANs are constrained networks deployed in three-level network architectures with a central server and several intermediate edge routers. A protocol and system are proposed in this paper for creating rules that can enforce application-layer semantics, both at a central server and at edge routers. With this system in place, multicasting wireless networks would be able to filter messages more efficiently, preserving network and node resources. Rules are implemented by means of the Constrained Application Protocol (CoAP) resources of the nodes, placed in an IPv6 over Low power Wireless Personal Area Network (6LoWPAN). Moreover, we discuss the applicability of the proposed solution by a performance and a security analysis.

Keywords: Constrained Application Protocol, Key management, Wireless sensor and actuator networks

1 Introduction

With the current enhancements of wireless communication and sensing technologies, many applications deployed in smart healthcare and homecare environments in ambient assisted living (AAL) systems are increasingly relying on sensor and actuator networks for collecting and processing information. Furthermore, wireless sensor and actuator networks (WSANs) are also becoming the norm due to the remarkably low installation and maintenance costs [1]. In general, these sensor and actuator devices (i.e., the end-nodes) are highly resource constrained with respect to the memory, processing power, and battery capacity.

One of the key properties of the automated AAL systems is that the operational and the control messages are transmitted to the end-nodes using an automatic ruling mechanism [2]. These rules can be general rules or triggered by a certain input based on a situation or an activity. Furthermore, the rules can be designated to a specific group of end-nodes or lead to some impacts on a particular set of end-nodes. There can be an application in home automation, for instance, that has the general rule to control the light bulbs (i.e., switch on/off)

S. Loshkovska and S. Koceski (eds.), *ICT Innovations 2015*,
Advances in Intelligent Systems and Computing 399,
DOI: 10.1007/978-3-319-25733-4_19

during a certain period of the day or based on the light intensity level in the environment or the number of people in the house. The rules for the light bulbs can be defined on local level (e.g., nodes in the same room) or global level (e.g., nodes in different rooms in the same house).

For the sake of maintaining the confidentiality, security, and privacy of the users, strong security properties have attained an utmost importance in these networks [3]. The efficient and secure key management would be one such way of obtaining high security for automatic ruling scenarios in the automation networks. The network architecture considered in this paper consists of three levels such as the central server (CS), the edge routers (ERs) acting as gateway, and the resource constrained sensor nodes.

Efficiency can be achieved in the first place by making a distinction between local and global level, such that the communication path can be shortened. In the second place, by applying multicast communication to the group of reacting nodes in an automatisation rule instead of unicast communication, less messages need to be sent through the network, which is important for energy efficient solutions in resource constrained networks.

Our main intention is to define the key management and communication scheme for automatic ruling based on secure multicasting in this given network architecture with a minimum overhead at the constrained devices (i.e., Level 3) and to discuss the implementation aspects. The remainder of the paper is organized as follows: Section 2 provides an overview about the related work. Section 3 explains the network architecture, key definitions, and assumptions. Section 4 and 5 respectively describe the proposed solution and the testbed implementation. Section 6 presents the performance Finally, Sect. 7 concludes the paper.

2 Related Work

During the past two decades, many works have been produced and many literature is available about securing low power wireless sensor networks. Key management is one of the major concerns of their security features [3], [4]. One of the well known security frameworks for sensor networks was presented in SPINS [5]. SPINS had two security building blocks such as SNEP for data confidentiality, two-party authentication, and data freshness, and μTESLA for authenticated broadcast. Zigbee [6] based approaches have been also widely proposed for securing wireless sensor networks.

Recently, these types of schemes are applied in the particular domain of home automation [7]. As explained in [7], all the information is securely sent by means of three types of keys such as a master key for end-to-end security, a link key for network security, and a session key. Although, group key management and key update mechanisms are described in many previous literature [4], none of them advocate automatic ruling for multicasting. In this paper, we extend the key management process in order to deal in an efficient way with the automation rules that can be operated in a local or a global environment with multicasting.

The key update mechanism used in the the proposed solution is inspired by the mobile object-based secret key distribution scheme explained in [8].

As mentioned in [9] and [10], the exploitation of multicast communication advocates the efficiency and low energy consumption. They use a secret sharing based approach to establish the key. Although the idea of multicasting is used in our setting, the participants do not need to know each others' involvement in order to proceed. Constrained Application Protocol (CoAP) [11] is a specifically designed web transfer protocol for resource constrained network devices. The implementation of the proposing protocol is mainly inspired by the implementation of rules in the CoAP protocol and the routing principals given in the IPv6 Routing Protocol for Low-Power and Lossy Networks (RPL) [12]. The system is based on IPv6 over Low power Wireless Personal Area Network (6LoWPAN) [13], implying great scalability, minimum access delays, and the least configuration and installation cost.

3 Network Architecture and Definitions

3.1 Notations

All the notations are described in Table 1. For the clarity of presentation, the respective i^{th} and j^{th} values are not specifically used to describe the protocol in Sect. 4.

Table 1. Notations used in the proposed security scheme.

Notation	Description
N_{ID}	Sensor/Actuator node N identity
R_{ID}	Rule number
$K_{ERN_i}/\ C_{ERN_i}$	Shared symmetric key/counter between ER and i^{th} node.
$K_{CSN_{ji}}/\ C_{CSN_{ji}}$	Shared symmetric key/counter between CS and i^{th} node under j^{th} ER domain.
$K_{CSER_i}/\ C_{CSER_i}$	Shared symmetric key/counter between CS and i^{th} ER.
K_{CS}	Group key for the ERs under the domain of CS.
K_{EG}	Group key for the end-nodes under the domain of ER.
K_{R_iCS}/C_{R_iCS}	Group key/counter for end-nodes of ith rule set by CS.
K_{R_iER}/C_{R_iER}	Group key/counter for end-nodes of ith rule set by ER.

3.2 Three-level Network Architecture

We consider the network architecture with three levels of participants as illustrated in Figure 1. Level 1 includes the sensor nodes and the actuators (i.e., known as end-nodes) that are functioning in 6LoWPAN networks which operate according to the IEEE 802.15.4 standard. The required security materials are pre-installed in these end-nodes by the administration during the deployment. The pre-installed security credentials include the node identity N_{ID}, the shared symmetric key with the edge router K_{ERN} and central server K_{CSN},

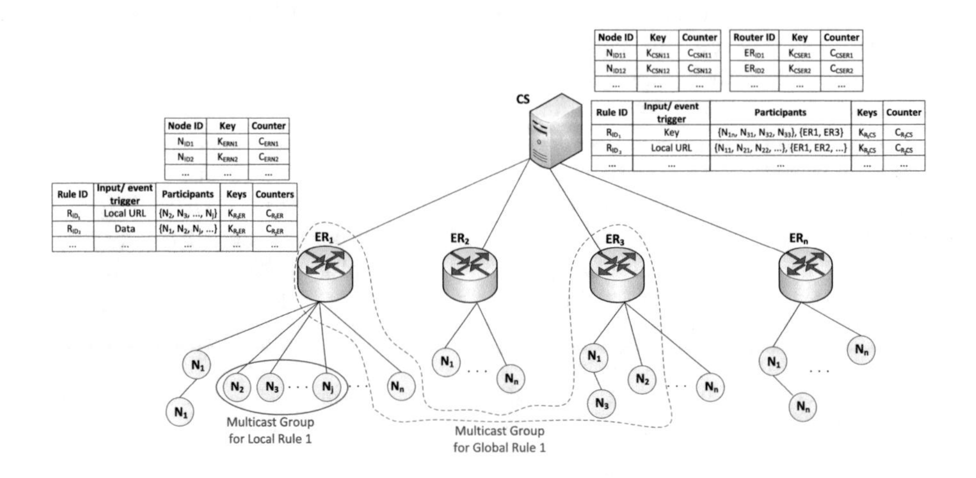

Fig. 1. Three-level network architecture.

and the corresponding counter values C_{ERN}, C_{CSN}. Except the node identity N_{ID}, other values are subjected to change according to the requirements of the network operations, as will be explained later.

In Level 2, the ERs are performing as gateways between the 6LoWPAN network and the IPv6 network, where the CS is deployed. Communication cost and transmission time between end-nodes and ER are much lower than those between end-nodes and CS. Therefore, it would be more efficient to take decisions based on sensed data for controlling sensor devices by the ER if all involved end-nodes belong to the same ER domain. The ER maintains two tables that include a list of active N_{ID}s under its domain with their keys and counters, and a list storing the information belonging to the local level rules. An ER can communicate with the end-nodes in its domain by direct or multiple hops. Level 3 is the CS that maintains three tables. Two tables are used for storing the identities of the ERs and the active end-nodes in the network along with their corresponding keys and counters. The third table contains the required information for the global level rules.

3.3 Local and Global Rules

During the pre-installation, rule tables are set in ERs and CS by the administrator and the corresponding rule activities are implemented in the end-nodes. Rules are triggered by ER or CS as a particular action for a certain set of inputs or events. As shown in the rule tables in Figure 1, each rule is defined with a unique identity (R_{ID_1}) and four parameters. First is the input or event that might trigger some activity (or activities) by the end-node(s). Secondly, there should be a list of participants who act according to the rule. Third and fourth are the group key and counter value exploited for the rule based multicast communication. Local rules are executed by ERs. The list of participants is a multicast group consisting of end-nodes from the same ER domain. The key K_{R_iER} and counter C_{R_iER} are exploited for the ith rule managed by the ER. Global rules are set by the CS whereas the participants include the recipient group of end-nodes and their respective ERs. The multicast group can be formed of end-nodes deployed under multiple ER domains. The respective ERs are also included in the list, due to the purpose of delivering messages to the end-nodes. The key K_{R_iCS} and counter C_{R_iCS} are exploited for the ith rule managed by the CS. Based on the receiving data from the end-nodes or other control messages, first, the ER checks whether the information is interested on a local level. If the information triggers a local rule, the ER will perform accordingly. Otherwise, the ER transfers the information to the CS to execute the appropriate global rule.

3.4 Assumptions

It is assumed that the wireless communication enabled environment assures the confidentiality, where the input of the node is secret (e.g., status of the alarm and presence detectors), and the authentication of the source and the content (e.g., attacker is not able to deactivate alarm sensors). We propose a system enabling these properties and resistant against altering, deleting or inserting data, and replay attacks by transmitting the data in an authenticated encryption mode, taking into account a refreshment countermeasure (i.e., a counter). Moreover, we consider additional resistance measurements such as intrusion detection systems against the interception of the data and the denial-of-service attacks outside the scope of this paper. In addition to that, we also assume the nodes are tamper resistent.

4 Proposed Solutions

4.1 Phase 1: System Initialization and Rules Establishment

During phase 1, all the end-nodes, ERs and the CS are embedded with the security credentials.

Initialization. Initially, all the end nodes are pre-installed with node identities (N_{ID}) and two keys with the corresponding counters (i.e., K_{ERN}/C_{ERN} and K_{CSN}/C_{CSN}) during the deployment phase by the network administrator. Those values are also stored in the databases of the respective ERs and the CS. Moreover, the ER shares a symmetric key and a counter with the CS as K_{CSER} and C_{CSER}. Later, the ER determines the group authentication key K_{EG} and securely delivers the key to each end node under its domain using the shared keys K_{ERN}s. Similarly, the CS derives and delivers the group authentication key K_{CS} to ERs using the respective K_{CSER}s.

Local and Global Rule Establishment. The administrator defines local rules for the ERs and global rules for the CS. These rules are supposed to be executed whenever there is a set of nodes that needs to be controlled based on an input or an event of one or a group of nodes. When an ER receives sensed data not corresponding to a local rule, the ER transfers it to the CS for further processing. The ER uses the shared symmetric keys K_{ERN_j}s (i.e., N_j belongs to the multicast group of the given local rule) in order to encrypt and unicast the local rule keys and counters (e.g., K_{R_iER} and C_{R_iER} for the ith local rule). Similarly, the CS uses the shared symmetric key K_{CSN_j}s (i.e., N_j belongs to the multicast group of the given global rule) in order to encrypt and unicast the global rule keys and counters (e.g., K_{R_iCS} and C_{R_iCS} for the ith global rule). Accordingly in both cases, the activities and corresponding conditions together with the keys and counter values of the rules, are implemented at the end-nodes as explained further in Sect. ??.

4.2 Phase 2: Message Flow for Multicasting

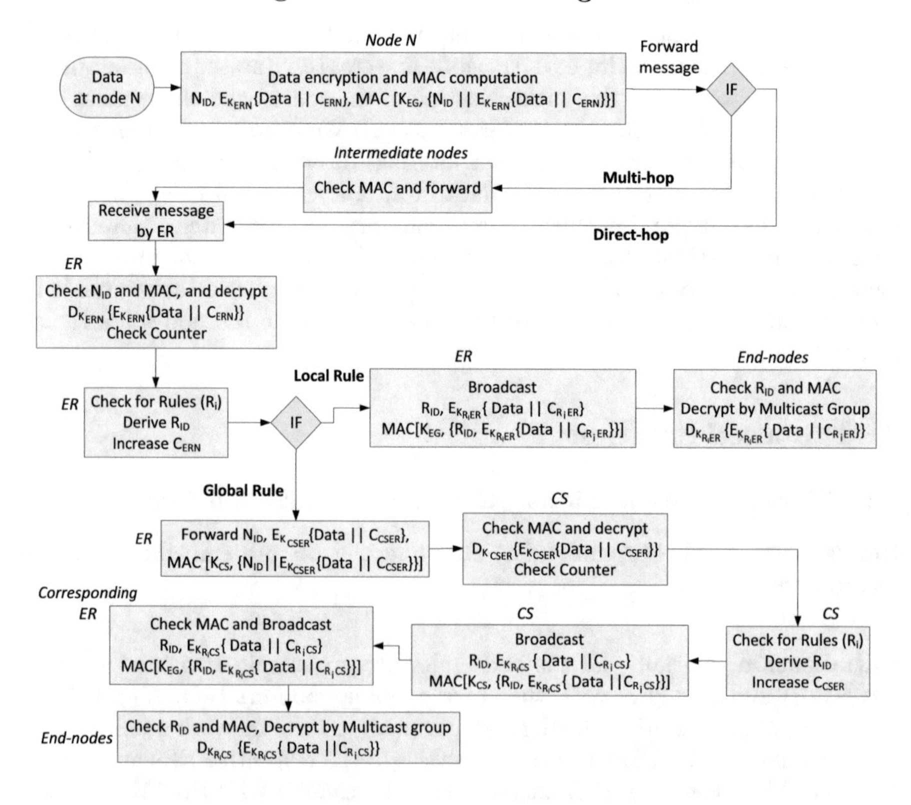

Fig. 2. Phase 2: Message flow Multicasting.

As shown in Figure 2, phase 2 explains in detail the communication flow of the execution of a rule activated by an end-node. Furthermore, the rules can also be activated by the ER or the CS according to the same flow, however, with less steps.

An end-node sends data to the ER either randomly (i.e., for every fixed period) or when the status is changed. Therefore, the following actions are executed. First, the corresponding counter C_{ERN} is increased. Second, it encrypts the data concatenated with the counter C_{ERN} by the unique symmetric key K_{ERN}. Finally, the node sends it along with the node identity N_{ID} and Message Authentication Code (MAC) value $(MAC[K_{EG}, \{N_{ID} \| E_{K_{ERN}}\{Data \| C_{ERN}\}\}])$. The end-node may access the ER by direct or multiple hops. In multi-hop situation, the intermediate nodes check the integrity of the message by comparing the MAC value calculated by K_{EG}, and forward the message only in case of success. Otherwise they discard the message.

Once the ER receives the message, it checks the sender's identity, and the message integrity by computing the MAC value. If it is successful, then, the ER decrypts the message using K_{ERN} and checks first whether the counter is in a certain threshold range. This threshold range is typically defined by the Packet Delivery Ratio (PDR) of the network. This ratio should be experimentally determined in the network. Next, the ER consults whether the received information triggers a local rule. If it does, the corresponding R_{ID} is derived by the ER. Next, the ER encrypts the input data with respect to that R_{ID}, and an increased value of the corresponding counter C_{R_iER}, by K_{R_iER}, and broadcasts this message, along with the MAC value. Only the list of participants in that particular rule can decrypt this message.

Otherwise, the ER encrypts the data concatenated with the corresponding increased counter value C_{CSER} by K_{CSER}, and forwards this message, together with the MAC value to the CS, $(MAC[K_{CS}, \{N_{ID}, E_{K_{CSER}}\{Data \| C_{CSER}\}\}])$. Upon receiving the message, the CS checks its integrity and decrypts it using K_{CSER}. If the counter is increased, the CS checks whether the received information triggers a global rule. If the global rule is initiated, first, the R_{ID} is derived by the CS. Then, the CS encrypts the corresponding data concatenated with an increased counter C_{R_iCS} by K_{R_iCS} (i.e., $[E_{K_{R_iCS}}\{Data \| C_{R_iCS}\}]$) and broadcasts the message (i.e., $R_{ID}, [E_{K_{R_iCS}}\{Data \| C_{R_iCS}\}$, and MAC) to the entire group. The ERs who receive this message will forward it to their groups only if it is assigned to the given R_{ID}. For security reasons, at random times a dummy message starting with a certain R_{ID} is send in the network.

4.3 Key and Counter update

The keys and the counters should be updated when they are frequently used for a long period in order to avoid some security vulnerabilities. For instance, if an adversary can compromise the keys K_{ERN} or K_{CSN}, he could decrypt all the exchanged messages, including these from the past, and if the counter reaches its limit and would be reset, replay attacks can be performed. Therefore, a counter is used for each communication with K_{ERN}, K_{R_iER}, K_{CSN}, and K_{R_iCS}. It is at the ERs and CS that a key and counter update should be initiated when the corresponding counter reaches its limit. The updates of keys K_{ERN}, K_{CSN}, and corresponding counters are performed by unicast communication whereas, the updates of keys K_{R_iER}, K_{R_iCS}, and corresponding counters are performed by multicast communication. These updates can also be implemented as rules. Moreover, at node addition or node removal in the network, the group authentication key K_{EG} should be updated. In addition, if the node is involved in rules (i.e., globally or locally), it is necessary to update the respective keys and counters by the ER or CS. Also the tables of ER and CS, containing the list of active nodes, require an update.

5 Network specifications

The Zolertia Z1 hardware platform running on Contiki-OS is substituted to the role of the end-nodes in the given network architecture. The hardware AES128 co-processor is used for the security operations along with the primitives from the CC2420 transceiver. Furthermore, the authenticated encryption mode of AES-CCM is applied. Each end-node is attached to different kinds of sensors and actuators, that need to be made available to the network. In order to provide the accessibility to the server with the information and control signals of the end-nodes, each end-node is accessed through a RESTful interface as described in the CoAP protocol [11].

The ERs are formed as dual systems by combining a Zolertia mote [14] together with a Beaglebone board [15]. The Linaro-ARM compiler has been used to compile the Linux kernel with all the necessary drivers. This compiler is also used for tunslip6 and other application running on the Beaglebone board. The application tunslip6 ensures the translation of IPv6 data-packets into 6LoWPAN data-packets. The dual system (i.e., Beaglebone with Zolertia attached) also stores the RPL-graph for packet routing inside the 6LoWPAN-network, a database containing the rules, and a list of active nodes together with the corresponding security material.

The CS maintains a database, which holds the necessary information about the entire network. The database is subdivided into five major parts: users/roles, rules, embedded nodes, security and locations. Users may access the system (with roles), and decide which actions to be taken on a given location, or a set of nodes or rules. The key management part is also integrated into the database. The database is interfaced through a Java-application on the CS. This application converts the information in the database into useful packets/requests (UDP/CoAP) towards the nodes.

6 Protocol Analysis

We compare our protocol with the intuitive approach of a unicast communication at CS and ERs, where all the automation rules are defined at CS level. We make distinction between the global rule and local rule mechanism.

Global rule mechanism. In this situation, the message flow is exactly the same in our protocol as in the intuitive approach. The gain of performance from our protocol is obtained by the multicast communication executed by the CS and the ERs. As a consequence, the more ERs and nodes involved in the rule activation, the larger the impact improving performance.

Local rule mechanism. When applying a local rule the message flow in our protocol is shorter than in the intuitive approach since the message does not need to go up to the central server. In order to demonstrate the impact of this, we discuss the difference between timing values for submission of data from end-node to ER (i.e., direct and one hop distance), and from end-node to CS.

In every test, three series of 100 packets of size 108 bytes are sent, and the round trip time (RTT) is measured. It was observed that the RTT for the direct-hop from an end-node to ER was 240 ms. Consequently, the reaction time of an end-node to an assignment of the ER would be around 120 ms. For the tests with one-hop distance between the end-node and ER, the RTT was slightly increased to 360 ms. Therefore, the reaction time would be approximately 180 ms. Comparing these values for RTT and reaction time, the time for encryption/decryption functionalities (i.e., <2ms) induce negligible impact on the total transmission time [7]. Another observation is that the extra hop in the communication path (i.e., one-hop) increases the transmission time (i.e., compared to direct-hop), due to the extra processing time at the intermediate nodes.

In our setting, we used 54 Mbps WiFi network to deploy the CS and ER. It was measured 50 ms and 71 ms RTT respectively for the communication between Zolertia and BeagleBone, and BeagleBone and CS. Consequently, since the ER is formed with Zolertia and Beaglebone, the total RTT between CS and ER is 121 ms. Therefore, two nodes (i.e., in the same ER domain and with direct-hop distance to ER) can communicate via the ER in 240 ms and via the CS in 361 ms (240 + 121 ms). Consequently, the communication cost via the CS would be around 33% higher than that via the ER.

Similar as in the case of the global ruling, additional performance gain is obtained from the multicast communication.

7 Conclusions

This paper proposes and discusses the implementation aspects of an automatic ruling for three-level network architecture based on efficient and secure multicasting mechanism. The security analysis shows the immunity of the scheme for well known security attacks. It is proven by the timing values, that the efficiency would be approximately 33% higher, if the rules are handled at the edge router level rather than by the central server. However, based on the application scenario, the rules can be executed at the ER or the CS. Ultimately, with this novel approach, multicasting wireless networks would be able to filter messages more efficiently, preserving network and node resources.

References

1. Rawat, P., Singh, K., Chaouchi, H., Bonnin, J.: Wireless Sensor Networks: A Survey on Recent Developments and Potential Synergies. The Journal of Supercomputing **68** (2014) 1–48
2. Gomez, C., Paradells, J.: Wireless Home Automation Networks: A Survey of Architectures and Technologies. IEEE Communications Magazine **48** (2010) 92–101
3. Zhou, Y., Fang, Y., Zhang, Y.: Securing Wireless Sensor Networks: A Survey. Communications Surveys Tutorials, IEEE **10** (2008) 6–28
4. Klaoudatou, E., Konstantinou, E., Kambourakis, G., Gritzalis, S.: A Survey on Cluster-Based Group Key Agreement Protocols for WSNs. IEEE Comm. Surveys Tutorials **13** (2011) 429–442
5. Perrig, A., Szewczyk, R., Tygar, J.D., Wen, V., Culler, D.E.: SPINS: Security Protocols for Sensor Networks. Wireless Networking **8** (2002) 521–534
6. Gill, K., Yang, S.H., Yao, F., Lu, X.: A Zigbee-based Home Automation System. IEEE Transactions on Consumer Electronics **55** (2009) 422–430
7. Smeets, R., Aerts, K., Mentens, N., Braeken, A., Segers, L., Touhafi, A.: Cryptographic key management architecture for dynamic 6LoWPAN networks. In: International Conference on Applied Informatics. (2014)
8. Kumar, P., Porambage, P., Ylianttila, M., Gurtov, A.: A Mobile Object-Based SECRET Key Distribution Scheme for Wireless Sensor Networks. In: IEEE Conference on Ubiquitous Intelligence and Computing and International Conference on Autonomic and Trusted Computing (UIC/ATC). (2013) 656–661
9. Harn, L., Lin, C.: Authenticated Group Key Transfer Protocol Based on Secret Sharing. IEEE Transactions on Computers **59** (2010) 842–846
10. Lee, C.Y., Wang, Z.H., Harn, L., Chang, C.C.: Secure Key Transfer Protocol Based on Secret Sharing for Group Communications. IEICE Transactions **94-D** (2011) 2069–2076
11. Shelby, Z., Hartke, K., Bormann, C.: Constrained Application Protocol (CoAP). RFC 7252 (2014)
12. T. Winter et. al.: RPL: IPv6 Routing Protocol for Low-Power and Lossy Networks. RFC 6550 (2012)
13. Shelby, Z., Bormann, C.: 6LoWPAN: The Wireless Embedded Internet. Wiley Publishing (2010)
14. : Zolertia Z1 Development Platform. (Zolertia Shop)
15. : BeagleBone Black. (CircuitCo..)

Toward 3D Avatar Visualization of Macedonian Sign Language

Boban Joksimoski, Ivan Chorbev, Katerina Zdravkova, and Dragan Mihajlov

Faculty of Computer Science and Engineering,
Rugjer Boshkovikj 16, P.O. Box 393,
1000 Skopje, Republic of Macedonia
{boban.joksimoski,ivan.chorbev,katerina.zdravkova
dragan.mihajlov}@finki.ukim.mk
http://www.finki.ukim.mk

Abstract. Sign language is the first language of many people suffering hearing impairments. They learn it before getting their initial skills in writing or communicating through other methods. Written and auditory expressions are more difficult for users of sign languages. For them, the output should be presented in the form of gestures, facial expressions and body language.

This paper presents a 3D visualization system that extensively uses animation and game concepts for accurately generating sign languages using 3D avatars. Every avatar is endowed with expressive range of gestures and anatomical features as subtle as muscles for mimicking the changes in facial expression. The platform works with transferring text into skeletal control parameters that are passed to the visualization engine. Transforms are calculated so the avatar rigging system can extensively utilize the skeletal and morph target controls for depicting visual signing.

Keywords: Sign Language; 3D Avatar; Virtual Signing; Visualization

1 Introduction

According to the World Health Organization statistics, updated in 2015 [1], there are around 360 million people worldwide that have disabling hearing loss. Hard of hearing and deaf people have difficulties integrating in the society and often are part of a different cultural and linguistic group. Deaf individuals often acquire a sign language as their first language and are most fluent and comfortable in this method of expression. Sign languages are complex and are classified as natural languages, expressing common language characteristics like dialects, vocabulary and grammar.

For deaf individuals, interfaces utilizing sign language are highly desirable. Sign languages are equally (or more) complex than ordinary written and auditory speech.

There are various methods of achieving input and output that is familiar to users of sign languages. The output usually is based on using predefined video

S. Loshkovska and S. Koceski (eds.), *ICT Innovations 2015*,
Advances in Intelligent Systems and Computing 399,
DOI: 10.1007/978-3-319-25733-4_20

sequences or using interactive avatar based virtual environment. In this paper we address the problems that are present in Macedonia, and propose an avatar based virtual signing system for the deaf and hard of hearing societies in Macedonia.

2 Sign Languages and Their Use

The process of expressing thoughts using visual cues, motion and body language is natural to the living beings. With the dawn of the information era and the development of assistive technologies, more sophisticated tools are developed for integration of people with disabilities. This creates opportunities for narrowing the gap between communities that primarily use non-verbal communication and the general population. A lot of research has been done in this subject, initially by linguists and later accompanied by computer scientists.

The preferred method of communication of the deaf society is by using gestures, facial expressions and body language. Sign languages, contrary to the common misconception, are various and differ in different countries and geographical areas. Standardized sign language is a sign language that has extensively defined vocabulary and grammar rules. The best known standardized sign language is the American Sign Language (ASL), initially researched by William Stokoe [2]. Stokoe is responsible for co-creating the best known dictionary of ASL, and also created the famous Stokoe notation [3], used to represent the written notation of the ASL.

Other written notations of sign languages have been created and used, including the Hamburg Notation System (HamNoSys) [4] and SignWriting [5]. A lot of work has been done for incorporating sign languages and computer sciences. Bigger communities tend to create better tools and have bigger incentives to experiment with various methods. The American Hard of Hearing Society has a big list of applications intended for sign language learning, interpretation and communication. Human-Computer Interaction approaches are used for promoting bilingualism between families that include both deaf and non-deaf members [6][7].

Other examples use special algorithms, usually based on Hidden Marcov Models or Neural Networks, for motion analysis of the video sequences and interpreting the sign language information [8]. Other efforts focus on video communication tailored for usage by hard of hearing people. For example, a special video encoding and compression algorithm is used along with image detection algorithms for video communication between speakers of ASL [9]. Robotics has also been used for teaching sign languages [10]. Parton [11] offers an overview of the topics and multidisciplinary research that has been done in the field of sign languages.

Signing avatars is a relatively young research area with two decades of active research. The articulatory representation of the Stokoe notation inspired the first attempts of creating a 3D signed avatar [12][13]. Two influential European projects, ViSiCAST and eSIGN used the HamNoSys notation for visual signing of the avatars [14], utilizing an articulatory approach. A more recent avatar project is called Paula with a number of interesting results for the synthesis of finger-spelling, nonverbal components and natural pose computation [15].

3 Challenges for the Macedonian Sign Language Users

The Macedonian Sign Language (MSL) poses a specific challenge. It is a relatively small language that has a valid but non-standardized grammar. Its vocabulary is estimated to be around 2800 signs, including the alphabet. As opposed to the ASL that has a much bigger vocabulary, MSL is relatively limited. As noted before, ne of the greater shortcomings of MSL is that it is not standardized and very few researchers in Macedonia are actively working on it.

The law introduced in 2009 recognizes MSL as an official method of communication, equal to the Macedonian spoken language. Although it is recognized as an official method of communication, the Macedonian society of deaf and hard of hearing people and its members face big obstacles in integrating into the Macedonian society. The organization estimates that there are more than 6000 people that are hard of hearing or completely deaf in Macedonia. As opposed to that, there are only 12 officially certified translators of MSL [16]. The small amount of translators cannot service all of the persons in need, although it is required by law. Also, the only official guide for learning Macedonian Sign Language is the "Dictionary of the Macedonian sign language", that gives an picture based overview of the language.

Also, the nature of the Macedonian speaking and written language poses additional problems for hard of hearing people. Macedonian language is heavily phonetic, and for every sound there is an accompanying letter from the alphabet. Given that the written representation of words is the actual spoken form, every phoneme has an accompanying sign language gesture 1.

The phonetic structure of the Macedonian language poses a great challenge for hard of hearing people, and especially children to become literate. The write as you speak, read as it is written concept of the Macedonian language poses a significant obstacle; because people that are hearing impaired from birth cannot grasp some of the phonological concepts.

Furthermore, the Macedonian Sign Language doesn't have a formal written notation, making it extremely difficult for hard of hearing people to express themselves in a form that is better suited for their needs.

All of the previously stated problems are the motivation for creating the Macedonian Sign Language Platform (MSLP). The primary focus of the platform is creating a service that will be available to all of the population, regardless of the medical conditions, culture and community. Along with the previous efforts for promoting and incorporating the Macedonian sign language, we are striving to achieve a multimedia platform that can be used for learning the sign language.

The MSLP platform is an approach different from the work done in the field by other researchers [17]. In their previous work, various applications have been created, ranging from interactive multimedia games that are tailor-made for learning the Macedonian sign language, content management system for storing

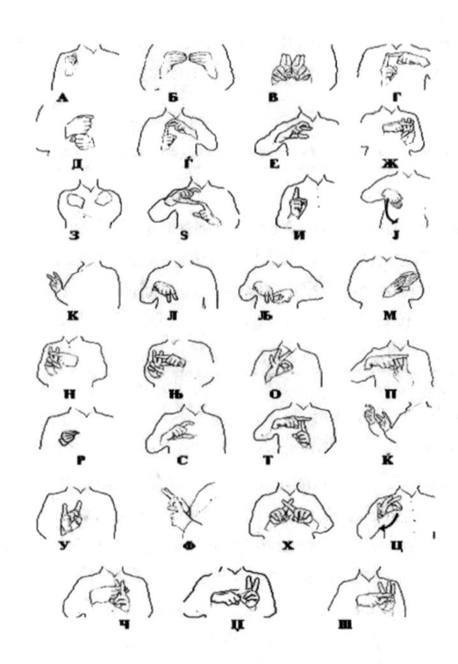

Fig. 1. One-handed alphabet of the Macedonian Sign Language.

and categorizing sign language visualizations and creating animated video and gif sequences for visualization of the sign language. The effort was mostly targeted at children. It has paved the way for more sophisticated methods through embracing the opportunities of the technology available.

4 System Overview of Sign Language Visualization

Having the commodity of a rather inexpensive and powerful hardware, broadband Internet connections, multiplatform open standards and advancements in the real-time rendering performance, it is finally achievable to implement ideas that have proven challenging in the past. Currently the Macedonian Sign Language Platform enables parsing of macedonian text and translating it to transformation parameters that are then animating the model. The input text can be supplied by the client side browser by submitting a request to the server side. After the server-side processing is done, it responds to the client by means of an avatar based animation. The whole platform is separated in two distinct parts, the frontend that is responsible for visualization and the backend, responsible for processing and interpreting the requests from the client (Fig. 2).

Distinct parts of the platform include:

Backend – Python web application stack, PostgreSQL database, text processing module and an Apache web server

Frontend – Client JavaScript Library, Three.js visualization library, Tween.js library for interpolation and the 3D avatar.

We are pursuing this type of separation for couple of reasons: providing the backend as a public web-service with a defined API; creating a separation of the visualization and the platform functionality; ease of access to distinct parts of the system and centralized control of the data. The modular approach promotes easier maintenance and upgrades.

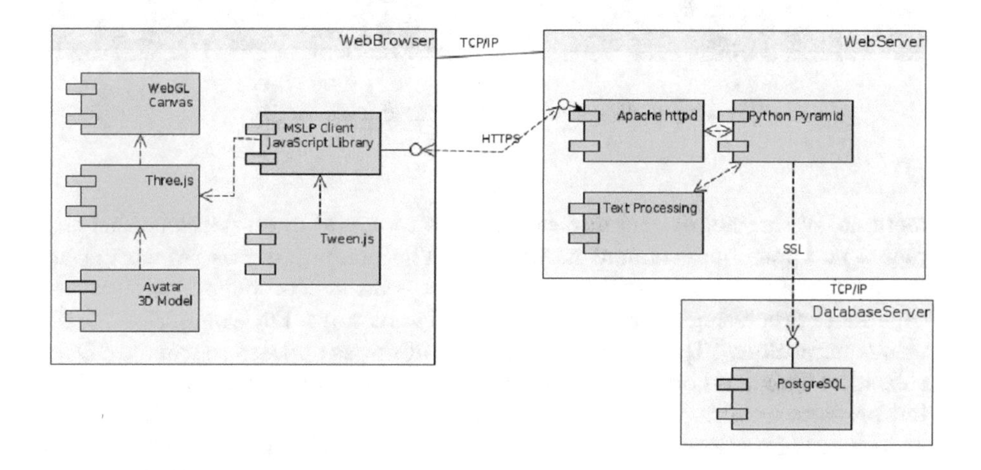

Fig. 2. General UML component diagram of the system.

The client side of the applications is made as a client web-application that can be embedded in web sites. For the purpose of creating the 3D environment, we are using the WebGL standard [18][19], accompanied by the Three.js [20] library. Using these technologies we are able to create a 3D real-time rendering engine that can load three dimensional models to represent avatars. Entering in full 3D domain enables creating fluid, real-time animations that are more suitable for conveying the intended message. The drawback is the exponential growth of the complexity of the system and initial lack of fine-grained control over the sign language visualization. As the system matures and through test scenarios, the fine-grained control will be regained as additional data will be available.

Currently adult male, female and child model have been developed (Fig. 3). Every avatar is designed with coordination with an experienced 3D character modeler, so that the models are anatomically correct and all of the muscle systems are present.

For controlling the avatar, we have utilized the support for skeletal animation that is widespread in the CGI industry. An identical bone system is created for every avatar and it is connected to the mesh using the smooth skin weights

Fig. 3. Wireframe and textured male model.

method. With smooth skinning we are able to create mesh deformations that represent upper body human movements. The bone system is responsible for movement of the torso, extremities and the head 4. The bone system is also capable of processing transformations in different ways. For example, we prefer using quaternions [21] for representing 3D rotations as opposed to standard Euler rotations. This method sacrifices human readability of the rotation parameters, but provides us with a gimbal lock escape mechanism that is crucial for creating wide range of motion without human supervision. Quaternion interpolation can be troublesome, but most of the problems like SLERP-ing, SQUAD-ing and orientation limits are algorithmically solvable.

Facial animation is another point that is addressed in the platform. Our We truly believe that there cannot be successful sign language visualization without accompanying facial expressions. Facial expressions and eye gaze are a hard topic for sign languages and the previous work of Cao [22] and Cassell [23] has shown that additional data is required. The current focus of the facial animation for MSLP is lip intonation. A significant constraint for creating an efficient facial rig is the lack of control that is present in standard animation software applications like Autodesk Maya and Blender. Our environment has to render the model and all animations in real-time, so we cannot rely on complex facial rigs. A workaround is using a technique called mesh morphing or mesh blending [24], that interpolates the polygonal mesh to specific target meshes. A target mesh is created for every phoneme and every expression that is of interest. This approach, albeit simple and efficient, creates duplicate mesh data for every morph target, thus increasing the avatar file size. The mesh targets that are created for visualization of the phonemes in the Macedonian language can be grouped into approximately 13 groups. Along with the data for the facial expressions, the model has over 30 different morph targets (Fig. 5).

In the Macedonian phonology, a lot of the phonemes can be left out and the phonetic rules for the behavior depend on the neighboring phonemes, minimal pairs, tone, segmental sounds, stresses and allophones [25]. We are also making

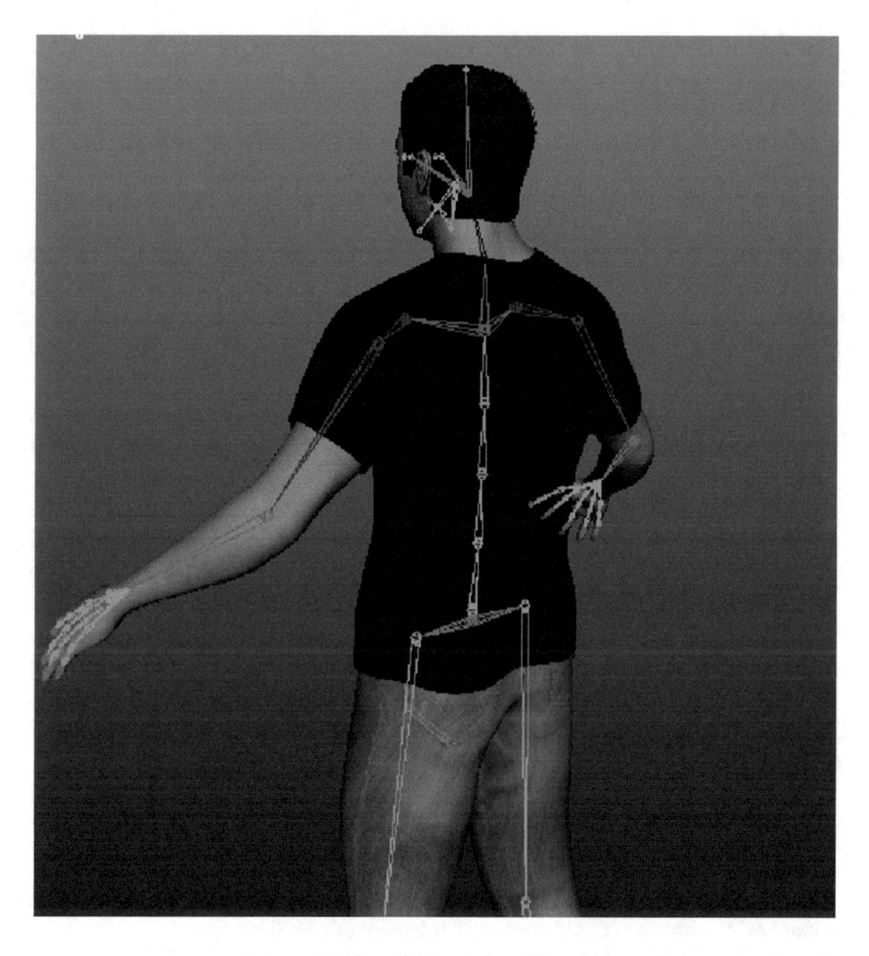

Fig. 4. Skeletal Rig of the 3D avatar.

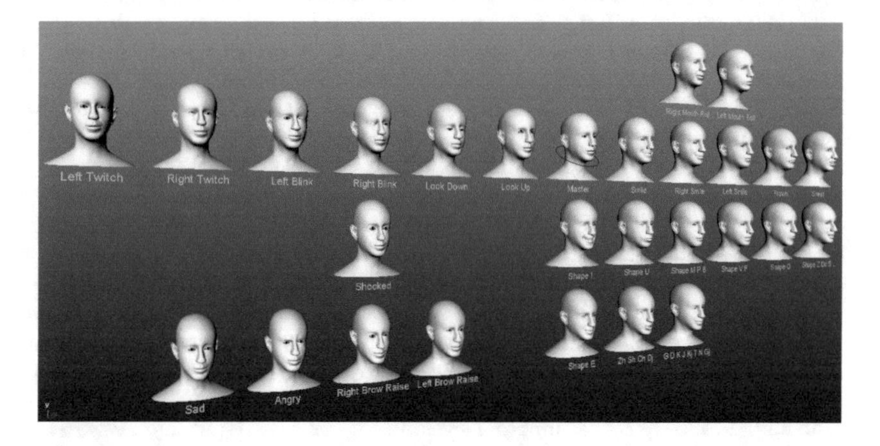

Fig. 5. Part of the morph targets for creating facial expressions and lip synchronization

an effort for translating the client as a web browser plugin for Google Chrome and Mozilla Firefox. With this, we can achieve better user experience and client side caching of the avatars.

5 Conclusion and Further Work

In this paper we have given an overview of the Macedonian Sign Language and the current state of its use among the deaf and hard of hearing people in Macedonia. In the context of digital tools for the Macedonian Sign Language, we have presented the use of the concept of 3D avatar models and their animation for enabling real-time gestures and lip intonation. The system is intended as a web service to the Internet users in Macedonia. Because of its client-server architecture, it can be used on any device that supports the WebGL standard and has an active internet connection.

The possible uses of this kind of systems are vast. The main objective is creating a digital database of all the gestures present in the Macedonian Sign Language, as a method to improve its standardization. The second focus is to create a game learning system based on the Macedonian sign language and to promote the learning and integration of the Macedonian sign language among non-deaf people, so they can actively communicate with their deaf colleagues and friends. Further improvement that we are working on is tuning the lip-sync features so that they can be used for learning and practicing lip-reading. The ultimate goal is to make a generic system suitable to all sign languages.

References

1. World Health Organization – Deafness and hearing loss, `http://www.who.int/mediacentre/factsheets/fs300/en/`
2. Stokoe, W.C.: Sign Language Structure: An Outline of the Visual Communication Systems of the American Deaf. Journal of Deaf Studies and Deaf Education, 10, 3-37 (2005)
3. Stokoe, W.C., Casterline, D.C., Croneberg, C.G.: A dictionary of American sign language on linguistic principles. Linstok Press, [Silver Spring, Md.] (1965)
4. Hanke, T.: HamNoSys-representing sign language data in language resources and language processing contexts. In: LREC. pp.16 (2004)
5. Sutton, V.: Sign writing. Deaf Action Committee (DAC) (2000)
6. Guimaraes, C., Antunes, D., Garcia, L., Guedes, A., Fernandes, S.: Conceptual meta-environment for Deaf children Literacy challenge: How to design effective artifacts for bilingualism construction. In: Sixth International Conference on Research Challenges in Information Science (RCIS) 2012, pp. 1–12 (2012)
7. Guimaraes, C., Antunes, D., Garcia, L., Peres, L., Fernandes, S.: Pedagogical Architecture Internet Artifacts for Bilingualism of the Deaf (Sign Language/Portuguese). In: 46th Hawaii International Conference on System Sciences (HICSS), pp. 40-49 (2013)
8. Marshall, I., Safar, E.: Grammar Development for Sign Language Avatar-Based Synthesis. In Proc. Of the 3rd International Conference on Universal Acess In Human-Computer Interaction (UAHCI), Las Vegas, (2005)

9. Chon, J., Cherniavsky, N., Riskin, E.A., Ladner, R.E.: Enabling access through real-time sign language communication over cell phones. In: Conference Record of the Forty-Third Asilomar Conference on Signals, Systems and Computers, pp. 588-592,IEEE Press, (2009)
10. Kose, H., Yorganci, R., Itauma, I.: Humanoid robot assisted interactive sign language tutoring game. In: IEEE International Conference on Robotics and Biomimetics (ROBIO) 2011, pp. 2247-2248 (2011)
11. Parton, B.S.: Sign Language Recognition and Translation: A Multidisciplined Approach From the Field of. Artificial Intellifence, Journal of Deaf Studies and Deaf Education, 11 (2005)
12. Gibet, S., Lebourque, T., Marteau, P.F.: High-level specification and animation of communicative gestures. Journal of Visual Languages & Computing, pp. 657–687, 12 (2001)
13. Losson, O., Vannobel, J.: Sign language formal description and synthesis. In: Proc. of 2nd European Conference on Disability, Virtual Reality & Assoc. Tech., Skvde, Sweden. (1998)
14. Kennaway, J., Glauert, J.R., Zwitserlood, I.: Providing signed content on the internet by synthesized animation. ACM Transactions on Computer-Human Interaction (TOCHI), 14, (2007)
15. Wolfe, R., McDonald, J., Davidson, M., Frank, C.: Using an animation-based technology to support reading curricula for deaf elementary schoolchildren. In: The 22nd Annual International Technology & Persons with Disabilities Conference. (2007)
16. Association of the Deaf and Hard of Hearing of Macedonia: List of translators, `http://www.deafmkd.org.mk/index.php?option=com_content&view=article&id=55&Itemid=80`
17. Ackovska, N., Kostoska, M., Gjurovski, M.: Sign language tutordigital improvement for people who are deaf and hard of hearing. ICT Innovations 2012, Web Proceedings ISSN 1857–7288 103, (2012)
18. WebGL Specification, `https://www.khronos.org/registry/webgl/specs/1.0/`
19. Parisi, T.: WebGL: Up and Running, OReilly Media (2012)
20. Three.js – JavaScript 3d library, `http://threejs.org/`
21. Kuipers, J.: Quaternions and Rotation Sequences: A Primer with Applications to Orbits, Aerospace and Virtual Reality. (Paperback). Princeton University Press (2002)
22. Cao, Y., Tien, W.C., Faloutsos, P., Pighin, F.: Expressive speech-driven facial animation. ACM Transactions on Graphics 24, pp. 1282-1302 (2005)
23. Cassell, J., Pelachaud, C., Badler, N., Steedman, M., Achorn, B., Becket, T., Douville, B., Prevost, S., Stone, M.: Animated Conversation: Rule-based Generation of Facial Expression, Gesture & Spoken Intonation for Multiple Conversational Agents. In: Proceedings of the 21st Annual Conference on Computer Graphics and Interactive Techniques. SIGGRAPH 94, New York, NY, USA, ACM, pp. 413–420, (1994)
24. Alexa, M.: Recent Advances in Mesh Morphing. Computer Graphics Forum 21, pp. 173–198, (2002)
25. Koneski, B., Vidoeski, B.: A historical phonology of the Macedonian language. C. Winter, Universitatsverlag, Heidelberg, (1983)

Using NLP Methods to Improve the Effectiveness of a Macedonian Question Answering System

Jasmina Jovanovska, Ivana Bozhinova, Katerina Zdravkova

Ss. Cyril and Methodius University, Faculty of Computer Science and Engineering, Skopje, Republic of Macedonia
{jasmina.armenska,katerina.zdravkova}@finki.ukim.mk
bozhinova.ivana@students.finki.ukim.mk

Abstract. The process of retrieving particular information from a huge amount of text significantly depends on the language specific features. The most imposing one for Macedonian language is the possibility of a word to have various derivational and inflectional suffixes. In this research we investiage how particular NLP tools influence the retrieval, putting special emphasis on the use of Part-of-Speech (PoS) tagging, word forms, and stemming. In absence of a stemming algorithm for Macedonian language, we used the Dice Coefficient and the single-link clustering in order to group words with a common base form. All these features were implemented in an already existing Macedonian Question Answering System (QAS). We tested different strategies for weighting terms in the documents (the queries), as well as different approaches for query expansion with word forms and words with the same stem. The experimental results show that the word variations strongly influence the retrieval, improving our system's accuracy.

Keywords: Question Answering, PoS tagging, Word forms, Stemming.

1 Introduction

Natural language, which is a tool that people use to communicate, has its own specific properties that reduce the effectiveness of textual retrieval systems. The most significant ones are the linguistic variation and ambiguity. The linguistic variation is a possibility to use different words and phrases to express the same idea (such as synonymy), whereas the linguistic ambiguity appears whenever a word or phrase has more than one interpretation (such as polysemy) [1]. Many problems also arise due to many variations occurring during word formation, which happens because of the usage of affixes, alternative spelling, transliteration, abbreviations and spelling errors. Natural language processing is a discipline studying the aforementioned features that demonstrate the complexity of the natural language. The achieved progress in this area over the years enabled widespread applications of NLP methods for processing huge amounts of textual information with an acceptable level of accuracy and

S. Loshkovska and S. Koceski (eds.), *ICT Innovations 2015*,
Advances in Intelligent Systems and Computing 399,
DOI: 10.1007/978-3-319-25733-4_21

efficiency. These methods are essential components in the modern search engines, in the tools for automated translation, as well as in the summary generators.

Macedonian has many language specific properties that have to be taken into account when retrieving particular information. The most imposing one is the possibility of a word to have different derivational and inflectional suffixes. For information retrieval purposes, this abudance of words with a common base form means a greater likelihood of mismatch between the retrieval query and the relevant documents. Hence, all those words should be conflated, which on the other hand we believe will increase the effectiveness of our QA system. At the time we did this research, neighter has a morphological analyzer been developed for Macedonian language, nor has the influence of word variants been evaluated for IR tasks.

Therefore, in the research reported here we investigate how particular NLP tools affect the retrieval of relevant information using textual documents written in Macedonian language (putting special emphasis on the use of PoS tagging, word forms, and stemming).

2 NLP in Question Answering Tasks

NLP methods are applicable and useful in all of the components of a general question answering system [2]:

- **Question processing.** The goal of the question processing phase is to extract two things from the question: a keyword query suitable as an input to an Information Retrieval (IR) system and the answer type, a specification of the kind of entity that would constitute a reasonable answer to the question. Many NLP techniques are suitable for this phase and generally perform the morpho-syntactic analysis of the question posed in a natural language.
- **Document (or passage) retrieval.** The query that was created in the previous phase is next used to identify the relevant parts of documents (passages) that are likely to contain the answer of the original question.
- **Answer processing.** The final stage of question answering is to extract a specific answer from the retrieved documents, matching the expected type, and to represent it to the user.

The two key approaches to natural language processing are statistical and linguistic approach. They differ considerably, but in practice IR and QA systems generally use a mixed approach, combining parts of both methodologies.

2.1 Linguistic Processing of Natural Languages

The linguistic approach is based on the application of different techniques and rules that explicitly encode linguistic knowledge. The most frequently used linguistic tools in IR (as well as QA) are PoS tagging, parsing, and lexical databases.

Part-of-speech (PoS) tagging is the process of assigning a Part-of-speech or other syntactic class marker to each word in a corpus. Many methods have been applied to this problem for English language and most of them have an accuracy of around 96-97% for simple tagsets. In some cases, PoS tagging algorithms work well without large modifications for many other languages [3]. But, when dealing with highly inflected or agglutinative languages, numerous changes have to be done. To overcome this problem, it is essential to extract the morphological information for each word. For Czech and the MULTEXT-East languages, Hajič and Hladká [4] and Hajič [5] use a fixed external dictionary for each language, which compiles out all the possible forms of each word, and lists possible tags for each word form. Some experiments were performed to define which PoS tags are highly important in the retrieval process. Kraaij and Pohlmann [6] found that the nouns are the best document content indicators, whereas Arampatzis et al. [7] found an improvement when using only the nouns during the retrieval phase, compared to using all stemmed words.

Syntactic analysis of text corpus is performed using Chunking and Shallow Parsing, which separates sentence words into basic phrases. One of the famous tools for semantic processing is the Princeton University's WordNet [8]. WordNet is a large lexical database of English, containing interlinked synsets by means of conceptual-semantic and lexical relations. It labels the semantic relations among words, enabling annotation of each word with its PoS tag, and accompanying it with its word forms and synonyms.

2.2 Statistical Processing of Natural Language

The huge amount of data available today has increased the importance of statistical approaches that are language independent and can at the same time deal with the data heterogeneity.

Statistical techniques are widely applicable in the pre-processing phase of QA systems, which decide what the most important document features (terms) are, and how they will be quantified. Almost all QAS remove the *stopwords* (function words and low-content words) before processing the documents and queries. This approach usually increases the systems performance.

Stemming is another method that is frequently used when the term index is created or during the process of analysing users queries. Its goal is to reduce the inflectional forms and sometimes derivationally related forms of a word to a common base form [9]. Many stemmers have been developed for a wide range of languages and their effectiveness across languages is varied and influenced by many factors. However, the research made to date shows that stemming in general doesn't deteriorate the retrieval results. It is important to emphasize that the stemming usually appears to improve the effectiveness of highly inflected languages more than of other languages [10,11,12]. The stemmer design is a labor work and what is particularly important is the necessity of a linguistic expertise in the language. Therefore, the statistical methods can be used to construct language-independent approaches for word conflation. Words that have the same form can be grouped based on various string-similarity measures. The most frequently used approaches include n-grams. Word groups can be formed from words that share the same initial n-grams or using a proportion of n-grams appearing in the words, and these results can be improved when a clustering technique is implemented. Using the Dice Coefficient for string similarity, Adamson and Boreham [13] successfully clustered a word sample into groups of semantically related words. Robertson and Willett [14] concluded that string similarity is an appropriate method to implement in environments where a large dictionary has to be searched. Statistical stemming has shown to be very effective for other languages as well, including Turkish [15] and Malay [16].

3 Our Approaches in the Question Answering Process

Our initial hypothesis is that the implementation of PoS tagging, derivation of word forms and stemming strongly influences the process of retrieving information written in Macedonian language. Therefore, we upgraded the existing Macedonian QA system with implementation of the aforementioned NLP tools [17].

3.1 The Macedonian Test Collection

The evaluation and strategy comparison in this research were performed over an existing test collection for Macedonian language [17]. We enriched the basic corpus of that collection (covering several topics from Information technology) with additional documents, which finally consists of around 145724 not unique words. This extension was performed towards the extraction of statistically significant results for words distribution in the corpus.

We also inherited a pool of 156 multiple-choice questions. For each question, four answers are provided, one of which is correct. It is important to emphasize that the questions are extracted from the document collection. Table 1 shows the question distribution according to their category. It can be noted that the most prevalent are the descriptive questions, constituting 53.2% of the entire question pool.

Table 1. Question distribution according to their category.

Category	Collection
Factoid Questions	50
Descriptive Questions	83
List Questions	23
Total	**156**

3.2 Creating Corpus Dictionary

The first step in the corpus dictionary creation is the determination of the PoS tag, that is achieved using previously developed annotated dictionary for Macedonian language [18]. It contains PoS tags for most of the frequently used Macedonian words. Our system works only with nouns, verbs, adjectives, adverbs and numbers. All other words in Macedonian are low-content and we decided not to take them into consideration (such as prepositions, conjunctions, particles, pronouns). According the performed PoS tagging, we created three different dictionaries:

- **Dictionary_1**. This is the corpus dictionary and contains 12570 unique words.
- **Dictionary_2**. Its entries are clusters of words (from Dictionary_1) representing a new form of the **same word** (word forms), and contains 7411 clusters.
- **Dictionary_3.** Its entries are clusters of words (from Dictionary_1) with a *common base form* (words with the same stem), and contains 5191 clusters.

The process of creating the last two dictionaries was performed as follows. In absence of a complete and concise stemming algorithm for Macedonian language, we used the Dice Coefficient as a measure for word-similarity and single-link clustering, so that we can group the words that have the derivational and inflectional suffixes implemented on the common base form [19]. Therefore, we represent each word as a sequence of bigrams and calculate the Dice Coefficient only for the words that start with the same bigram. In each step of the hierarchical clustering we conflate two clusters with the maximum Dice coefficient. The clustering stops when the highest coefficient has a value lower than 0.7.

We decided not to quantify the similarity of words starting with different bigram. There are examples of such words in Macedonian language with high Dice coefficient that have completely different meanings. Therefore, we extended this similarity measure by assigning a value of 0 for those dictionary words.

Afterwards, the clusters were divided into a number of sub-clusters according to the PoS tag of the constituting words. We also made minor manual changes to further split the clusters containing words with the same tag, which are not word forms of a same word, but words with a same stem. Thus, for each dictionary word we have a cluster containing its word forms appearing in the documents, as well as the clusters containing words with the same stem. Since our document collection is domain specific, we did not have PoS tags for some of the corpus dictionary words. Therefore, we had to label those words manually.

Below is given an example of several words appearing in our corpus dictionary, which have a common base form (смета):

- The nouns: сметач (computer), сметачи (computers), сметање (calculation)
- The verbs: смета (calculate), сметаше (calculated), and
- The adjective: сметачка (computational).

The Table 2 shows the entries they form in each of the three dictionaries.

Table 2. An example of dictionaries' entries.

Dictionary	Number of entries	Entries
Dictionary_1	6	{сметач}; {сметачи}; {сметање}; {смета}; {сметаше}; {сметачка}
Dictionary_2	4	{сметач, сметачи}; {сметање}; {смета, сметаше}; {сметачка}
Dictionary_3	1	{сметач, сметачи, сметање, смета, сметаше, сметачка}

3.3 Term and Cluster Weighting

Previous studies for answering questions posed in Macedonian language have already shown that the well-known Vector Space Model (VSM) with cosine similarity measure gives superior results compared to Probabilistic and Language Model [17]. This was the main reason to use VSM in this research as well. According to the created dictionaries, we built three different indexes to get the best retrieval approach we will further implement.

- **Index_1.** It consists of all the terms appearing in *Dictionary_1*. The *tf-idf* weighting scheme is used to measure the term importance in each of the documents from the collection (named *Weighting Strategy 1*).
- **Index_2.** It consists of the entries (clusters) from *Dictionary_2*, for which we tested two different weighting strategies:

– *Weighting Strategy 2.1*. Cluster's weight is equal to the average *tf-idf* weight of all its constituting terms.
– *Weighting Strategy 2.2*. Cluster's weight is equal to the maximum *tf-idf* weight of its constituting terms.

- **Index_3.** It consists of the entries (clusters) from *Dictionary_3*, for which we tested the same two aforementioned weighting strategies (named *Weighting Strategy 3.1* and *Weighting Strategy 3.2*, accordingly).

4 Experiments and Results

In this section we present the results from the practical implementation of the PoS tagging, word forms and stemming in our QAS, applied to the upgraded Macedonian test collection. The system functionalities are implemented as a collection of Python modules. The queries are represented as Python dictionaries, which are suitable data structures in terms of memory efficiency and performance.

4.1 Document Retrieval Process

In order to examine the impact of the created indexes in the retrieval process, we tested numerous techniques to determine the best one for our system.

Testing Index_1. In all of the methods we tested using the *Index_1*, the weights of the query terms are calculated with the *tf-idf* weighting scheme.
Method 1: No query expansion. The retrieval query is formulated only with the words appearing in the question. This approach is used as a baseline for our system and its accuracy is 83.97%.
Method 2: Query expansion with word forms regardless of the query length. In order to analyze the importance of incorporating word forms, we search for the words in the question that do not appear in the document under consideration in the exact same

form. For each of these words we look for its most similar word form (using the Dice coefficient) and append it to the query. The most similar word form can be retrieved from:

1. the entire corpus,
2. the document we compare with the question.

Query expansion using the entire corpus does not modify the effectiveness at all. However, the second approach leads to an improvement of the accuracy to 87.18%.
Method 3: Query expansion with word forms if the query length is less than four words. Considering that the longer questions already contain enough information necessary to find the relevant document, there is no need to expand them with additional terms. Thus, we conducted further testing for each of the two above-mentioned approaches. Firstly, we restricted the algorithm to append the word forms only if the query contains less than four words. When looking for the word form in the entire corpus the accuracy slightly increased to 84.62%. When appending the most similar word form from the document we compare with the question, we got the highest accuracy improvement (89.10%) of our system.

Testing Index_2. In all of the methods we tested using *Index_2* and *Index_3*, the weights of the query terms are calculated as follows:

- *Query Weighting Strategy A.* Simply using the term frequency in the query.
- *Query Weighting Strategy B.* Using modified *tf-idf* weighing, where *tf* is the term frequency in the query, while its *idf* is calculated according to one of the following approaches:

– *Query Weighting Strategy B.1.* The average *idf* of the words constituting the cluster to which the query term belongs, if the *Weighting Strategy 2.1 (3.1)* is used for the clusters,
– *Query Weighting Strategy B.2.* The maximum *idf* of the words constituting the cluster to which the query term belongs, if the *Weighting Strategy 2.2 (3.2)* is used for the clusters.

With *Method 1*, the system's performance grew to 87.82% for both *Weighting Strategy 2.1* and *Weighting Strategy 2.2.* We also tested each of the previously explained methods for query expansion. For *Method 2*, the best results (91.03%) were obtained with *Weighing Strategy 2.1* for the documents, and *Query Weighting Strategy B.1*, while appending word forms from the entire corpus. The highest accuracy (91.03%) for *Method 3* was achieved by the same combination of techniques as for the *Method 2*.

Testing Index_3. The examination steps for this index were exactly the same as the ones used for the previous index. The *Method 1*, combined with *Weighing Strategy 3.1* for the documents and *Query Weighting Strategy A*, slightly increased the system's accuracy compared to the baseline (85.26%). Using *Query Weighting Strategy B*, we got the highest accuracy for this method and index (85.90%). For

Methods 2 and *3,* the best results (87.82%) were obtained employing *Weighing Strategy 3.1* for the documents combined with *Query Weighting Strategy B*, while adding the most similar word form from the entire corpus.

Figure 1 represents a summary for the best results achieved using the three indexes, different weighting schemes and different methods for the query expansion.

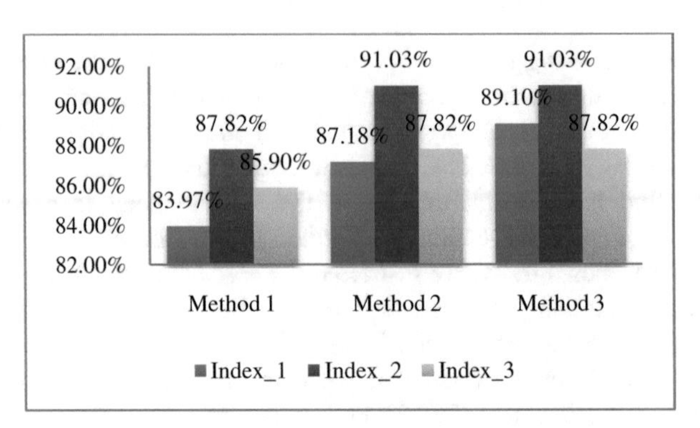

Fig. 1. The best system accuracies using different methods and indexes

4.2 Question Answering Process

The next step after retrieving the relevant document for a given question is to determine the exact answer. In this phase the documents are split into paragraphs, using the new line character as a delimiter.

We tested two different approaches for finding the right answer to a question.

- The first approach creates four queries for a given question. Each query consists of the question terms, as well as the terms from one of the given answers. Then, cosine similarity is calculated between each paragraph in the relevant document (the document specified in the IR phase), and each of the four queries. The combination that would have the maximum similarity is supposed to contain the right answer.
- The second approach creates five queries, one for the question and one for each of the answers. Then, by finding the highest similarity between the question query and the paragraphs in the document, the system retrieves the most relevant paragraph. Next, an equivalent comparison is made between that paragraph and each of the answer queries. The answer query with the highest similarity is presumed to contain the exact answer.

Our previous analysis confirmed that the latter approach is significantly more effective. Therefore, we used it for the QA phase in our system [20].

The baseline for the QA phase (combining the *Method 1* and *Index_1*) gives an accuracy of 59.62%. In this phase we also implemented the transliteration module due to the numerous foreign words appearing in the Macedonian test collection [20]. This feature increased the system's accuracy to 62.18%. Using the best resulting combination of techniques for *Index_2* and *Index_3* (defined in the IR phase), system's performance is improved to 64.74% and 63.46%, accordingly.

Figure 2 represents an overview of the best results for the QA phase.

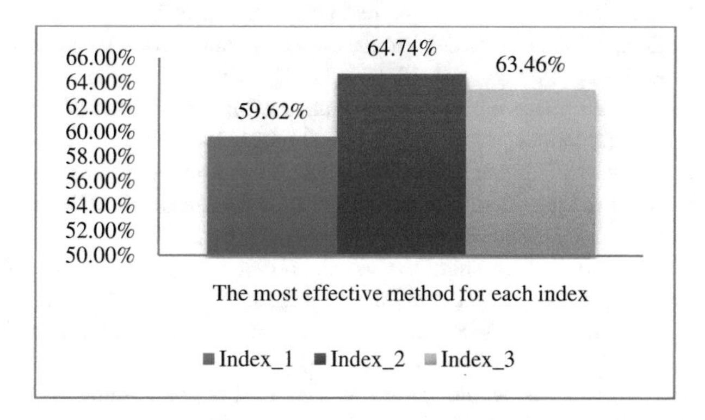

Fig. 2. QA phase accuracy using different indexes.

5 Conclusion and Future Work

The creation of a question answering system for a particular natural language is a rather exhaustive process that demands understanding of the language specific elements. The main goal of the research presented in this paper was to examine the importance of using word forms and stemming in the process of Information Retrieval and Question Answering, when working with Macedonian language. In order to achieve this, we tested three different dictionaries from which we built three different indexes. Afterwards, aiming to reach superior results, we examined various methods in order to incorporate them into our existing QA system [17].

Based on the conducted analyses, we can conclude that when it comes to design a NLP QA system for Macedonian language, the best practice is to include different word forms of the query terms in the retrieval process. We also founded that even the query expansion with the words deriving from the same stem gives better results than using only the query words.

The NLP area is rich with challenging and unexplored problems, whose solutions could possibly bring great benefits to the QA systems. In the future research, we intend to develop more language-specific modules for the Macedonian language in order to increase system's understanding of the language.

Analyzing the final results from our system, we found that 40.96% of the descriptive questions, 28% of the factoid and 39.13% of the list questions are wrongly answered. To achieve better results for answering descriptive and list questions, our future steps are directed towards the use of collocations, co-occurring words and synonyms, which on the other hand will improve the overall effectiveness of our system. In order to do this successfully, we also need to further enrich the Macedonian test collection with additional documents and questions.

References

1. Vallez, M., Pedraza-Jimenez, R.: Natural Language Processing in Textual Information Retrieval and Related Topics. In: Hipertext.net, num. 5 (2007)
2. Sanjay K Dwivedi and Vaishali Singh. Research and reviews in question answering system. Procedia Technology, vol. 10, pp. 417–424. Elsevier (2013)
3. Brants, T.: TnT: a statistical part-of-speech tagger. In: Proceedings of the sixth conference on Applied natural language processing, pp. 224–231. Morgan Kaufmann, Seattle (2000)
4. Hajič, J., Hladká, B.: Tagging inflective languages: Prediction of morphological categories for a rich, structured tagset. In: Proceedings of COLING-ACL Conference, vol. 1, pp. 483-490. Association for Computational Linguistics, Montreal (1998)
5. Hajič, J.: Morphological tagging: Data vs. dictionaries. In: Proceedings of ANLP-NAACL Conference, pp. 94-101. Association for Computational Linguistics, Seattle (2000)
6. Kraaij, W., Pohlmann, R.: Viewing stemming as recall enhancement. In: Proceedings of SIGIR-96, pp. 40–48. ACM, New York (1996)
7. Arampatzis, A., van der Weide, Th.P., Koster, C.H.A., van Bommel, P.: Text Filtering using Linguistically-motivated Indexing Terms. Technical Report CSI-R9901, Computing Science Institute, University of Nijmegen, Nijmegen, The Netherlands (1990)
8. WordNet, A lexical database for English, http://wordnet.princeton.edu/
9. Jivani, A. G.: A Comparative Study of Stemming Algorithms. International Journal of Computer Technology and Application 2(6), 1930-1938 (2011)
10. Pirkola, A.: Morphological typology of languages for IR. Journal of Documentation 57(3), 330-348 (2001)
11. Popovic, M., Willett, P. The effectiveness of stemming for natural-language access to Slovene textual data. JASIS. 43(5), 384-390 (1992)
12. Bijal, D., Sanket, S.: Overview of Stemming Algorithms for Indian and Non-Indian Languages. International Journal of Computer Science and Information Technologies 5(2), 1144-1146 (2014)
13. Adamson, G.W., Boreham, J.: The use of an association measure based on character structure to identify semantically related pairs of words and document titles. Information Storage and Retrieval, 10(7-8), pp. 253-260. Elsevier (1974)
14. Robertson A.M., Peter Willett, P.: Applications of n-grams in textual information systems. Journal of Documentation 54(1), pp. 48 – 67 (1998)
15. Ekmekcioglu, F.C., Lynch, M.F., Willett, P.: Stemming and n-gram matching for term conflation in Turkish texts. Information Research 2(2) (1996)
16. Sembok, T.M.T., Bakar, Z.A.: Characteristics and Retrieval Effectiveness of n-gram String Similarity Matching on Malay Documents. In: Proceedings of WSEAS International Conference on Applied Computer and Applied Computational Science, pp. 165-170. ACM Press, Wisconsin (2011)
17. Armenska, J., Zdravkova, K. Comparison of information retrieval models for question answering. In: Proceedings of the Fifth Balkan Conference in Informatics, pp. 162–167. ACM, New York (2012)
18. Multext-East Resources, http://nl.ijs.si/ME/V4/
19. Frakes, W. B.: Stemming algorithmas. In: Information retrieval: Data structures and algorithms, Frakes, W.B., Baeza-Yates, R. (eds.), pp.132-139. Englewood Cliffs, NJ: Prentice Hall (1992)
20. Bozhinoski, S., Bozhinova, I., Armenska, J.: Information retrieval over Macedonian test collection using word forms and transliteration. In: 12th International Conference on Informatics and Information Technologies, Bitola (2015)

Medical Image Retrieval for Alzheimer's Disease Using Data from Multiple Time Points

Katarina Trojacanec, Ivan Kitanovski, Ivica Dimitrovski, Suzana Loshkovska, for the Alzheimer's Disease Neuroimaging Initiative*

"Ss. Cyril and Methodius" University
Faculty of Computer Science and Engineering, Skopje
"Rugjer Boshkovik" 16, PO Box 393, Skopje, Macedonia

{katarina.trojacanec,ivan.kitanovski,ivica.dimitrovski, suzana.loshkovska}@finki.ukim.mk

Abstract. This paper presents medical image retrieval for Alzheimer's disease based on information extracted from multiple time points. The aim is to analyze the retrieval performance directed by the information combination of different time points and from each time point separately. For each subject, Magnetic Resonance Images (MRI) of four consecutive time points are obtained from ADNI database. Measurements of cortical and subcortical brain structures, including volumes and cortical thickness of the brain regions are used as features. The feature selection is performed aiming to select the most relevant features and reduce redundant and possibly noisy data.

The evaluation is based on ten scenarios defined for separate time points and for a combination of multiple time points. According to the obtained results, it can be concluded that the retrieval performance gets better while using the time point that is more temporally distant from the baseline, regarding the scenarios in which separate information at any time point is used. When using a combination of multiple time points, the retrieval performance is improved only in the case when more than two time points are available and the selected features from the last two points are used. Feature selection algorithm leads to better results in all cases, while significantly reduces the feature vector dimensionality. The selected features are known as significant markers for Alzheimer's Disease.

Keywords: Alzheimer's Disease, MRI, Medical image retrieval, Feature extraction, Longitudinal data, Feature selection, ADNI

---

* Data used in preparation of this article were obtained from the Alzheimer's Disease Neuroimaging Initiative (ADNI) database (adni.loni.usc.edu). As such, the investigators within the ADNI contributed to the design and implementation of ADNI and/or provided data but did not participate in analysis or writing of this report. A complete listing of ADNI investigators can be found at: http://adni.loni.usc.edu/wp-content/uploads/how_to_apply/ADNI_Acknowledgement_List.pdf

S. Loshkovska and S. Koceski (eds.), *ICT Innovations 2015*,
Advances in Intelligent Systems and Computing 399,
DOI: 10.1007/978-3-319-25733-4_22

1 Introduction

Alzheimer's Disease (AD) is a progressive neurodegenerative disorder and one of the most common form of dementia in older adults nowadays. Currently, there is no known therapy that slows the progression of this disorder [1]. Finding relevant biomarkers, early diagnoses of the disease, monitoring disease progression or treatment reaction, identifying the patients that are most probable to develop AD are open research topics.

Vast amount of data related to AD investigations is continuously generated in the clinical and research centers. Among this data, medical images are very beneficial and contain extremely important information. They provide precise and consistent markers for diagnosis, understanding the disease pathology and monitoring the disease progression [2]. Magnetic Resonance Imaging (MRI) is found to be a powerful imaging technique for diagnosis of AD. MRI clearly reflect the structural brain changes imposed by the disease progression such as thinning of the cerebral cortex, ventricular enlargement, hippocampus shrinkage etc. [3,4,5,6,7,8]. This leads to enormously increased number of images stored in the medical databases that need to be efficiently organized and retrieved. Being able to retrieve images from the large medical databases with similar Volume of Interest (VOI)/pathology/disease might be very useful in the clinical and research centers mainly in two directions: (1) providing clinically relevant information to the physicians at right moment, thus supporting the diagnosis process and improving its quality and efficiency [9,10], and (2) for educational purposes [11].

There is a very critical open question in this domain: the lack of addressing the temporal factor and using longitudinal information in the retrieval process. Thus, the main goal of this paper will be to include these longitudinal data, namely MRI information from multiple time points and to perform analysis on the retrieval performance in several scenarios: (1) considering each time point separately, and (2) using a combination of different successive time points. Although the analysis based on the baseline scans is important to provide additional information at the moment of the first visit, inclusion of the longitudinal data has several more benefits. Considering image retrieval problem, addressing of the temporal information might provide more information on the disease progression and/or treatment reaction. The main contribution of this paper is towards including temporal information for AD in the retrieval process. The influence of the temporal factor on the retrieval performance is identified. The present research contributes in two main directions: (1) to provide clinically relevant information to the physicians at right moment in a way that will enable deeper understanding of the disease/disease progression/treatment reaction and (2) for educational purposes.

The paper is organized as follows. The related work is given in Section 2, while the experimental setup is described in Section 3. Experimental results are presented in Section 4. Section 5 provides concluding remarks and future directions.

2 State of the Art and Related work

Several research studies were performed on medical image retrieval for AD [12,13,14,15]. The main concern in this research domain is the lack of direct inclusion of longitudinal information in the retrieval process. Images for multiple time points are usually available for each patient. Using the information extracted from such longitudinal scans might be very beneficial in the retrieval process.

At general, the retrieval process involves generating feature vector that represents the image given as a query, and then comparing it with other feature vectors stored in the system [10]. Some of studies apply Intensity Histograms, Local Binary Pattern (LBP) and Gradient Magnitude Histograms to the middle slice [12], other apply Discrete Cosine Transform, Daubechie's Wavelet Transform and LBP to a subset of slices selected by radiologists [13,14]. The usage of Laguerre Circular Harmonic Functions expansions together with Bag-of-Visual-Words approach is investigated in [15]. There are two main critical aspects that need a special attention here: (1) the feature extraction is performed only in a slice by slice manner on one/several slice/s, and (2) dimensionality of the feature vector. The first one might result in exclusion of possibly significant spatial information, while the second one might lead to a high computational complexity. Examples include 256 features in each descriptor in [12], 13312 features for 3D Grey Level Co-occurrence Matrices, 1920 for 3D Wavelet Transforms, 9216 for Gabor Transforms and 11328 for 3D LBP per volume [16], all applied to brain MRI. To overcome this, alternative approach can be applied [17]. It uses domain knowledge and is based on the structural MRI measures sensitive to the pathology, such as volumes of the brain structures and cortical thickness. Among others, hippocampus, inferior lateral ventricle, amygdala, entorhinal cortex are significant biomarkers closely related to AD. This method directly involves information for structural brain changes into the retrieval process. Consequently, it is used to generate image representation in this work.

3 Experimental Setup

3.1 Participants and Inclusion Criteria

Data used in the preparation of this study were obtained from the Alzheimer's Disease Neuroimaging Initiative (ADNI) database (adni.loni.usc.edu). The ADNI was launched in 2003 as a public-private partnership, led by Principal Investigator Michael W. Weiner, MD. Investigation on whether serial magnetic resonance imaging (MRI), positron emission tomography (PET), other biological markers, such as cerebrospinal fluid (CSF) markers, APOE status and full-genome genotyping via blood sample, as well as clinical and neuropsychological assessments can be combined to measure the progression of mild cognitive impairment and early Alzheimer's Disease has been the primary goal of ADNI. The database contains data about cognitively normal individuals, adults with early or late Mild Cognitive Impairment, and people with early AD with different follow up duration of each group, specified in the protocols for ADNI-1, ADNI-2, and ADNI-GO (see http://www.adni-info.org).

In this research, a total of 162 different subjects from ADNI-1 were selected by using the following criteria: (1) each subject belongs to AD or Normal Controls (NL) group; (2) for each subject scans at baseline, and 6, 12, and 24 months later are available; (3) each subject has overall quality control (QC) outcome "pass" according to the Center for Imaging of Neurodegenerative Diseases, UCSF [18].

3.2 Feature Generation

We represent the information extracted from the MRI volumetric data by measurements of the brain structures such as cortical thickness and volumes of the separate brain regions [17]. To obtain such measurements, cortical reconstruction and volumetric segmentation was performed with the FreeSurfer image analysis suite. Shortly, the pipeline includes removal of non-brain tissue using a hybrid watershed/surface deformation procedure, automated Talairach transformation, segmentation of the subcortical white matter and deep grey matter volumetric structures, intensity normalization, tessellation of the grey matter white matter boundary, automated topology correction, and surface deformation following intensity gradients. After the completion of the cortical models, registration to a spherical atlas follows which utilizes individual cortical folding patterns to match cortical geometry across subjects. Then the parcellation of the cerebral cortex into units based on gyral and sulcal structure is performed [18].

In this study, information from scans obtained at multiple time points is considered. When this information is available for the subjects, scans from multiple time point undergo the longitudinal stream to extract reliable volume and thickness estimates. This is also available in the FreeSurfer package. A within-subject template space and image, unbiased toward the chronological scan order, is created using robust, inverse consistent registration. Several processing steps, such as skull stripping, Talairach transforms, atlas registration as well as spherical surface maps and parcellations are then initialized with common information from the within-subject template, significantly increasing reliability and statistical power [18]. A thorough visual QC is then performed by UCSF, providing all data and QC outcome for ADNI data users.

Due to the main goal of the paper, analysis of the retrieval performance is performed at different time point and when mutual temporal information is used. For this purpose the strategy used in [17] is modified/extended, leading to these ten retrieval scenarios:

- Case 1: For each time point (baseline (TP0), and 6 (TP6), 12 (TP12), and 24 (TP24) months later) a combination of cortical thickness and volumes of the separate brain structures is used for representation. A total number of features in the feature vector is 112 (44 volumes and 68 cortical thickness, 34 for each hemisphere)
- Case 2: For each time point after the baseline, an extended version of the representation used in case 1 is tested. Namely, a different combination of the measures at that point and its predecessors is used. Following this strategy, for the second time point, a combination of the measurements at TP0 and TP6 (224 features in total) is used. For the third time point, two combinations are analyzed, a combination of TP0, TP6 and TP12 measures (336 features in total), and a combination of TP6 and TP12 measures (224). The strategy is similar for the fourth time point, leading to three more scenarios, (1) a combination of measurements at TP0, TP6, TP12 and TP24 (448 features), (2) measurements at TP6, TP12 and TP24 (336 features), and (3) measures at TP12 and TP24 (224 measures in total).

3.3 Feature Selection

For each image representation scenario, the feature subset selection method was applied. This step aims to further reduce the feature vector dimensionality and to improve the retrieval performance by selecting the most relevant features. We have used the Correlation-based Feature Selection (CFS) method [19]. It evaluates subsets of features taking into account the usefulness of individual features for predicting the class along the degree of intercorrelation among them, meaning that good feature subsets contain features highly correlated with the class, yet uncorrelated to each other [19]. Considering the application domain, features sensitive to the disease are expected to be selected more frequently. To be able to examine this, we recorded the inclusion rate, i.e. how frequent each feature is selected.

3.4 Retrieval

Taking into consideration the number of subjects used in this study, we have performed leave-one-out strategy (each subject representation was used as a query against all other representations stored in the database). Euclidean distance was used as a similarity measurement. Influence of the temporal information on the retrieval performance was evaluated for each scenario by using mean average precision (MAP). These scenarios include retrieval using scan information at each time point (TP0, TP6, TP12 and TP24), and a combined information from multiple time points. To get an unbiased result, the feature selection was performed independently of the query subject. Thus, the specific feature subset for each query subject was obtained.

4 Experimental Results

Experimental results using the proposed strategy for analysis of temporal factor influence on the retrieval performance are presented in this subsection. Table 1 contains values of MAP for each scenario with and without feature selection.

Table 1. Evaluation of the MAP-based retrieval performance by using information obtained from scans at different time points and a combined information for 1.5T MRI ADNI dataset with and without feature selection (classes: NL, AD).

MAP	Without feature selection	With feature selection
TP0	0.705	0.783
TP6	0.723	0.8
TP0 + T6	0.717	0.798
TP12	0.732	0.81
TP0 + TP6 + TP12	0.723	0.81
TP6 + TP12	0.73	0.811
TP24	**0.758**	0.822
TP0 + TP6 + TP12 + TP24	0.736	0.818
TP6 + TP12 + TP24	0.743	0.821
TP12 + TP24	0.748	**0.825**

According to the obtained MAP-based results, several conclusions can be derived. Regarding the cases where separate information of different time points are used, the retrieval performance gets better, as information is used from the time point that is more temporally distant from the baseline (starting from the value of MAP of 0.705 using baseline information, the value is continuously growing to 0.758, obtained at TP24). The situation is the same when feature selection is applied in these cases (the value of MAP at baseline is 0.783 with continuous improvement to the value of 0.822 at TP24). Considering these results from the medical point of view, this is very logical because as the disease progresses, the changes of the brain structure in the relevant brain regions are bigger in subjects with AD. As a result, it can be concluded that this kind of representation provides meaningful reflection of the disease progression. Considering the scenario where the combined information is used, it can be concluded that adding additional information not always leads to better results.

The feature vector containing the information from the first two points results in a value of MAP between the appropriate values at TP0 and TP6 (with and without feature selection). Regarding the combinations that include the third time point information, only a slight improvement is achieved in a TP6 and TP12 combination with feature selection. In the scenarios where TP24 is included, the only combination that leads to better results is the combination of TP12 and TP24 with feature selection. This MAP value of 0.825 is the best value obtained in all cases in general. According to the performed study for the examined ADNI subset, when more than two time points exist, the combination of the last two time points with feature selection will lead to the best MAP.

Table 2. Inclusion rate for features (scenarios in which each time point is treated separately)

Volumes	**TP0**	**TP6**	**TP12**	**TP24**	**Cortical thickness**	**TP0**	**TP6**	**TP12**	**TP24**
RightPutamen	0	0	0	1	LeftBankssts	146	146	146	146
LeftAccumbensArea	0	146	0	0	LeftEntorhinal	146	146	146	146
LeftAmygdala	146	146	4	0	LeftFrontalPole	0	0	0	1
LeftCerebralCortex	0	2	3	110	LeftInferiorParietal	10	146	146	113
LeftHippocampus	146	146	146	146	LeftInferiorTemporal	146	1	13	28
LeftInferiorLateralVentricle	146	146	146	146	LeftIsthmusCingulate	0	128	58	143
LeftPutamen	0	0	0	124	LeftMedialOrbitofrontal	140	0	4	145
RightAccumbensArea	0	0	146	146	LeftMiddleTemporal	18	146	145	124
RightAmygdala	0	0	10	145	LeftParahippocampal	145	122	125	140
RightCerebralCortex	0	0	0	1	LeftParsOrbitalis	0	0	0	1
Csf	0	0	0	1	LeftPosteriorCingulate	0	0	1	0
RightHippocampus	26	137	84	137	LeftSuperiorTemporal	0	1	3	0
RightInferiorLateralVentricle	0	0	0	3	RightBankssts	0	0	132	124
					RightEntorhinal	146	130	146	145
Cortical thickness					RightInferiorParietal	0	146	0	8
RightParahippocampal	9	0	35	9	RightInferiorTemporal	136	0	0	1
RightPosteriorCingulate	0	0	0	33	RightIsthmusCingulate	0	0	3	0
RightPrecuneus	8	0	83	0	RightLateralOccipital	0	0	0	145
RightRostralAnteriorCingulate	0	0	1	0	RightLingual	146	0	1	0
RightRostralMiddleFrontal	0	0	0	1	RightMedialOrbitofrontal	0	1	1	0
RightSupramarginal	0	0	0	1	RightMiddleTemporal	10	0	98	128
RightInsula	3	1	0	0	RightMiddleTemporal	10	0	98	128

From the point of view of the influence of the feature selection on the retrieval performance, it can be concluded that in all cases this step significantly improves the results. To be able to analyze the stability of the selected features, the inclusion rate for each feature in the experiments was additionally calculated. It is given in the Table 2 and Table 3.

Table 3. Inclusion rate for the features used in the scenarios in which a combination of the information at each time point is considered

Volumes	**TP0+6**	**TP0+6+12**	**TP6+12**	**TP0+6+12+24**	**TP6+12+24**	**TP12+24**	**Cortical thickness**	**TP0+6**	**TP0+6+12**	**TP6+12**	**TP0+6+12+24**	**TP6+12+24**	**TP12+24**
LeftAmygdala_0	22	5		13			LeftLingual_6	0	0	1	0	0	
LeftHippocampus_0	146	146		146			LeftMiddleTemporal_6	146	88	9	2	1	
RightHippocampus_0	3	0		1			LeftParahippocampal_6	3	0	0	0	1	
LeftAccumbensArea_6	146	0	0	0	0		RightEntorhinal_6	139	82	6	0	0	
LeftAmygdala_6	140	139	140	129	140		RightInferiorParietal_6	146	133	59	0	0	
LeftCerebralCortex_6	0	0	3	0	0		RightIsthmusCingulate_6	1	1	0	0	0	
LeftHippocampus_6	146	146	146	146	146		RightMedialOrbitofrontal_6	1	1	0	0	0	
LeftInferiorLateralVentricle_6	146	128	124	130	125		RightParahippocampal_12		114	59	0	0	0
OpticChiasm_6	0	0	0	1	1		RightPrecuneus_12		18	9	0	0	0
RightHippocampus_6	113	16	18	24	99		RRostralAnteriorCingulate12		1	1	1	1	1
LeftAmygdala_12		0	0	0	0	1	LeftBankssts_12		146	146	146	146	146
LeftCerebralCortex_12		1	1	0	0	0	LeftEntorhinal_12		146	146	142	142	142
LeftHippocampus_12		146	146	131	144	146	LeftInferiorParietal_12		146	146	146	146	146
LeftInferiorLateralVentricle_12		32	22	18	23	146	LeftInferiorTemporal_12		20	9	3	2	3
RightAccumbensArea_12		146	146	134	132	110	LeftIsthmusCingulate_12		3	70	0	0	0
RightHippocampus_12		8	3	8	8	9	LeftMedialOrbitofrontal_12		1	5	0	0	0
RightPutamen_24				1	1	1	LeftMiddleTemporal_12		145	145	128	130	103
LeftCerebralCortex_24				125	127	126	LeftParahippocampal_12		16	101	13	6	15
LeftHippocampus_24				146	146	146	LeftSuperiorTemporal_12		1	1	1	1	2
LeftInferiorLateralVentricle_24				1	1	1	RightBankssts_12		143	128	128	120	112
LeftPutamen_24				26	0	1	RightEntorhinal_12		60	133	0	1	105
RightAccumbensArea_24				146	146	146	RightInferiorTemporal_12		1	1	0	0	0
RightAmygdala_24				16	16	143	RightIsthmusCingulate_12		6	4	3	1	0
RightCerebralCortex_24				1	1	1	RightMedialOrbitofrontal_12		1	1	1	0	1
Csf_24				1	2	1	RightMiddleTemporal_12		102	102	1	0	5
RightHippocampus_24				0	0	5	RightParahippocampal_24				83	16	2
							RightPosteriorCingulate_24				14	4	12
Cortical thickness							RRostralMiddleFrontal_24				1	1	1
RightParahippocampal_0	4	2		0			RightInsula_24				0	1	0
RightInsula_0	2	2		3			LeftBankssts_24				146	146	146
LeftBankssts_0	145	29		1			LeftEntorhinal_24				146	146	146
LeftEntorhinal_0	3	0		0			LeftFrontalPole_24				1	1	1
LeftInferiorTemporal_0	146	1		0			LeftInferiorTemporal_24				1	1	1
LeftMedialOrbitofrontal_0	146	19		0			LeftIsthmusCingulate_24				142	142	146
LeftParahippocampal_0	143	127		97			LeftMedialOrbitofrontal_24				145	145	143
LeftPosteriorCingulate_0	0	1		0			LeftMiddleTemporal_24				39	28	113
RightEntorhinal_0	3	0		0			LeftParahippocampal_24				36	127	129
RightInferiorTemporal_0	21	3		0			LeftParsOrbitalis_24				1	1	1
RightLingual_0	146	144		146			RightBankssts_24				31	24	30
RightInsula_6	1	1	1	1	1		RightEntorhinal_24				145	145	107
LeftBankssts_6	1	0	0	0	0		RightInferiorParietal_24				8	13	68
LeftEntorhinal_6	144	3	0	0	0		RightInferiorTemporal_24				1	1	1
LeftInferiorParietal_6	146	100	83	132	130		RightLateralOccipital_24				142	137	132
LeftIsthmusCingulate_6	22	110	55	0	0		RightMiddleTemporal_24				126	127	42

Only the features that were selected by the algorithm at least once are depicted. It can be concluded that the stability of the selected features in most of the cases is very high in both cases, when the information is considered separately for each time point, and in the case of concatenation. The most frequently selected features are reported in the literature as relevant biomarkers for Alzheimer's disease, including volume of the hippocampus, inferior lateral ventricle, amygdala, cortical thickness of the entorhinal cortex etc. [1]. This makes the inclusion of feature selection step even more meaningful.

Considering the feature vector dimensionality in the case of separate representation at any time point, between 11 and 18 features were selected in most of the cases after the feature selection was applied. In the case of concatenation, the feature selection method reduced the dimensionality to 15-22 features in the case of concatenation of two time points in most of the cases, 19-23 in the case of three time points, while in the case of four time points, subsets of 25 features were selected in most of the cases. Significant dimensionality reduction is evident in the examined scenarios, while improving the retrieval performance.

5 Conclusion

The aim of this work was to evaluate the retrieval performance directed by the temporal information. Measures of the brain structures such as volumes and cortical thickness were used to construct the feature vectors. Considering the representation, ten scenarios were defined on the bases of combination of different time points and each time point separately. The advantage of using feature selection to select the most relevant features was also studied. Regarding the representation based on separate time points, it can be concluded that the retrieval performance gets better while using the time point that is more temporally distant from the baseline. This means that the feature extraction procedure used in this work properly reflects the disease progression.

Considering the scenarios where combination of multiple time points is used, the enhancement with additional information not always leads to better results. If more than two time points exist, the combination of the measures obtained from the last two measures leads to the best MAP with the feature selection procedure included in the retrieval process. The feature selection improved the results in all cases in general, while significantly reducing the feature vector dimensionality. To be able to analyze the stability of the selected features, the inclusion rate for each feature was also provided. According to the results, high stability of the features can be noticed. Additionally, most of the selected features are known as significant markers for AD.

The examination performed in this work is beneficial in the direction of providing right and clinically relevant information contained in the result of the retrieval process to the clinicians, researchers or students. Besides that, including longitudinal information in the retrieval also enables analysis of the treatment reaction and monitoring the progression of the disease that is of great importance for this application domain.

In future, we plan to perform further longitudinal analysis and combined information extracted from multiple time points including atrophy rate, cortical thinning rate, and rate of change. Additionally, an accent will be put on the analysis of the follow-up period of the group of patients, converters from NL to AD.

Acknowledgement. Data collection and sharing for this project was funded by the Alzheimer's Disease Neuroimaging Initiative (ADNI) (National Institutes of Health Grant U01 AG024904) and DOD ADNI (Department of Defense award number W81XWH-12-2-0012). ADNI is funded by the National Institute on Aging, the National Institute of Biomedical Imaging and Bioengineering, and through generous contributions from the following: Alzheimer's Association; Alzheimer's Drug Discovery Foundation; BioClinica, Inc.; Biogen Idec Inc.; Bristol-Myers Squibb Company; Eisai Inc.; Elan Pharmaceuticals, Inc.; Eli Lilly and Company; F. Hoffmann-La Roche Ltd. and its affiliated company Genentech, Inc.; GE Healthcare; Innogenetics, N.V.; IXICO Ltd.; Janssen Alzheimer Immunotherapy Research & Development, LLC.; Johnson & Johnson Pharmaceutical Research & Development LLC.; Medpace, Inc.; Merck & Co., Inc.; Meso Scale Diagnostics, LLC.; NeuroRx Research; Novartis Pharmaceuticals Corporation; Pfizer Inc.; Piramal Imaging; Servier; Synarc Inc.; and Takeda Pharmaceutical Company. The Canadian Institutes of Health Research is providing funds to support ADNI clinical sites in Canada. Private sector contributions are facilitated by the Foundation for the National Institutes of Health (http://www.fnih.org). The Northern California Institute for Research and Education is the grantee organization. The study is coordinated by the Alzheimer's Disease Cooperative Study at the University of California, San Diego. ADNI data are disseminated by the Laboratory for Neuro Imaging at the University of Southern California.

Authors also acknowledge the support of the European Commission through the project MAESTRA - Learning from Massive, Incompletely annotated, and Structured Data (Grant number ICT-2013-612944).

References

1. Weiner, M. W., Veitch, D. P., Aisen, P. S., Beckett, L. A., Cairns, N. J., Green, R. C., Harvey, D. et al.: The Alzheimer's Disease Neuroimaging Initiative: a review of papers published since its inception. Alzheimer's & Dementia 9(5), e111-e194, (2013)
2. Ye, J., Wu, T., Li, J., Chen, K.: Machine learning approaches for the neuroimaging study of Alzheimer's disease. Computer 44(4), 99-101 (2011)
3. Velayudhan, L., Proitsi, P., Westman, E., Muehlboeck, J. S., Mecocci, P., Vellas, B. et al.: Entorhinal cortex thickness predicts cognitive decline in Alzheimer's disease. Journal of Alzheimer's Disease 33(3), 755-766 (2013)
4. Dai, D., He, H., Vogelstein, J. T., Hou, Z.: Accurate prediction of AD patients using cortical thickness networks. Machine vision and applications 24(7), 1445-1457 (2013)
5. Nho, K., Risacher, L. S., Crane, P. K., DeCarli, C., Glymour, M.M., Habeck, C., Kim, S. et al.: Voxel and surface-based topography of memory and executive deficits in mild cognitive impairment and Alzheimer's disease. Brain imaging and behavior 6(4), 551-567 (2012)

6. Lötjönen, J., Wolz, R., Koikkalainen, J., Julkunen, V., Thurfjell, L., Lundqvist, R. et al.: Fast and robust extraction of hippocampus from MR images for diagnostics of Alzheimer's disease. Neuroimage 56(1), 185-196 (2011)
7. Sabuncu, M. R., Desikan, R. S., Sepulcre, J., Yeo, B. T. T., Liu, H., Schmansky, N. J, et al.: The dynamics of cortical and hippocampal atrophy in Alzheimer disease. Archives of neurology 68(8), 1040-1048 (2011)
8. Gerardin, E., Chételat, G., Chupin, M., Cuingnet, R., Desgranges, B., Kim, H. S. et al.: Multidimensional classification of hippocampal shape features discriminates Alzheimer's disease and mild cognitive impairment from normal aging. Neuroimage 47(4), 1476-1486 (2009)
9. Oliveira, M. C., Cirne, W., de Azevedo Marques, P. M.: Towards applying content-based image retrieval in the clinical routine. Future Generation Computer Systems, 23(3), 466-474 (2007)
10. Akgül C. B., Rubin, D. L., Napel, S., Beaulieu, C. F., Greenspan, H., Acar, B.: Content-based image retrieval in radiology: current status and future directions. Journal of Digital Imaging. 24(2), 208-222 (2011)
11. Rosset A., Muller H., Martins M., Dfouni N., Vallée J.-P., Ratib O.: Casimage project - a digital teaching files authoring environment. Journal of Thoracic Imaging. 19(2), 1-6. (2004)
12. Akgül, C. B., Ünay, D., Ekin, A.: Automated diagnosis of Alzheimer's disease using image similarity and user feedback. In: Proceedings of the ACM International Conference on Image and Video Retrieval, p. 34. ACM (2009)
13. Agarwal M., and Mostafa J.: Image Retrieval for Alzheimer's Disease Detection. Medical Content-Based Retrieval for Clinical Decision Support. Springer Ber. Heid.. 49-60 (2010)
14. Agarwal, M., & Mostafa, J.: Content-based image retrieval for Alzheimer's disease detection. In: Content-Based Multimedia Indexing (CBMI), 2011 9th International Workshop on, 13-18 (2011)
15. Mizotin, M., Benois-Pineau, J., Allard, M., Catheline, G.: Feature-based brain MRI retrieval for Alzheimer disease diagnosis. In: Image Processing (ICIP), 2012 19th IEEE International Conference on, 1241-1244 (2012)
16. Qian, Y., Gao, X., Loomes, M., Comley, R., Barn, B., Hui, R., Tian, Z.: Content-based retrieval of 3D medical images. In: eTELEMED 2011, The Third International Conference on eHealth, Telemedicine, and Social Medicine, 7-12 (2011)
17. Trojacanec K., Kitanovski I., Dimitrovski I. and Loshkovska S.: Content Based Retrieval of MRI Based on Brain Structure Changes in Alzheimer's Disease. In: Proceedings of the In-ternational Conference on Bioimaging, 13-22. DOI: 10.5220/0005182200130022 (2015)
18. Reuter, M., Schmansky, N.J., Rosas, H.D., Fischl, B.: Within-Subject Template Estimation for Unbiased Longitudinal Image Analysis. Neuroimage 61 (4), 1402-1418 (2012)
19. Hall, M. A., Holmes, G.: Benchmarking attribute selection techniques for discrete class data mining. IEEE Transactions on Knowledge and Data Engineering 15(6), 1437-1447 (2003)

Generic Face Detection and Pose Estimation Algorithm Suitable for the Face De-identification Problem

Aleksandar Milchevski[1], Dijana Petrovska-Delacrétaz[2] and Dejan Gjorgjevikj[3]

[1] Faculty of Electrical Engineering and Information Technologies, Skopje, Republic of Macedonia
milchevski@gmail.com
[2] TELECOM SudParis, Évry, France
dijana.petrovska@telecom-sudparis.eu
[3] Faculty of Computer Science and Engineering, Skopje, Republic of Macedonia
dejan.gjorgjevikj@finki.ukim.mk

Abstract. In this work we tackle the problem of face de-identification in an image. The first step towards a solution to this problem is the design of a successful generic face detection algorithm, which will detect all of the faces in the image or video, regardless of the pose. If the face detection algorithm fails to detect even one face, the effect of the de-identification algorithm could be neutralized. That is why a novel face detection algorithm is proposed for face detection and pose estimation. The algorithm uses an ensemble of three linear SVM classifiers. The first, second and the third SVM classifier estimate the pitch, yaw and roll angle of the face and a logistic regression is used to combine the results and output a final decision. Second, the results of the face detection and a simple space variant de-identification algorithm are used to show the benefits of simultaneous face detection and face de-identification.

Keywords: De-identification, Nonfrontal face detection, Pose estimation, Classifier fusion, SVM, Logistic regression

1 Introduction

The issue of privacy protection in video surveillance has drawn a lot of interest lately. There are different levels of privacy protection schemes that can be applied. Regarding the resolution of the video, people silhouettes or faces need to be protected. In order to be efficient, frontal as well as nonfrontal faces need to be protected. Both of them require a face detection algorithm in order to localize the region that needs to be hidden, encrypted, etc. There are already some research efforts of privacy protection solutions to hide distinguishing frontal facial information and to conceal identity. The available face detection algorithms work well, however the problem of nonfrontal face detection needs to be further studied. In order to be efficient, privacy protection schemes for nonfrontal faces have to be studied also. However, such research efforts are still lacking.

S. Loshkovska and S. Koceski (eds.), *ICT Innovations 2015*,
Advances in Intelligent Systems and Computing 399,
DOI: 10.1007/978-3-319-25733-4_23

2 Previous Work

2.1 Face Protection

The problem of face privacy protection can be in general defined as finding a way for the protection of the identity of the subject in the image or the video, while keeping the usability of the image or the video. The image is transformed in such a way that the subject cannot be identified by face recognition algorithm or a human observer. The definition of the usability of the video and the answer to the question: "Why not just delete the face region?" depend on the specific area of use of the de-identified video e.g. usually, it is important for the de-identification algorithm to retain the facial expressions.

In [1] the k-same algorithm is proposed. The algorithm determines the similarity between faces based on a distance metric and replaces the face with a new face which is an average of components of several faces. However, in order for the algorithm to be successful all of the used faces should be from different subjects. Several experiments are done using Eigen face recognition algorithm. The experiments made show that the naïve approaches, such as blurring, pixelization, adding noise, etc. although produce results from which a human observer cannot identify the subject, they do not provide good protection against face recognition algorithms.

The de-identification algorithm presented [2] is based on the k-same algorithm previously described; however, an AAM (Active Appearance Model) is first fitted for the face which is de-identified. The result of this improvement is that the output of the de-identification algorithm is without artifacts i.e. with better visual quality. The experiments made are also with an Eigenfaces recognition algorithm and they show successful de-identification. The experiments are limited on frontal faces and the AAM ground-truth is manually established.

In [3] a scrambling technique as a solution to the face de-identification problem. The sign of the H.264 transformed image is pseudo randomly flipped. The advantages of this approach are the low computational cost and the full reversibility of the applied modification of the image. The authors also provide an alternative of the algorithm by using permutation of the coefficients instead of sign change.

In [4] an algorithm for privacy protection in video surveillance is proposed which uses geometric warping of the face region. Several experiments are made using the OpenCV's Viola Jones implementation for the face detection and FLDA (Fisher Linear Discriminant Analysis) for the face recognition.

In [5] a system for automatic face replacement in images is proposed. First the pose of the face is estimated and a face with similar face is found from a large data set (yaw and pitch angles differ by no more than 3° from the yaw and pitch of the original face). After that, the face is replaced while keeping some of the original features. The new face is than color and light adjusted. The algorithm is fully automated and produces highly plausible results.

2.2 Face Detection

Face detection is probably one of the most researched problems in the areas of computer vision and image processing. There are a vast number of published algorithms and different approaches, but the most revolutionary is the Viola – Jones algorithm.

Very good survey on face detection algorithms is done in [6]. In the following text the most important and recent algorithms are summarized.

In the work of [7] a multi-view face detector is proposed using a detector pyramid. They use coarse-to-fine approach to deal with the out of plane rotations of the head. The full range of possible rotations is partitioned into several partitions, and ranges are narrowed as the level of the pyramid increases. The detector at the top of the pyramid is very simple with a main task to reject as much of the non-face images. They deal with in-plane rotations by applying their detector on rotated test images with 30° and -30° rotations.

In [8] the authors present a simple solution for the multi-view face detection problem. They use the Viola – Jones framework, however they modify it by using LUT as weak classifiers instead of the stump weak classifier, used in the original work of Viola – Jones.

Human faces are divided into several categories and a cascade is trained for every category individually. For the yaw axis there are 5 categories with the following intervals: [-90°, -50°], [-50°, -20°], [-20°, +20°], [+20°,+50°], [+50°, +90°].

The authors in [9] build a low dimensional face manifold parameterized by the pose of the face. They train a convolution network and use Energy Minimization Framework to map the face images onto the face manifold and non-face image far away from the face manifold. The authors elaborate that the multi-view face detection and pose estimation are very closely related so they should not be done separately. The authors claim that the system is highly reliable, and runs in real time on standard hardware.

In [10] the authors present a method for simultaneous face detection, pose estimation and landmark localization. For the landmark localization they use mixtures of trees with a shared pool of parts, instead of densely-connected elastic graphs. They used HOG (Histogram of Oriented Gradients) as a feature descriptor. The authors claim that their method is better or comparable to the state-of-the-art algorithms in all three categories. The presented results clearly show that the algorithm works well, however, they have limited the test set to images with relatively big faces, so that the landmarks are clearly visible.

In [11] the authors motivated by the success of [10] propose a face detection algorithm which uses part models. However, they propose Cascade Deformable Part Models, arguing that the use of Tree Structure Model in [10] is suboptimal for face detection, because is too slow and limited to high resolutions. The presented results show that the algorithm works well even on the AFLW data set. The average detection time reported for the method is 0.52s, compared to 26.06s for the TSM algorithm published in [10].

3 Proposed Algorithm for Generic Face Detection

Usually the HOG is described as a feature descriptor with great descriptive power, but also as very computationally expensive. In FDDB [12] the best scoring algorithm uses HOG for the feature description. Many authors have suggested simplification or ways to compute the feature in a faster way. Because of the superior descriptive power the HOG is chosen as a feature descriptor.

Almost all of the reviewed work on unconstrained face detection have treated the face detection and pose estimation as a combined problem. However, the two problems are conflicting: the pose estimation tries to find differences between the several view groups and the face detection tries to find similarities in all view groups. Nevertheless, because of the difficulty of the multi-view face detection problem (large number of possible variations) the pose information should be used even if the detection of the face is of main concern.

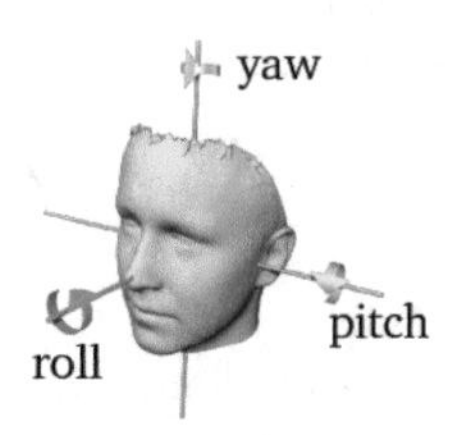

Fig. 1. The three rotation angles used to describe the pose of the head

Analyzing the data set used for training (AFLW) [13] and the effect of the observed face on the image regarding the three axis (Fig. 1), three separate classifiers are proposed:

1. SVM – pitch (nodding)
   An SVM (Support Vector Machine) classifier using four classes: three classes corresponding to faces with values for the pitch in the intervals of [-90°,-12°), [-12°,12°], (12°,90°] and a fourth class corresponding to images without faces.
2. SVM – roll (In plane rotation)
   An SVM classifier using six classes: five classes corresponding to faces with values for the roll in the intervals of [-90°,-30°), [-30°,-12°), [-12°,12°], (12°,30°], (30°,90°] and a sixth class with images without faces. (An assumption is made that the maximum in-plane rotation of the face is 90°. The detector can be applied on a 180° rotated test image if other faces are expected).
3. SVM – yaw (out of plane rotation)
   An SVM classifier using eight classes: seven classes corresponding to faces with values for the yaw in the intervals of [-108°,-60°), [-60°,-30°), [-30°,-12°), [-12°,12°], (12°,30°], (30°,60°], (60°, 108°] and a eighth class corresponding to images without faces.

The three classifiers are trained individually with different training sets. Next, the decisions and the probability estimates from all three classifiers are combined to form a final decision if the tested image should be classified as face or not. An ensemble of classifiers is also used [14]. First, they use three independently trained SVM classifiers for frontal, profile and semi-profile face. The probability estimations of the three classifiers are combined using SVM regression which should output 4 distinct values representing the four classes.

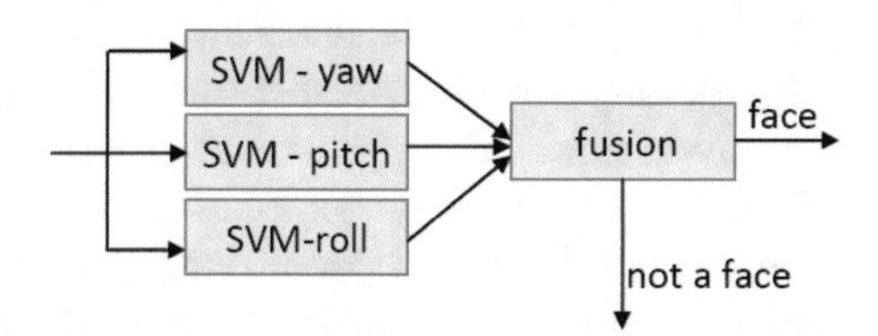

Fig. 2. Block diagram of the proposed face detection algorithm

In Fig. 2 a block diagram of the proposed algorithm is given. A block with a size of 36x36 is first extracted and the HOG is calculated. The feature vector is inputted to the three linear SVM classifiers. The outputs of the three classifiers are given to the decision fusion block which gives the final decision of the tested block being a face or not.

3.1 Training

Training of the linear SVM classifiers (for roll, pitch and yaw). For the first step of the proposed algorithm three linear SVM are trained using the LIBLINEAR [15] library.

Training set with images containing faces. For the training of the three linear SVM classifiers the training sets with images containing faces are created using the AFLW data set. The first 15.000 images of the data set are used for this step of the training. The face region is extracted from the image using a square region that contains all of the landmarks provided with the data set. The square region is than scaled to a block of size 36x36. Every block is than mirrored in order to increase the number of face images. The obtained training set is split into three equal subset, which are used for the separate SVM classifiers. In this way, every SVM classifier is trained with approximately 10.000 independent positive face samples.

Training set with images without faces. In order to obtain independent training and to find difficult samples without faces, the ILSVRC[16] data set is used. Only the images which are labeled that do not belong to the person class are used. A large number of feature vectors are extracted from one image in the following way: First, a feature vector is calculated for every 36x36 block in the image without overlapping. Than the image is scaled by a factor of two and again a feature vector is extracted for every

36x36 block. The procedure is repeated until the rescaling of the image produces an image with height or width less than 36.

The training set with images without faces is divided into several subsets which will be used separately. A subset of the training set is created by extracting feature vectors from the data set until the total number of negative feature vectors is above 20.000.

Choice of C- parameter for the SVM. The choice of the C -parameter controls the trade-off between complexity of decision rule and frequency of error [17]. If the parameter is too large, a high penalty for nonseparable samples is introduced and there is an increased chance of overfitting. If the parameter is too small, there is an increased chance of underfitting. That is why the choice of the C parameter is analyzed.

In order to obtain an optimal value for C parameter a grid search has been performed using 5-fold cross-validation. The C parameter is varied exponentially in the range from 2^{-5} to 2^{3}, a value 10 is also included as a value other authors used. An independent subset of the training set with images without faces is formed for the training of every SVM as explained above.

The results from the cross-validation show that a value of 2^{-4} is a good choice for every SVM classifier.

Mining for hard samples without faces. An iterative procedure in order to obtain hard samples without faces was performed. An SVM classifier is trained with a new subset of the training set with images without faces. After the training is completed the classifier is tested on the training set and the falsely classified as faces are extracted. This procedure is repeated for 15 iterations. After all of the iterations are finished the SVM is trained using another subset of the training set with images containing faces and all of the extracted hard samples.

Decision Fusion. Two methods are tested to fuse the final decision, using an SVM and using LR.

With the first method a final decision about the block is obtained with a new kernel – SVM classifier. As a feature vector the outputs of all of the three linear SVM are used. Every multiclass SVM is implemented using several binary SVM classifiers with the one-against-all approach. The margins outputted from every binary SVM are used as features for the training of the SVM, which will output the final decision. In that way, the feature vector has 18 features now, because 4 margins are obtained from the first linear SVM, 6 from the second and 8 from the third.

A separate training set is created using 5000 images from the AFLW data set, with the mirroring of the images about 10000 independent positive samples are obtained. (These images are exclusively used for the training of the fusing algorithm.)

Choice of Cost (C) parameter and gamma for the SVM. The optimal values of the C parameter and the gamma value for the radial basis function are estimated using grid search and 5-fold cross-validation. The range for the gamma parameter is from 2^{-15} to

2^3, and the range for the C parameter is the same as for the previous tests. The cross-validation accuracy yielded highest accuracy for C=4 and gamma=2^{-3}.

Mining for hard samples without faces. A similar iterative procedure is used to obtain hard samples without faces for the training of the kernel-SVM classifier. A new subset of the training set without faces is created for each iteration with a size of about 20.000 samples. The classifier is tested on the training set and the samples falsely classified as face are saved. Additionally, the support vectors corresponding to images without faces are also saved. The procedure is repeated for 50 iterations.

Final Training. After all the iterations have been finished a new classifier is trained with all of the extracted hard samples and a new subset of training set with images without faces. Again a grid-search and cross-validation is done in order to test the accuracy of the system for different values.

Another way to fuse the decision was also tested. For this alternative the decision is fused using logistic regression. The logistic regression classifier was trained using the same training set as for the final kernel-SVM classifier.

3.2 Classifier Test

In order to compare the two ways to fuse the decision an experiment has been made using the remaining images from the AFLW data set (about 2000 independent images, 4000 in total using mirroring) and a new subset from the ILSVR data set. The ROC curve is plotted and shown on Fig. 3. It can be seen that decision fusion with logistic regression outputs superior results compared to the kernel-SVM.

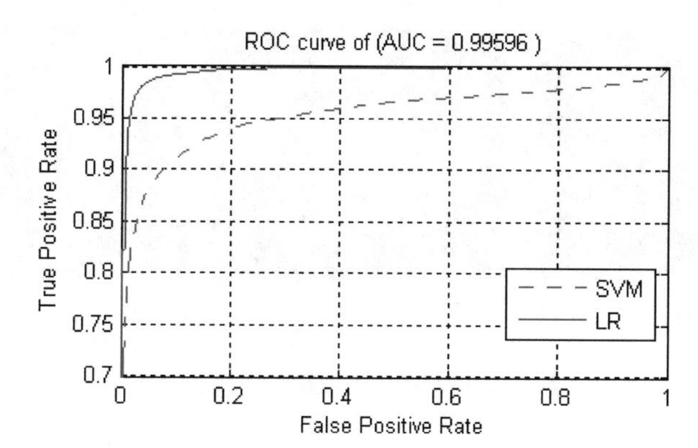

Fig. 3. ROC curve comparison of the two ways to fuse the decision

3.3 Face Detection

The face detection algorithm is implemented using a standard window-sliding approach. Every 36x36 block is tested in the image using the designed system.

The image is than rescaled and the same procedure is repeated. The regions with a probability higher than a threshold (0.99) are than grouped to form the final detection result. A probability map for the whole image is also created by adding all of the outputted probabilities. The procedure is repeated until the height or the width of the rescaled image is less than 36.
Fig. 4.a) and Fig. 5.a) show the output of the algorithm for the selected images. The images are part of the FDDB.

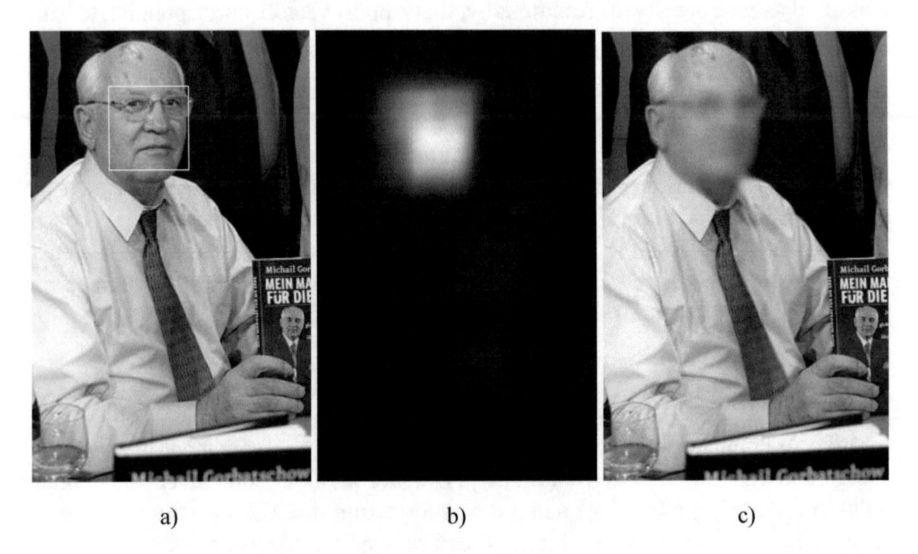

a) b) c)

Fig. 4. Results obtained with the proposed method

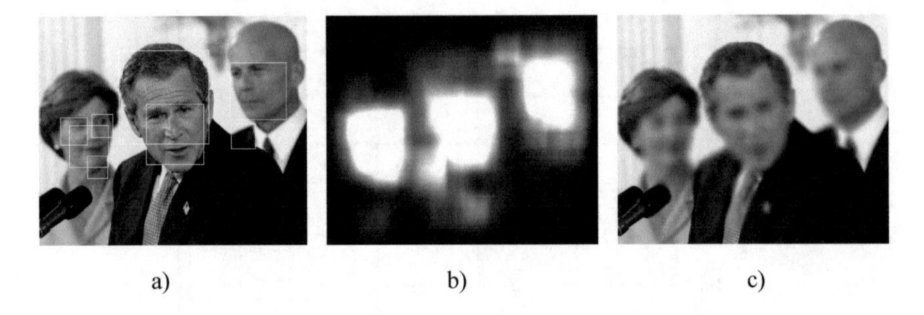

a) b) c)

Fig. 5. Results obtained with the proposed method

3.4 Face De-identification

The output of the designed face detection algorithm can be used for de-identification. To show that, the probability map outputted from the face-detection is used to do a space variant blurring. The blurring is not a good choice for a de-identification algorithm and serves only to show the benefits of simultaneous face detection and face de-identification. The PSF (Point Spread Function) is chosen to be Gaussian window with a size of 19x19 and a standard deviation proportional to the output from the probability map. Fig. 4.c) and Fig. 5.c) show the output of the de-identification algorithm.

4 Conclusion

In this paper a review of the most important and most recent algorithms for face detection algorithm was done. A new approach was proposed for face detection in unconstrained condition using an ensemble of linear SVMs. The algorithm was tested on several images in order to evaluate the performance. Two algorithms were tested for the fusion of the decisions, SVM and logistic regression. The fusion with the logistic regression yield better results. The tests also showed that a better way should be used to group the rectangles with high probability of face. The accuracy of the detection of profile faces should also be improved.

The output of the algorithm is used to implement a simple de-identification algorithm. Although the proposed de-identification method is simple it shows the benefits of the simultaneous solution of the face detection and the face de-identification problem.

Acknowledgements. This work was partially done during a STSM (Short Term Scientific Mission), supported by the COST Action IC1206, hosted by TELECOM SudParis and prof. Dijana Petrovska-Delacrétaz.

References

1. Sweeney, E.N.L., Malin, B.: Preserving privacy by de-identifying facial images (2003)
2. Gross, R., Sweeney, L., Torre, F.D.l., Baker, S.: Model-based face deidentification. In: Computer Vision and Pattern Recognition Workshop, 2006. CVPRW'06. Conference on. pp. 161-161. IEEE (2006)
3. Dufaux, F., Ebrahimi, T.: A framework for the validation of privacy protection solutions in video surveillance. In: Multimedia and Expo (ICME), 2010 IEEE International Conference on. pp. 66-71. IEEE (2010)
4. Korshunov, P., Ebrahimi, T.: Using warping for privacy protection in video surveillance. In: Digital Signal Processing (DSP), 2013 18th International Conference on. pp. 1-6. IEEE (2013)
5. Bitouk, D., Kumar, N., Dhillon, S., Belhumeur, P., Nayar, S.K.: Face swapping: automatically replacing faces in photographs. ACM Transactions on Graphics (TOG) 27(3), 39 (2008)
6. Zhang, C., Zhang, Z.: Boosting-based face detection and adaptation. Synthesis Lectures on Computer Vision 2(1), 1-140 (2010)
7. Li, S.Z., Zhu, L., Zhang, Z., Blake, A., Zhang, H., Shum, H.: Statistical learning of multi-view face detection. In: Computer Vision ECCV 2002, pp. 67-81. Springer (2002)
8. Wu, B., Ai, H., Huang, C., Lao, S.: Fast rotation invariant multi-view face detection based on real adaboost. In: Automatic Face and Gesture Recognition, 2004. Proceedings. Sixth IEEE International Conference on. pp. 79-84. IEEE (2004)

9. Osadchy, M., Cun, Y.L., Miller, M.L.: Synergistic face detection and pose estimation with energy-based models. The Journal of Machine Learning Research 8, 1197-1215 (2007)
10. Zhu, X., Ramanan, D.: Face detection, pose estimation, and landmark localization in the wild. In: Computer Vision and Pattern Recognition (CVPR), 2012 IEEE Conference on. pp. 2879-2886. IEEE (2012)
11. Orozco, J., Martinez, B., Pantic, M.: Empirical analysis of cascade deformable models for multi-view face detection (2013)
12. Jain, V., Learned-Miller, E.G.: Fddb: A benchmark for face detection in unconstrained settings. UMass Amherst Technical Report (2010)
13. Kostinger, M.,Wohlhart, P., Roth, P.M., Bischof, H.: Annotated facial landmarks in the wild: A large-scale, real-world database for facial landmark localization. In: Computer Vision Workshops (ICCV Workshops), 2011 IEEE International Conference on. pp. 2144-2151. IEEE (2011)
14. Yan, J.: Ensemble svm regression based multi-view face detection system. In: Machine Learning for Signal Processing, 2007 IEEE Workshop on. pp. 163-169. IEEE (2007)
15. Fan, R.E., Chang, K.W., Hsieh, C.J., Wang, X.R., Lin, C.J.: Liblinear: A library for large linear classification. The Journal of Machine Learning Research 9, 1871-1874 (2008)
16. Russakovsky, O., Deng, J., Su, H., Krause, J., Satheesh, S., Ma, S., Huang, Z., Karpathy, A., Khosla, A., Bernstein, M., et al.: Imagenet large scale visual recognition challenge. International Journal of Computer Vision pp. 1-42 (2014)
17. Cortes, C., Vapnik, V.: Support-vector networks. Machine learning 20(3), 273-297 (1995)

RS-fMRI Data Analysis for Identification of Changes in Functional Connectivity Networks of Bi-polar Patients

Tommy Boshkovski[1], Ilinka Ivanoska[1], Kire Trivodaliev[1], Slobodan Kalajdziski[1], Pablo Villoslada[3], Magi Andorra[3], Vesna Prčkovska[3], and Ljupco Kocarev[1,2]

[1] Faculty of Computer Science and Engineering, Ss. Cyril and Methodius University
tommy@manu.edu.mk,
{ilinka.ivanoska,kire.trivodaliev,slobodan.kalajdziski}@finki.ukim.mk
[2] Macedonian Academy of Sciences and Arts, Skopje, Macedonia
lkocarev@ucsd.edu
[3] Center of Neuroimmunology, Institut d'Investigacions Biomdiques August Pi i Sunyer (IDIBAPS), Barcelona, Spain
{pvilloslada,andorra,vprchkov}@clinic.ub.es

Abstract. One third of the world's population suffers from some kind of neurological disorder. The development of technology allows us to analyze, model and visualize these disorders in order to help MDs in further treatments. Resting state fMRI is one of the most common ways for investigating the functional connectivity of the brain, which produces time series data of activation of the brain's regions when subjects are in resting state. In this paper we show that changes occur in the Default Mode Network of bipolar patients by statistically analyzing time series data from their resting state fMRI. We discover several differences in the functional connectivity of these subjects compared to a control group. We then use clustering algorithm in order to find the clusters of active regions during the rs-fMRI, i.e. the groups of regions with similar time series data.

Keywords: Resting state fMRI, Functional Connectivity, Default Mode Network, Clustering, Bi-polar Patients.

1 Introduction

The brain is an incredibly complex organ which simultaneously processes information from our body, It controls organs; generates wishes and feelings; stores data and calls memories, and controls movements. In fact this organ make us humans, it gives us sense for art, languages. All of this leads us to the question: "How the brain works?". To answer this question one needs to look inside the brain, and observe the brain connectivity i.e. how the neurons are interconnected and organized on different levels. There are 3 types of brain connectivity [1]: structural (anatomical), functional, and effective connectivity. Structural

S. Loshkovska and S. Koceski (eds.), *ICT Innovations 2015*,
Advances in Intelligent Systems and Computing 399,
DOI: 10.1007/978-3-319-25733-4_24

connectivity refers to network of physical (anatomical) links between neural elements i.e. from connections of local circuits between groups of cells (neurons) to connections of large-scale networks that link different regions in the brain (interregional pathways). If these connections are observed in a short period of time then they are relatively static, but there may be changes while observing them in a long period of time, which shows the dynamic nature of structural connectivity [1,2]. Functional connectivity aims to the statistical analysis of the dependence of time series obtained from anatomically distributed brain regions. This can be measured using correlation between time series [1,2,3]. Effective connectivity can be described as union of structural and functional connectivity. It studies the network of causal effects between neural elements, which can be inferred through time series analysis [1,2].

Using resting state fMRI one can observe the default mode network (DMN) that operates while subjects are in resting state and don't participate in any activity or task [4]. If subjects start making any activities or tasks then this network deactivates. Areas of the brain included in the DMN include the medial temporal lobe, the medial prefrontal cortex, and the posterior cingulate cortex, as well as the ventral precuneus and parts of the parietal cortex [5].

Because the DMN is most active at rest and because of the structures involved, some people have inferred that it is associated with activities like daydreaming or retrieving memories [6,7]. Others, in contrast, have suggested that the activity could be related to physiological processes unrelated to any particular activity [8]. Changes in the DMN have been linked to a wide number of different diseases e.g. Alzheimer's disease, autism, schizophrenia, bipolar disorder, depression [8,9]. Diseases can cause either too little or too much activity of the brain regions of DMN.

Currently, the resting state data processing methods include seed-based approaches, amplitude of low frequency fluctuation (ALFF) and fractional amplitude of low-frequency fluctuation (fALFF) independent component analysis (ICA), regional homogeneity (ReHo) analysis etc [10]. A different research was made in [11,12] for demonstrating a hierarchical structure of resting state activity in only the healthy brain using clustering algorithms. [13] investigates the difference of small-world properties of brain functional network derived from resting-state fMRI between healthy subjects and patients with Bipolar disorder (BD).

To investigate the functional connectivity we use fMRI acquisition technique which measures the blood oxygen level depended (BOLD) signal [14,15,16]. Öngür et al. [17] highlight the differences in DMN by using resting state fMRI data analysis. They used independent component analysis (ICA) to identify the DMN component in bipolar disorder and schizophrenia and compared them using ANOVA test.

In this paper we compute functional connectivity as correlation between time series of the BOLD signal. Then the functional connectivity is normalized and in contrast of using ICA for identifying DMN and ANOVA for comparation, we use one sample t-test to denote the DMN and two sample t-test to find the differences in DMN. Next, we use network analysis to identify the differences on network level. We make a comparison between the bipolar disorder group and the control group using a clustering algorithm. With this we aim to show that the bipolar disorder group will have reduced activity in the DMN i.e will have weaker correlations as compared to the control group.

In Sect. II we describe the data structure, along with the methods and techniques used to preprocess, compute and examine functional connectivity. In Sect. III we identify the differences between the bipolar disorder (BD) and the Control group. Finally, a conclusion is made in virtue of the results obtained from the experiment (Sect. IV).

2 Materials and Methods

2.1 Resting State Data

We analyzed 7 BD patients (5m, 2f, 57.57 $\pm$ 8.32 years) and 21 healthy controls (9m, 12f, 28.19 $\pm$ 3.04 years). At the date of scanning all of the BD patients were in the depressive phase with HDRS score $>$22. Brain images were acquired on a 3 Tesla TrioTim scanner (Siemens, Erlangen, Germany) using the 8-channel phased-array head coil supplied by the vendor. A custom-built head holder was used to prevent head movement, and earplugs were used to attenuate scanner noise. Functional resting state data (of 14 minutes) was acquired using a gradient-echo echo-planar pulse sequence sensitive to blood oxygenation level-dependent (BOLD) contrast (TR/TE=2000/30ms, FA=85°, 3.0mm^3 isotropic voxels, 3.0mm thick, no gap between slices) [18]. Because of the artifacts in the fMRI scans, that incurs as a consequence of scanner malfunction, head movements during the scan, breathing, and heart rate, which can cause problems in the further analysis, the data need to be preprocessed.

To preprocess the imaging data we use the DPARSF toolbox [19]. First, the images of each subject were corrected for slice timing and realigned (motion corrected)[20]. Then the images were normalized [21] using MNI EPI template with affine registration followed by nonlinear transformation. Next we smoothed the images using Gaussian Kernel of 4mm Full Width at Half Maximum and detrend the signal to remove any noise that may remain from previous steps. Lastly, the signal need to be preprocessed to get low frequent fluctuation (0.01 0.08 Hz).

2.2 Statistical Analysis of Functional Connectivity of the Brain

Functional connectivity of the resting state subjects [22] between 34 Regions of Interest (ROIs) (Table 3 in Appendix) in the brain, is computed by using Pearson linear correlation, defined as:

$$\rho = \frac{\sum_{i=1}^{n} (x_i - \bar{x}) \cdot (y_i - \bar{y})}{\sqrt{\sum_{i=1}^{n} (x_i - \bar{x})^2 \sum_{i=1}^{n} (y_i - \bar{y})^2}} \quad (1)$$

where ρ is Pearson correlation coefficient, $\bar{x}$ and $\bar{y}$ are mathematical expectations based on measurements of time series of random variables x_i and y_i of length n. This process is repeated for every subject to get r maps (the Pearsons correlation map between the ROIs) which we transform to z maps using Fisher z score transformation (eq. 2). This transformation increases normality of the distribution of correlations in the sample.

$$z = \frac{1}{2}(\ln(1 + r) - \ln(1 - r)) \quad (2)$$

Next, one sample t test were performed for each group with H: $p < 0.001$, where p is activation threshold, in order to obtain the significant patterns in functional connectivity for the group, and two sample t-test to highlight the differences in functional connectivity between the groups. To make all of these measurements we used the REST toolkit [23] which is one of the widespread used toolkits for analysis of resting state functional connectivity.

3 Results

After obtaining functional connectivity of both groups and performing one sample t - test we discovered that for BD group z threshold is 3.090232 and the minimal cluster size is 135 voxels according to activation threshold on voxel level (p=0.001). For the control group with the same threshold on voxel level we get the same z threshold but the minimal cluster size was 309 voxels. The functional connectivity in the control group [24] is greater in the following regions: *Temporal_Mid_R*, *Cerebelum_Crus1_R*, *Thalamus_R*, *Frontal_Sup_R* (Table 1). On the other hand, using the hypothesis H: BD > C the obtained differences in functional connectivity in the BD group is greater in the following regions: *Lingual_L*, *Insula_L*, *Postcentral_R*, *Rolandic_Oper_R* and *Postcentral_R* (Table 2). The differences are visualized on Fig. 1.

Table 1. Differences in functional connectivity in control group using hypothesis H:C>BD i.e. regions that have greater activation in control subjects than the activation in the subjects with bipolar disorder

Id	Voxels	X	Y	Z	z-score	Label
1	529	66	−36	−12	6.5425	*Temporal_Mid_R*
2	915	12	−84	−24	6.3979	*Cerebelum_Crus1_R*
3	208	6	−9	12	6.2728	*Thalamus_R*
4	123	24	27	60	5.9291	*Frontal_Sup_R*

Table 2. Differences in functional connectivity in control group using hypothesis H:C<BD i.e. regions that have greater activation in control subjects than the activation in the subjects with bipolar disorder

Id	Voxels	X	Y	Z	z - score	Label
1	2634	−9	−81	−3	−7.5209	*Lingual_L*
2	1651	−36	−9	12	−7.5228	*Insula_L*
3	127	66	−3	18	−5.624	*Postcentral_R*
4	88	42	−9	15	−5.9227	*Rolandic_Oper_R*
5	700	39	−30	66	−9.4618	*Postcentral_R*

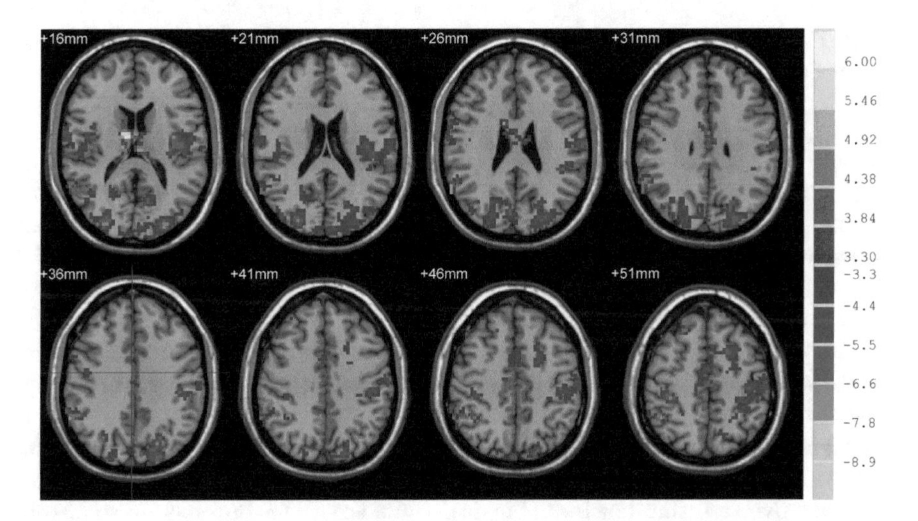

Fig. 1. Visualized differences between control and BD group. Positive t-score refer to Table 1 and negative to Table 2

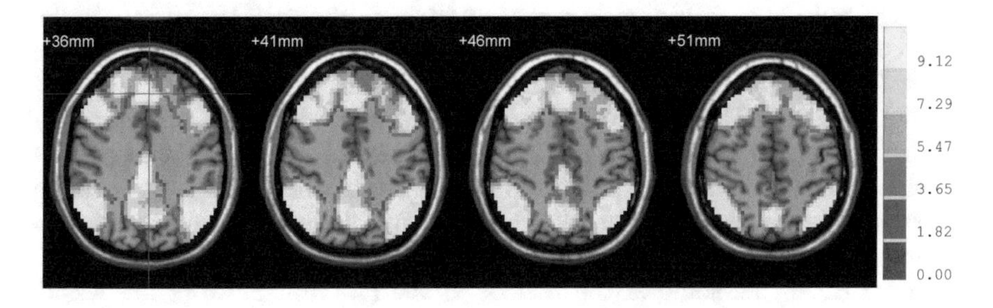

Fig. 2. Functional connectivity of the Control group that represent the activation of the brain regions during resting state i.e. DMN of the Control group

Interestingly the activation of the brain regions that are part of the DMN of BD group (Fig. 2) is lower than the activation in the Control group (Fig. 3).

Using Louvian's clustering algorithm, we clustered the nodes of the network or in our case we cluster the cross-correlation matrices. In the algorithm first we

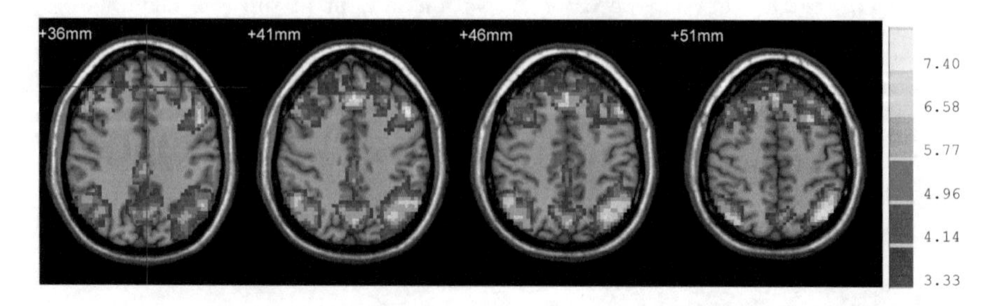

Fig. 3. Functional connectivity of the BD group that represent the activation of the brain regions during resting state i.e. DMN of the BD group

calculate the optimal community structure and modularity using Louvain algorithm [25,26]. The optimal community structure is a subdivision of the network into non-overlapping groups of nodes in a way that maximizes the number of within-group edges, and minimizes the number of between-group edges. In the study of networks the modularity is a quality measure that quantifies the degree to which the network may be subdivided into such clearly delineated groups. The clusters obtained by optimizing the modularity are visualized by first reordering the vertices so that same-cluster nodes are next to each other and then clusters along the diagonal are outlined [27].

On Fig 4 we can see the visualized clustering results of the cross-correlation matrices of the control subjects (Fig. 4a) and the BD group (Fig. 4b). The control group has five resulting clusters (Fig.4a), while the BD group has four clusters (Fig.4b). The differences between the two groups can be seen even at this level of details, and are even more obvious after performing some additional visualizations. We reconstruct the networks for both groups using spherical ROIs, as shown in Table 1, with radius = 5 mm and edges with weights corresponding to the correlation between the ROIs. Next, we threshold the edges' weight using t = 0.2 and obtain the clustered ROI correlation networks for the control (Fig. 4c) and the BD (Fig. 4d) group. As we can see from the new clustered ROI network visualizations aPFC_L,aPFC_R, frontal_R,pCC, mCC, antthal_R, precun_L, precun_R and IPS belong to different clusters which is an essential difference between the control and BD group. Additionally there are less connections in the BD ROI network (Fig. 4d) as compared to the control ROI network (Fig. 4c) i.e some of the regions in the BD ROI network are weakly correlated or uncorrelated in contrast to the same regions in the control ROI network which can be formally seen by comparing averaged node degree which in the control ROI network is 7.1176 ± 4.0733 while in the BD ROI network is 5.7059 ± 3.5379.

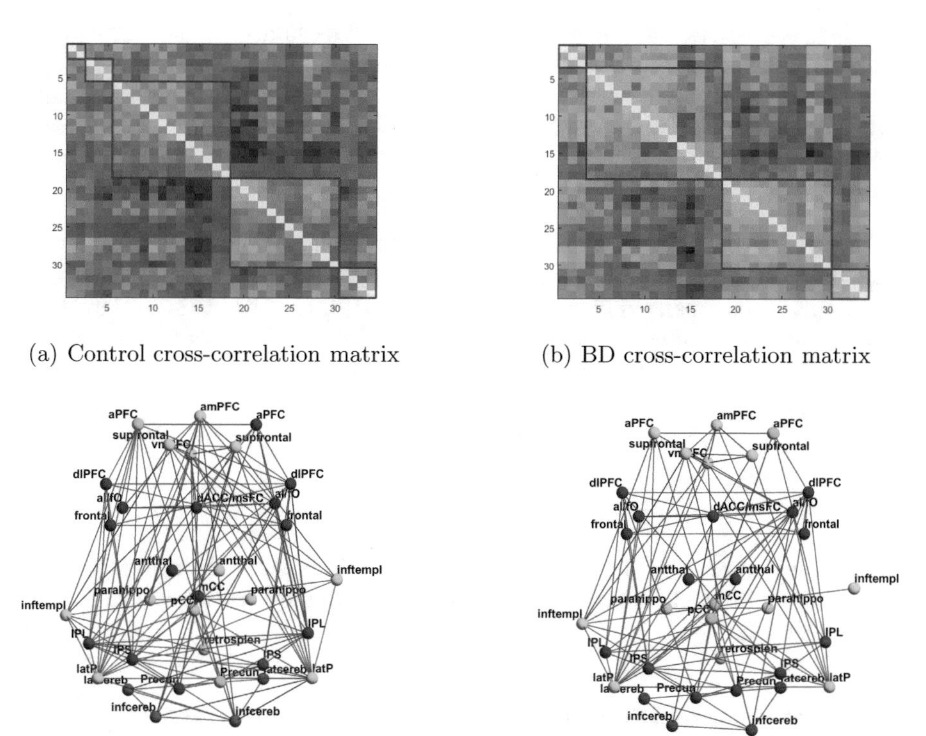

(a) Control cross-correlation matrix

(b) BD cross-correlation matrix

(c) Network for control subjects

(d) Network for BD subjects

Fig. 4. Cross-correlation matrices (a and b) and corresponding networks (c and d) of the control and BD group. The edges of both networks (c and d) were thresholded with threshold=0.2 to discard week connections

4 Conclusion

In this paper we investigate the concepts of functional connectivity in the brain and default mode network. We analyze a set of 28 subjects resting state data divided in a control group (21 subjects) and a bipolar disorder group (7 subjects). After some initial preprocessing we used a ROI based approach to determine the functional connectivity. By running two sample t test we discovered several differences between the control and BD group when the patients are experiencing a depressive episode. Namely, the activation in DMN in subjects with bipolar disorder is lower than the activation in healthy subjects (without history of any neurological disorder). We peformed clustering on the data using a modularity optimization algorithm which produced different results for the control group (4 clusters) and the BD group (5 clusters), having same nodes (ROIs) belonging to different clusters in the two groups. These differences are highlighted when edges' weights are thresholded and connections existing in the control group are lost in the BD group. This is a pilot study with small number of subjects and therefore the results should be taken with care. The limitation is that controls and patients are not matched by age, gender, and number. However, we believe that our findings could serve in the creation of potential markers for bipolar disorder. Future work involves replicating these results on a larger population of patients with BD.

Acknowledgment. This work was supported by the Computational Analysis of Human Brain Data project, FCSE, UKIM, Macedonia, the Center for Neuroimmunology, Department of Neurosciences, IDIBAPS, Hospital Clinic, Barcelona, Spain, and the Fundación La Marató TV3. Vesna Prchkovska's fellowship during this work was supported by the FP7 Marie Curie Intra-European Fellowship, project acronym: ConnectMS, project number: 328060

Appendix

In this section we show the coordinates of the ROIs used for the computation of functional connectivity (Table 3). Besides the coordinates the labels of the ROIs are also shown.

Table 3. Coordinates of the ROIs in MNI space (a 3-dimensional coordinate system of the human brain suggested by Montreal Neurological Institute), between is computed functional connectivity

id	X	Y	Z	Label
1	-45.04	28.22	31.49	dlPFC
2	47.88	28.55	29.87	dlPFC
3	-42.77	8.23	35.67	frontal
4	45.82	8.54	34.13	frontal
5	1.57	-26.27	31.6	mCC
6	-53.35	-49.24	41.54	IPL
7	56.91	-43.97	45.86	IPL
8	-31.63	-57.05	48.66	IPS
9	34.24	-59.26	44.4	IPS
10	-7.88	-71.31	44.07	Precun
11	12.67	-67.84	45.61	Precun
12	-29.21	57.15	7.07	aPFC
13	30.32	57.13	15.02	aPFC
14	-36.76	16.73	-0.01	aI/fO
15	39.92	19.03	-2.67	aI/fO
16	0.56	16.87	45.28	dACC/msFC
17	-11.77	-13.81	4.83	antthal
18	12.01	-13.63	5.52	antthal
19	2.19	61.08	12.87	amPFC
20	-2.39	42.72	-11.01	vmPFC
21	-13.52	47.22	49.24	supfrontal
22	19.98	46.28	48.76	Supfrontal
23	-64.94	-35.43	-16.78	Inftempl
24	71.12	-17.93	-20.83	Inftempl
25	-22.84	-27.94	-19.36	Parahippo
26	27.96	-27.55	-18.02	Parahippo
27	-0.46	-33	40.15	pCC
28	-48.96	-66.24	43.14	latP
29	59.07	-65.86	41.27	latP
30	4.6	-51.94	9.44	retrosplen
31	-33.67	-71.87	-29.4	latcereb
32	34.38	-66.31	-31.11	latcereb
33	-19.63	-85	-32.82	infcereb
34	20.35	-86.99	-33.3	infcereb

References

1. O. Sporns, Networks of the brain, London, England: The MIT Press, Cambridge, 2010.
2. O. Sporns, Discovering the human connectome, London, England: The MIT Press Cambridge, 2012.
3. R. C. C. e. al, "Imaging human connectomes at the macroscale," Nature, vol. 10, no. 6, pp. 524-539, 2013.
4. Greicius, Michael D., et al., "Functional connectivity in the resting brain: a network analysis of the default mode hypothesis," Proceedings of the National Academy of Sciences, vol. 100, no. 1, pp. 253-258, 2003.
5. Honey, C. J., et al., "Predicting human resting-state functional connectivity from structural connectivity," Proceedings of the National Academy of Sciences, vol. 106, no. 6, pp. 2035-2040, 2009.
6. R. Kaplan, "Endogenous fMRI default mode network fluctuations both positively and negatively correlate with individual transfer of learning," Frontiers in systems neuroscience, vol. 8, 2014.
7. Sestieri, Carlo et al. "Episodic Memory Retrieval, Parietal Cortex, and the Default Mode Network: Functional and Topographic Analyses," The Journal of neuroscience: the official journal of the Society for Neuroscience 31.12 (2011): 44074420.
8. Jones, D. T., et al. "Age-related changes in the default mode network are more advanced in Alzheimer disease." Neurology 77.16 (2011): 1524-1531.
9. Sheline, Yvette I., et al., "The default mode network and self-referential processes in depression," Proceedings of the National Academy of Sciences, vol. 106, no. 6, pp. 1942-1947, 2009.
10. Liu CH, Ma X, Wu X, Fan TT, Zhang Y, et al. "'Resting-state brain activity in major depressive disorder patients and their siblings". J Affect Disord 149 (2013): 299306.
11. Megan H. Lee, et al. "Clustering of Resting State Networks." PLoS ONE 7(7) (2012).
12. Wang Y, Li T-Q. "Analysis of Whole-Brain Resting-State fMRI Data Using Hierarchical Clustering Approach." PLoS ONE 8(10) (2013).
13. S. Teng, et al. "Small-world Network for Investigating Functional Connectivity in Bipolar Disorder: A Functional Magnetic Images (fMRI) Study" 13th International Conference on Biomedical Engineering, IFMBE Proceedings Volume 23 (2009): 726-729.
14. Zuo, X.-N., Di Martino, A., Kelly, C., Shehzad, Z. E., Gee, D. G., Klein, D. F.,, "The oscillating brain: complex and reliable.," Neuroimage, vol. 49, no. 2, pp. 1432-1445, 2010.
15. Nikos K. Logothetis, Brian A. Wandell, "Interpreting the BOLD Signal," Annual Review in Phisiology, vol. 66, pp. 735-769, 2003.
16. ML Schlvinck et al., "Neural basis of global resting-state fMRI activity," Pnas, vol. 107, no. 22, pp. 10238-10243, 2010.
17. Öngür, Dost, et al. "Default mode network abnormalities in bipolar disorder and schizophrenia." Psychiatry Research: Neuroimaging 183.1 (2010): 59-68.
18. Tenev A. et al. "Connectomics based multi-modal graph measures in bipolar disorder", Poster at 20th Annual Meeting of the Organization for Human Brain Mapping (OHBM) (2014): 2163.
19. C. Z. Y. Yan, "DPARSF: A MATLAB Toolbox for Pipeline Data," Frontiers in systems neuroscience, vol. 4, no. 13, 2010.

20. Power, J. D., Barnes, K. A., Snyder, A. Z., Schlaggar, B. L., Petersen, S. E, "Spurious but systematic correlations in functional connectivity MRI networks arise from subject motion," Neuroimage, vol. 59, no. 3, pp. 2142-2154, 2012.
21. J. Ashburner, K.J. Friston, "Nonlinear spatial normalization using basis functions", Humman Brain Mapping, vol. 7, pp. 254-266, 1999.
22. A Di Martino, A Scheres, DS Margulies et al., "Functional connectivity of human striatum: a resting state FMRI study," Schizophrenia Bull, vol. 40, no. 3, pp. 653-664, 2014.
23. Song X-W, Dong Z-Y, Long X-Y, Li S-F, Zuo X-N, et al., "REST: A Toolkit for Resting-State Functional Magnetic Resonance Imaging Data Processing," PLoS ONE, vol. 6, no. 9, 2011.
24. Damoiseaux, J. S., et al., "Consistent resting-state networks across healthy subjects," Proceedings of the national academy of sciences, vol. 103, no. 37, pp. 13848-13853, 2006.
25. Blondel, Vincent D., et al. "Fast unfolding of communities in large networks." Journal of Statistical Mechanics: Theory and Experiment 2008.10 (2008).
26. Reichardt, Jorg, and Stefan Bornholdt. "Statistical mechanics of community detection." Physical Review E 74.1 (2006): 016110.
27. Rubinov, Mikail, and Olaf Sporns. "Complex network measures of brain connectivity: uses and interpretations." Neuroimage 52.3 (2010): 1059-1069.

ECG Signal Compression Using Adaptive Hermite Functions

Tamás Dózsa[1] and Péter Kovács[2]

[1] Eötvös L. University, Pázmány Péter stny. 1/C, 1117 Budapest, Hungary
tamasdzs@gmail.com
[2] Department of Numerical Analysis, Eötvös L. University, Pázmány Péter stny. 1/C, 1117 Budapest, Hungary
kovika@inf.elte.hu

Abstract. In modern medical science evaluation of electrocardiogram (ECG) has proven to be an important task for doctors. These signals contain valuable information on the patients' condition; however analysis of them has encountered numerous challenges, such as storage of long-term recordings, filtering, and segmentation of signals. Resolving these problems is important to ensure a high quality diagnosis. In this paper we propose an ECG analysis method which provides adequate solutions to all of these challenges. The proposed method is based upon the approximation theory in Hilbert spaces. Namely, using the affine transforms of orthonormal Hermite systems, the approach optimizes two free parameters. This is done in order to achieve the best approximation of the ECG signal using a fixed number of Fourier coefficients. The process of optimization is done using Particle Swarm Optimization (PSO), Nelder–Mead (NM) simplex method, and Monte Carlo (MC) algorithm which are embedded into a matching pursuit framework. The former procedure guarantees both good compression ratio and high accuracy, while the latter segments the heartbeats. As it is shown by experiments, the proposed method achieves better results than previously known approaches.

Keywords: e-Health, Biomedical signal processing, ECG data compression, filtering, segmentation.

1 Introduction

Electrocardiogram is a physiological signal which represents the electrical activity of the heart. There are different types of signals with respect to the electrodes which are used in the measurements. For instance, one of the most common configurations are the so-called limb leads where the electrodes are located on each arm and the left leg. Each lead provides a signal which examined by medical experts. Sometimes the diagnosis requires detailed multichannel measurements with long-term recordings (i.e., Holter monitoring). In this case, compression of the ECG data is inevitable, especially for real-time transmission such as telemedicine. Removing data redundancy can also be useful in modern clinical applications as well as automatic detection of abnormalities. The first step of

S. Loshkovska and S. Koceski (eds.), *ICT Innovations 2015*,
Advances in Intelligent Systems and Computing 399,
DOI: 10.1007/978-3-319-25733-4_25

these classification tasks is to extract relevant features of the heartbeats while preprocessing of the signal also includes filtering and segmenting procedures.

Techniques that solve these problems are divided into three major categories: parameter extraction methods, direct time-domain algorithms and transform-domain procedures (see e.g., [1]). The last class is based upon the approximation theory in Hilbert spaces. Namely, the original data is considered as a function which is transformed to certain subspaces of the Hilbert space. Then, the coefficients of the transformed signal are stored instead of the samples of the original ECG data. There is a wide range of algorithms which use the same idea including Fourier, Walsh, discrete cosine, Karhunen-Loeve, wavelet transform, etc. For detailed overview of these methods we are referring to the surveys [2] [3].

The paper is organized as follows. First we recite the achievements of previous research in this area, and clarify the advancements described in this paper in Section 2. In Section 3-4 we are recalling the theory of approximation in Hilbert spaces along with Hermite functions. Then, we are extending the original algorithm by using affine transforms in Section 5. We also introduce a matching pursuit algorithm for the purpose of segmentation in Section 6. and compare the proposed method via experiments in Section 8. Finally, we provide a summary of conclusions and future plans.

2 Related Works

The proposed method is based on formerly known ECG compression techniques utilizing the so-called Hermite functions. Due to the shape similarity between these basis functions and the ECG signal one can expect a compact and accurate representation of the electrocardiogram. Other classical orthogonal polynomials such as Chebyshev and Legendre polynomials were also used in modeling the QRS complex [4]. This concept was extended by L. Sörnmo et al. [5] which was used later by R. Jané et al. [6] to construct an adaptive ECG compression method. Then, a recent study improved this model by using discrete Hermite systems [7]. We note that, all of these methods are based on the observation that an orthogonal system is invariant under affine transforms, i.e. translation and dilation. However, the previously mentioned algorithms use only dilations of these systems and utilize segmentations of the heartbeats as a preprocessing step. For this reason, the results are highly dependent on the efficiency of the segmentation. In contrast, the proposed method uses both translation and dilation parameters. Additionally, we apply Monte Carlo, Nelder–Mead and PSO [8] simulations in order to find the best parameters of the system, therefore minimizing the error of the representation. It is worth mentioning that the original method [7] did not apply such optimization, i.e. only one dilation parameter was used for the whole lead which was determined experimentally. In order to compare the proposed method with the original one we performed tests on real data provided by the PhysioNet ECG database [9].

3 Approximation in Hilbert Spaces

Processing ECG signals proposes a number of problems. These include the storage of data in case of long-term recordings, and filtering of noisy signals. Both of these challenges can be resolved sufficiently if we present the approximation as the linear combination of a finite number of smooth functions $\Phi_0, \ldots, \Phi_n$ where the system $(\Phi_k \in \mathcal{H}, n \in \mathbb{N})$ forms an orthogonal basis in some Hilbert-space $\mathcal{H}$. The best approximation of the signal $f \in \mathcal{H}$ is expressed as a projection

$$S_n f = \sum_{k=0}^{n} \langle f, \Phi_k \rangle \Phi_k \,, \tag{1}$$

where $\langle .,. \rangle$ denotes the inner product of $\mathcal{H}$. Furthermore, the square of the difference between the signal and the best approximation can be written as

$$\|f - S_n f\|^2 = \|f\|^2 - \sum_{k=0}^{n} |\langle f, \Phi_k \rangle|^2 \,. \tag{2}$$

On one hand, an ECG can be represented within a given tolerance (error) by storing only a certain number of Fourier coefficients $\langle f, \Phi_k \rangle$ $(k = 0, \ldots, n)$. On the other hand, this approximation can also be used for noise-filtering of the signal. In order to implement the approximation method for ECG signals, classical orthogonal Hermite systems proved to be a good choice as was illustrated in [5]. This is due to a number of useful properties discussed in the next section.

In this paper, as it was sufficient, the Euclidean space $\mathcal{F}$ of piecewise continuous square integrable functions was used instead of the Hilbert space $\mathcal{H}$ of square integrable functions (with respect to the Lebesgue measure). In $\mathcal{F}$ the inner product and the norm are expressed as

$$\langle f, g \rangle = \int_{-\infty}^{\infty} f(t)g(t)\mathrm{d}x \,, \quad \|f\| := \sqrt{\langle f, f \rangle} \quad (f, g \in \mathcal{F}) \,. \tag{3}$$

Furthermore, the orthonormal Hermite functions

$$\Phi_n(x) = H_n(x) e^{-x^2/2} / \sqrt{\pi^{1/2} 2^n n!} \tag{4}$$

constitute a (complete) orthonormal system in $\mathcal{F}$:

$$\langle \Phi_n, \Phi_m \rangle = \delta_{nm} \quad (n, m \in \mathbb{N}) \,, \quad \lim_{n \to \infty} \|f - S_n f\| = 0 \,, \tag{5}$$

where H_n denotes the classical Hermite polynomials [10].

4 Hermite Functions

The choice of Hermite functions is supported by a number of good properties. It is worth mentioning that the shapes of the basis functions are highly correlated with the original ECG data. For this reason, they were successfully applied in biomedical signal processing, including compressing, filtering [6] of ECGs and classification [11,12] of heartbeats. Furthermore, these functions possess the following useful qualities:

- The Φ_n system is closed (complete) in the space $\mathcal{F}$.
- The system can be used to detect the main lobes of the ECGs. This is due to the fact that the basis functions are well localised in time. More precisely, the functions $\Phi_n(x)$ quickly converge to 0 as $|x| \to \infty$:

$$|\Phi_n| \leq M_n e^{-x^2/4} \leq M_n \quad (x \in \mathbb{R}, n \in \mathbb{N}) . \tag{6}$$

- The functions $\Phi_n(x)$ can be calculated with stable second degree recursion:

$$\Phi_0(x) := \frac{e^{-x^2/2}}{\pi^{1/4}} , \quad \Phi_1(x) := \frac{\sqrt{2}xe^{-x^2/2}}{\pi^{1/4}} ,$$
$$\Phi_n(x) := \sqrt{\frac{2}{n}} x\Phi_{n-1}(x) - \sqrt{\frac{n-1}{n}} \Phi_{n-2}(x) .$$

5 Affine Transforms of Orthonormal Systems

The representation of signals depends on the position of 0 on the time scale, and the length of the unit. In practice these parameters are often chosen in an arbitrary fashion. For example in paper [7] the authors chose the same parameters for every heartbeat of a specific lead. As a result the error of the approximation is usually not optimal, and in some cases this yields completely wrong results (for example in the cases of abnormal, sick heartbeats). Thus the problem of how to choose these parameters uniquely for each approximation to minimize the error arises. The precision of the approximation can be improved, if instead of the classical orthonormal Hermite systems we use their affine transforms:

$$\Phi_n^{a,\lambda} := \Phi_n(\lambda x + a) \quad (x, a \in \mathbb{R}, \lambda > 0) . \tag{7}$$

Note that the new system $\sqrt{\lambda}\Phi_n^{a,\lambda}$ is also orthonormal and complete in $\mathcal{F}$. In this case, the best approximation of a signal f can be rewritten in the form

$$S_n^{a,\lambda} f = \sum_{k=0}^{n} \langle f, \Phi_k^{a,\lambda} \rangle \Phi_k^{a,\lambda} \quad (x, a \in \mathbb{R}, \lambda > 0) . \tag{8}$$

Additionally, the error of the approximation is given as a function of the translation a and the dilatation λ :

$$D_n^2(a, \lambda) := \|f\|^2 - \sum_{k=0}^{n} \langle f, \Phi_k^{a,\lambda} \rangle \Phi_k^{a,\lambda} . \tag{9}$$

By optimizing these two free parameters one can get a better approximation of the signal without having to increase the number of Fourier coefficients. Finding the minimum of function D_n is equivalent to find the maximum of

$$F_n(a, \lambda) := \sum_{k=0}^{n} |\langle f, \Phi_k^{a,\lambda} \rangle|^2 . \tag{10}$$

We note that numerical experiments have shown that the optimal parameters probably exist. Providing formal proof of the existence of these extreme values however remains a challenge of the future (as the domain of each function is not a compact set).

6 Matching Pursuit Algorithm

In order to achieve a more adaptive representation we identify several dilatation and translation parameters for each heartbeat. The original signal $f \in \mathcal{F}$ is therefore approximated as:

$$S_{\mathbf{n}}^{\mathbf{a},\boldsymbol{\lambda}} := \sum_{i=1}^{N} \sum_{k=0}^{n_i} \langle f^{a_i,\lambda_i}, \Phi_k \rangle \Phi_k \quad (a_k \in \mathbb{R}, \lambda_k > 0)\,. \tag{11}$$

where $\mathbf{a} = (a_1, \ldots, a_N)$ and $\boldsymbol{\lambda} = (\lambda_1, \ldots, \lambda_N)$ are vectors of the applied translation and dilatation parameters. The number of coefficients used in each approximation is represented by $\mathbf{n} = (n_1, \ldots, n_N)$. ECG signals consist of three main segments (P, QRS, T), therefore in this case $N = 3$. Based on the results in [5], we have applied $\mathbf{n} = (7, 6, 2)$. It is worth mentioning that in practice instead of using the original equation in Eq. (8) we have used the transforms f^{a_i,λ_i} of the signal. Note that, this step does not restrict the original problem; however affine transforms can be applied much faster on f rather than every member of the systems $\Phi_{n_i}^{a_i,\lambda_i}$.

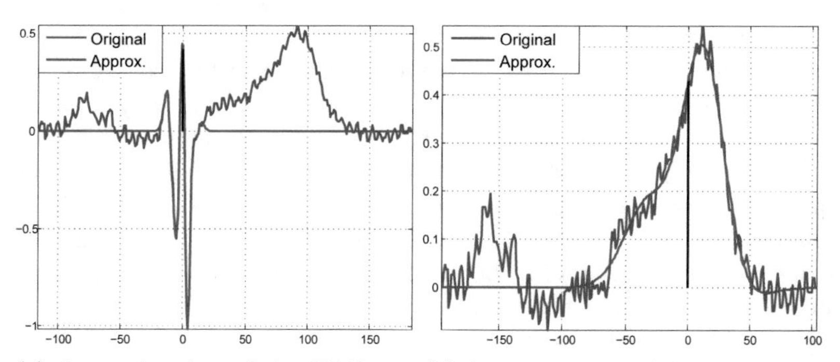

(a) Approximation of the QRS segment. (b) Approximation of the T segment.

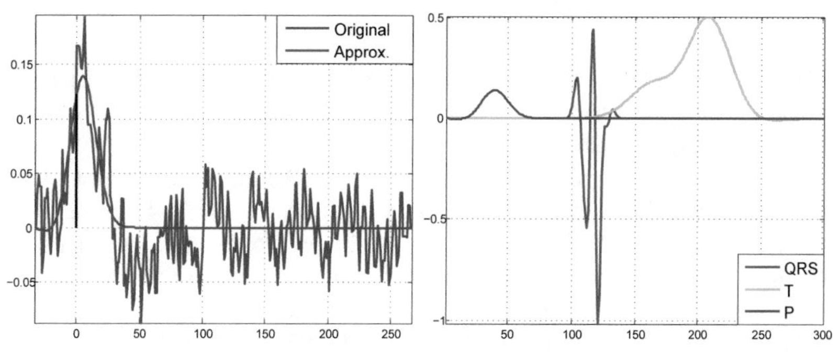

(c) Approximation of the P segment. (d) Segmentation of the heartbeat.

Fig. 1. Phases of the MP algorithm.

We independently optimize each pair of the parameters (a_i, λ_i). This however does not guarantee that the algorithm will generate separate approximations for the P, QRS, and T segments. In order to resolve this problem, we utilized the well known Matching Pursuit (MP) approach [13]. We use this algorithm to maximize the function $F_n(a_i, \lambda_i)$ defined in Eq. (10). The ith step of the algorithm is written as:

$$s^{(i)} = s^{(i-1)} + S_{n_i}^{a_i,\lambda_i} R^{(i-1)} \quad (1 \leq i \leq N), \tag{12}$$

where $R^{(i)} = f - s^{(i)}$ is the residuum signal. This method is able to segment the ECG signals, since in every iteration, a different orthonormal system is used. In the original method [6], the segmentation was done by a separate algorithm. As a consequence, the error of the approximation was highly dependent on the precision of the segmentation algorithm. The proposed method resolves this problem, even in the case of high signal-to-noise ratio. A visualization for each phase of the approximation is provided in Fig. 1.

7 Quantization

Along with the Fourier coefficients, the optimized parameters found by the MP algorithm also require storage. The representation of the signal is highly dependent on the translation parameters $\mathbf{a}$. This means that storage of $\mathbf{a}$, must be done on as many bits as is necessary for the accurate reconstruction of the signal. To achieve this $b = \log_2(\max_i |a_i|) + 1$ bits were used to store the vector of translation parameters. In contrast, the vector of dilatation parameters $\boldsymbol{\lambda}$ and Fourier coefficients $\mathbf{c} = (c_n^{(i)})\,(0 \leq n \leq n_i, 1 \leq i \leq N)$ can be stored with rounded values. To determine the optimal number of bits needed to store these, we used the method of linear quantization. Let $c_{\max}$ and $c_{\min}$ denote the maximal and the minimal element in the vector $\mathbf{c}$ respectively. Now it is possible to define linear quantization as:

$$Q(\mathbf{c}) := c_{\min} + \operatorname{sign}(\mathbf{c}) \cdot \Delta \cdot \left(\frac{|\mathbf{c} - c_{\min}|}{\Delta} + \frac{1}{2} \right), \tag{13}$$

where $\Delta = \frac{|c_{\max} - c_{\min}|}{2^b - 1}$. We store the $Q(\mathbf{c})$ rounded values on b bits, taking into consideration the position of $\mathbf{c}$ with respect to 0. The optimal value of b is determined via experiments.

In order to set the optimal value of b we have tested our method on the records of the PhysioNet MIT-BIH ECG database [9]. This contains 48 records, which are 30 minutes long each. The records contain two leads each resulting in 360 samples per second, which are stored on 11 bits. In our experiments, we used the records 118, 119, 201 and 213. The reason behind this is that records 118 and 119 contain more abnormal heartbeats, while 201 and 213 consist of mostly regular signals. This provides an opportunity to test the method under a wide range of circumstances. During the experiment we first determined the optimal dilatation and translation parameters for each heartbeat of these records. Then, we proceed with quantization of $\mathbf{c}$ and $\boldsymbol{\lambda}$ using a decreasing number of b bits, while taking into consideration the precision of the approximation (PRD) and the compression ratio (CR). We provide precise definition for these quantities in Section 8. In our first experiment, we have tested the method on the abnormal records 118, 119 only, which contain $1528 + 1979$ heartbeats. The results show that increasing b above 7 bits yields no significant increase in the precision of the approximation, therefore we have tested the method using $b = 7$. The results of these tests are shown in Fig. 2.

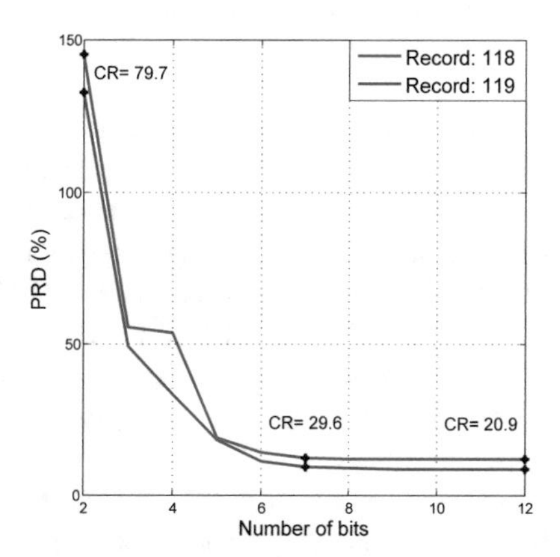

Fig. 2. Accuracy of the approximation using quantized parameters of the records.

8 Tests and Results

We tested our algorithm on the records $118, 119, 201$ and 213 of the MIT-BIH ECG database of PhysioNet [9]. Namely, we measured the performance of the proposed method on 9444 heartbeats which means a 120 minutes long ECG recording. In order to compare the method with previous approaches, we have repeated the tests on the original algorithm [6] as well. In both cases, we used the same number of coefficients, specifically we set $\mathbf{n} = (7, 6, 2)$. We evaluated the results by looking at the percentage root mean square difference (PRD) and the compression ratio (CR) of each method:

$$\text{PRD} := \frac{\left\|S_{\mathbf{n}}^{\mathbf{a},\boldsymbol{\lambda}} - f\right\|_2}{\left\|f - \overline{f}\right\|_2} \times 100\,, \quad \text{CR} := \frac{\text{size of original ECG}}{\text{size of compressed ECG}} \times 100\,. \tag{14}$$

Here $\overline{f}$ is the mean value of the original signal. Note that the PRD is the equivalent of the relative error of the approximation in the ℓ^2 norm. The comparison of different methods is difficult, because one has to take into account both the PRD and CR. In order to cure this problem a new unit of measure, the Quality Score (QS) was introduced in [14]. QS is defined as the quotient of CR and PRD, so the approach with the higher QS is considered better. We arranged our results in Tab. 1.

Table 1. Experimental results of 120 minutes long real ECG data (9444 heartbeats).

	PRD (%)				**CR** $(1:X)$				**QS** (CR : PRD)			
Rec.	**Orig.**	**NM**	**PSO**	**MC**	**Orig.**	**NM**	**PSO**	**MC**	**Orig.**	**NM**	**PSO**	**MC**
118	19.83	17.79	16.65	18.88	24.34	22.30	22.30	22.30	1.22	1.25	**1.33**	1.18
119	14.27	8.76	10.20	12.92	27.89	25.55	25.55	25.55	1.95	**2.91**	2.51	1.97
201	13.51	12.17	12.17	13.32	28.21	25.35	25.35	25.35	**2.08**	**2.08**	**2.08**	1.90
213	19.92	18.28	17.60	19.46	17.08	15.64	15.64	15.64	0.85	0.85	**0.88**	0.80
Avg.	16.85	14.25	14.16	16.15	24.38	22.21	22.21	22.21	1.53	**1.77**	1.70	1.46

One can see that, although the Monte Carlo method had slightly better PRD results, our approach achieved worse quality scores than the original algorithm. Despite this, it still can be used well for segmentation, even in the case of irregular heartbeats. As it was shown, the proposed method yields a considerably better quality score with the Nelder–Mead and PSO optimizations. With either optimization our algorithm has achieved better PRD scores, although the CR was higher with the original method. This however could be expected as our algorithm also has to store three additional translation and dilation parameters for each heartbeat.

Note that in case of record 119, we have found that the error of the approximation was remarkably lower than the error of the original method. In order to understand this phenomenon we have to look at Fig. 3, which shows the first heartbeat of the record in question. It is clear, that the former method [6] has an unacceptable large error near the T segment. This is, in part, due to the error of the segmentation algorithm, and also to the lack of the translation parameter. The original (Orig.) method approximates the signal in the intervals between the green dots in the figure, and the Hermite functions are positioned in the middle of these (black lines) segments. Since this particular heartbeat has an asymmetric T wave, the error of the approximation cannot be lowered solely by dilatation. Due to the free parameters (a_i, λ_i) and the optimization, our method has been able to tackle this problem and achieve about two times better PRD score. Note that the number of Fourier coefficients was the same in both cases, as our approach only adds a single translation parameter for each segment. This example also helps to understand the reason behind the fact that on the rest of the records, the original method achieved almost the same performance (QS) as the proposed algorithm. More precisely, if the signal consists of normal heartbeats, and therefore the segmentation algorithm operates with a small error, then the optimal translation is close to the middle of the intervals. This however cannot be guaranteed in reality, therefore the creation of our method is justified, and is more applicable in practice.

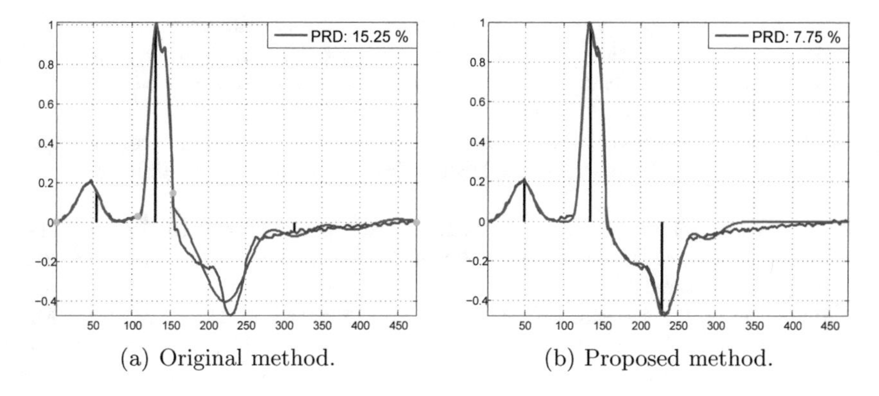

(a) Original method.

(b) Proposed method.

Fig. 3. Approximation of an asymmetric ECG signal.

9 Conclusion

We have shown that affine transforms of Hermite functions can be successfully used to compress, filter, and segment ECG signals. In addition our method is independent from the quality of the ECG, as the segmentation and compression yield good results even in the case of irregular heartbeats. The application of the method for regular heartbeats achieves similar quality scores to the original approach. This is due to the fact, that the optimal value of the translation parameter for regular heartbeats is close to 0. This means that in the case of healthy signals, the positive properties of Hermite functions are exploited. The difference in the performance of our method depending on whether it was applied with the Monte Carlo, PSO, or Nelder–Mead algorithms which justifies future trials of different optimizations. Providing formal proof of the existence of extreme values of F_n in Eq. (10) also remains an important challenge.

References

1. Jalaleddine, S.M.S., Hutchens, C.G., Strattan, R.D., Coberly, W.A.: ECG data compression techniques – a unified approach. IEEE Transactions on Biomedical Engineering **37**(4) (1990) 329–343
2. Addison, P.S.: Wavelet transforms and the ECG: a review. Physiological Measurement **26**(5) (2005) 155–199
3. Castells, F., Laguna, P., Sörnmo, L., Bollmann, A., Roig, J.M.: Principal component analysis in ECG signal processing. EURASIP Journal on Advances in Signal Processing **2007** (2007) 1–21
4. Georgiev, G., Valova, I., Gueorguieva, N., Lei, L.: QRS complex detector implementing orthonormal functions. Procedia Computer Science **12** (2012) 426–431
5. Sörnmo, L., Börjesson, P.L., Nygårds, M.E., Pahlm, O.: A method for evaluation of QRS shape features using a mathematical model for the ECG. IEEE Transactions on Biomedical Engineering **28** (1981) 713–717

6. Jané, R., Olmos, S., Laguna, P., Caminal, P.: Adaptive Hermite models for ECG data compression: Performance and evaluation with automatic wave detection. In: Proceedings of the International Conference on Computers in Cardiology, New York, IEEE press (1993) 389–392
7. Sandryhaila, A., Saba, S., Püschel, M., Kovacevic, J.: Efficient compression of QRS complexes using Hermite expansion. IEEE Transactions on Signal Processing **60**(2) (2012) 947–955
8. Kennedy, J., Eberhart, R.C.: Particle swarm optimization. In: Proceedings of IEEE International Conference on Neural Networks. Volume 4., New York, IEEE press (1995) 1942–1948
9. Goldberger, A.L., Amaral, L.A.N., Glass, L., Hausdorff, J.M., Ivanov, P.C., Mark, R.G., Mietus, J.E., Moody, G.B., Peng, C.K., Stanley, H.E.: PhysioBank, PhysioToolkit, and PhysioNet: Components of a new research resource for complex physiologic signals. Circulation **101**(23) (2000) 215–220
10. Szegő, G.: Orthogonal polynomials. 3rd edn. AMS Colloquium Publications, New York (1967)
11. Lagerholm, M., Peterson, C., Braccini, G., Edenbrandth, L., Sörnmo, L.: Clustering ECG complexes using Hermite functions and self-organizing maps. IEEE Transactions on Biomedical Engineering **47**(7) (2000) 838–717
12. Haraldsson, H., Edenbrandt, L., Ohlsson, M.: Detecting acute myocardial infarction in the 12-lead ECG using Hermite expansions and neural networks. Artificial Intelligence in Medicine **32** (2004) 127–136
13. Mallat, S.G., Zhang, Z.: Matching pursuit in time-frequency dictionary. IEEE Transactions on Signal Processing **41**(12) (1993) 3397–3415
14. Fira, C.M., Goras, L.: An ECG signals compression method and its validation using NNs. IEEE Transactions on Biomedical Engineering **55**(4) (2008) 1319–1326

Automated Ambient Open Platform for Enhanced Living Environment

Rossitza Goleva[1], Rumen Stainov[2], Alexander Savov[3], Plamen Draganov[4], Nikolay Nikolov[5], Desislava Dimitrova[6], Ivan Chorbev[7]

[1] Department of Communication Networks, Technical University of Sofia
Kl. Ohridski blvd. 8, 1756, Sofia, Bulgaria
rig@tu-sofia.bg

[2] Applied Computer Science Department, University of Applied Sciences, Leipziger Strasse 123, 36039 Fulda, Germany
rumen.stainov@informatik.hs-fulda.de

[3,4] Comicon Ltd., Sofia
Mladost 4, Roman Avramov blvd., Bitov kombinat, et.2, 1715, Sofia, Bulgaria
{comicon,plamen}@comicon.bg

[5] Security Solutions Institute Ltd., 8, Munich str., Hi Tech Bussness Center, Sofia, Bulgaria
nikolay.nikolov@pimprima.eu

[6] ETH Zurich, Switzerland
desislava.dimitrova@inf.ethz.ch

[7]Faculty of Computer Science and Engineering, University of Ss Cyril and Methodius, "Rugjer Boshkovikj" 16, P.O. Box 393, 1000 Skopje, Former Yugoslav Republic of Macedonia
ivan.chorbev@finki.ukim.mk

Abstract. The aim of this paper is to present an idea of open platform for Enhanced Living Environment that will allow flexible and reliable use of the cloud-computing, sensor, mobile, local and body area networks for highly customized services and applications. The platform architecture consists of sensors working on IEEE 802.15.4, ZigBee, EnOcean standards, Ethernet, GSM, 3G, 4G, 5G gateways to the cloud, and peer port application at application layer. The personal enhanced living environment consists of body area and home automation network. They are key factors for patient's comfortable living outside hospitals. Home, car, working place, park automation support the mobility, social integration, possibility of the patients to manage everyday tasks. Experiments carried out using sensors and holter prove the vitality of the solution partially. The platform is capable of working independently from the network connectivity, uses multi-homing, is open to migration to new technologies.

Keywords: Sensor-to-Cloud, ZigBee, IEEE 802.15.4, Peer-Port, Cloud Computing, 3G, 4G, 5G, Body Area Network, Internet of Things, Opportunistic Environments, Quality of Service

S. Loshkovska and S. Koceski (eds.), *ICT Innovations 2015*,
Advances in Intelligent Systems and Computing 399,
DOI: 10.1007/978-3-319-25733-4_26

1 Introduction

The aim of this work is to demonstrate an idea for Automated Ambient Open Platform for Enhanced Living Environment (AAPELE). The network consists of different domains starting from Body Area Network (BAN), ZigBee home/ business automation part, 3G, WiFi, LAN access, application server, Storage Area Network (SAN) in the cloud combined with peer-port implementation. The platform is presented as a hierarchical layered model and compared with OSI Reference Model. The main end-device scenarios are presented for better understanding and application testing. Performance analyses are based on the life experiment of ZigBee network, holter transmission and simulation model for result verification. The importance of the problem is obvious nowadays, when not only the patients but all citizens try to use information and communication technologies for reading, monitoring, control, storing important health related information. Furthermore, the society prefers to be informed about all processes concerning way of living and health conditions. Preventive behaviour and measures are becoming crucial for better life.

2 State of the Art

The topic of ambient assisted living is explored last 30 years intensively and the result is that many electronic devices are becoming part of the doctors' and patients' equipment [1,2]. The lack of flexible platform that will support devices' interoperability results in the separated data gathering, lack of reliable data collection [3], isolated data analyses, evaluation and interpretation. The patients could not understand results and be consulted easily. They could not collect proper information for their living environment; stay at home isolated; could not perform everyday tasks. Internet, cloud, LAN, mobile and wireless technologies allow interconnection between different types of devices over the single platform in a fully virtualized and reliable way. The idea how to apply ZigBee network for home automation is developed in [4]. Authors aim to show how web based solution could be used for building monitoring. Security problems and multicasting in the Ambient Assisted Living (AAL) platform could be seen in [5]. Distributed nature of the network is analyzed in [6]. Tests for specialized devices like holter could be found in [7,8]. The idea to propose near-real-time ZigBee solution is similar to the proposed solutions in [9,10,11]. The middleware application is shown in [12]. Peer-port application could be found in [13,14,15]. There could be differences in peer port implementation based on the type of the transport and application protocols applied. The application layer protocol could be "go back n" or "selected repeat" whereas the transport could be TCP or UDP [3]. Network design for building automation is also presented in [16]. Interoperability to the Bluetooth technology could be seen in [17]. Legislation issues are summarized in [18,19]. Data collection and appropriate acquisition is seen in [20,21]. Direct mapping to delay tolerant networks is presented in [22]. Details on the content and end-user interests is shown in [23]. Platform requirements could be seen in [24]. Software-defined networks processing times are analyzed in [25].

Open platform design for health services is shown in [26]. Middleware is presented in [27].

This paper is organized by presenting the architecture at the beginning; going next through use-case scenarios applied, test bed setup for ZigBee network, lab experiments, simulations settings, performance analyses and evaluation.

3 Platform Architecture

Open Automated Enhanced Living Environment platform is an end-to-end fully virtualized and reliable solution for personal environment data measurement, collection, analyses, transmission, evaluation and presentation. It is fully integrated with storage technologies. The reliability of the system is supported by different levels of redundancy. Sensor and other end-device data is collected locally and could be transferred through at least two different ports and at least two different network access points. The connectivity to the application server uses peer port that allows data buffering in many network devices simultaneously [3]. Fig. 1 presents the overall idea of the platform. It consists of BANs, sensors for car, home, working place, park, garden, shop, public buildings, and street automation. They form specific and personalized environment for the citizens. The variety of types of sensors, coordinators, gateways, controllers, application servers, applications, are supported through open platform that will be capable to ensure interoperability between fixed and mobile network domains. BAN consists of sensors, specialized devices for measurement and smart phone as a gateway to the cloud and is transient ad hoc network [1].

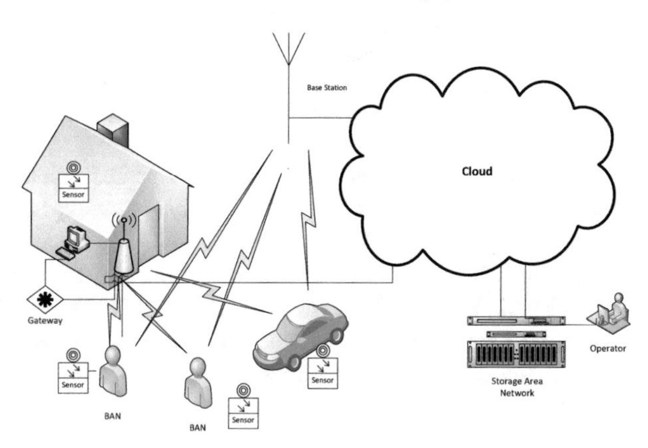

Fig. 1. Overall structure of the access network for automated Enhances Living Environment

People could wear special clothes for body index, temperature, blood sugar, pulse, activity measurement. They also could have more specialized devices as [7, 8] holter that could generate more complicated patterns of data irregularly (Fig. 2). Very important parts of the platform are information services and their support in the cloud. The amount of devices, people, homes, that will use the platform could generate big data and traffic. The lack of single point of failure could lead to the servers and storage overload. The reliability of the data transmitted is gained by the use of peer port technology. Data could be stored locally at home server and then replicated continuously or on demand to the centralized server and storage. Novel compression and information encoding algorithms should allow fast data processing and high-level evaluation and interpretation. In case the application is programmed to raise a flag to the general purpose doctor or cardiologist, or nurse there might be also possibility to avoid false alarms. The application should be capable to adapt to the behavior of the customer after training. Most of the customers do not change the behavior often. From the other side the application should be capable to detect dangerous situation and raise alarms at necessary levels of authority on time.

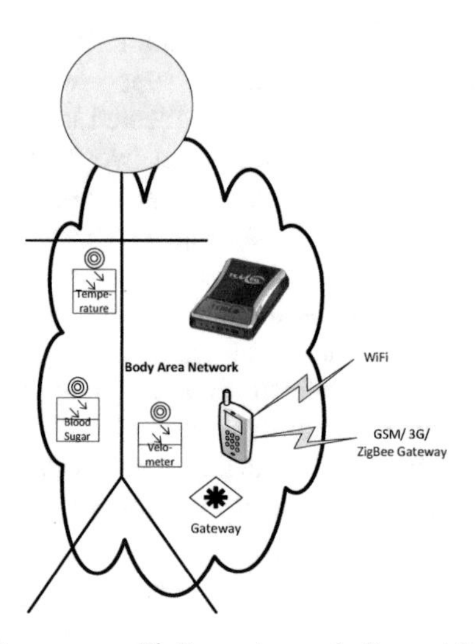

Fig. 2. BAN using ZigBee sensors, ZigBee gateway, holter and 3G access to the cloud

On Fig. 3, we show hierarchical model for the open automated enhanced living environment platform compared to the OSI RM [26]. It is based on ZigBee, 3G and cloud technologies but could be extended with other technologies [7]. Depending on the sensor technology, the sensor behavior will be different. End-users could choose between smart phone with 3G connectivity, GSM, fixed WiFi or LAN. There might be a local home server for storing data. Cloud technologies could be client/ server, peer-to-peer, virtual machine-based, middleware-based or mixed [25], [27].

4 Use-case Scenario

In this section, we present parts of the application use-case scenarios that will allow us to develop the information and communication services. Fig. 4 presents a typical periodical scenario when the holter in BAN is trying to send data including electrocardiography (ECG) to the general-purpose doctor or cardiologist every 5 minutes. Periodical data sending uses TCP protocol. The holter could collect the data locally for up to 48 hours [8]. It could also send data on demand.

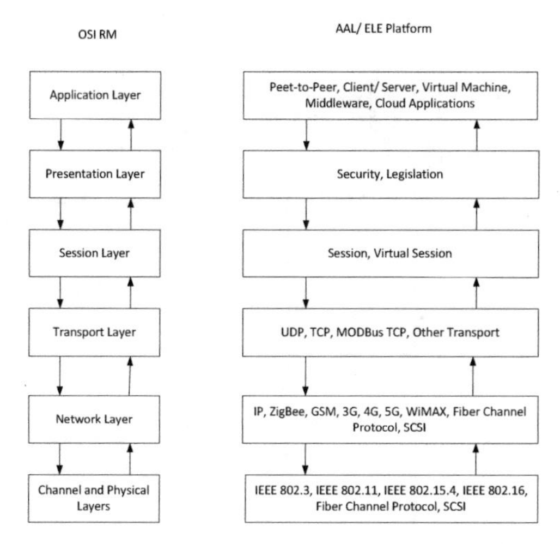

Fig. 3. Automated Enhances Living Environment open platform hierarchical model

The device could support user interface for additional information like general condition, complains, status reports, blood pressure, body weight, laboratory tests, radiographies, topographies. This data could be provided from the application via smart phone, tablet, or desktop computer. As a result, every person could obtain electronic file with monitored condition and evaluation. The data could be a matter of further examination by the authorized physician. It is also for personal use. On Fig. 5, datagram exchange between sensors and application server is shown. The measurements could be scheduled regularly in time using timer, could be reported according to the predefined timetable, and on demand. We also present sensor report through the gateway and without use of the local home server. The lower use-case scenario on this picture is data polling by the coordinator and reports. The network could work without coordinator and in condition of high radio interference. On Fig. 6 EnOcean sensor technology is presented. It is energy aware technology and is reporting only. Every data is reported 3 times and there is no acknowledgements. The data is collected by the network coordinator and stored at the home server. After pre-processing, the data is transmitted to the remote application server using the peer port technology [3]. Peer port shown on Fig. 6 could be based on UDP or TCP. The same idea could be implemented using the home network. In this case, the protocol between sensor gateway and the router is MODBus TCP and it is transferred to regular TCP between router and remote application server. The scenario is useful in all cases for regular status reports from patients and devices at user environment like temperature, humidity, light, activity, breading, etc. In all scenarios, the level of interoperability and redundancy is important. Some of the sensors could apply multi-homing. Some of the sensors could use star topology and have direct connection to the coordinator. Whereas the sensors have no direct access to the coordinator they could use mesh topology and retransmit data through sensor peers. In this case, the routing protocol is important for formation of the ad hoc network. Measurements shown below are in the lab experiment with sensors without coordinator.

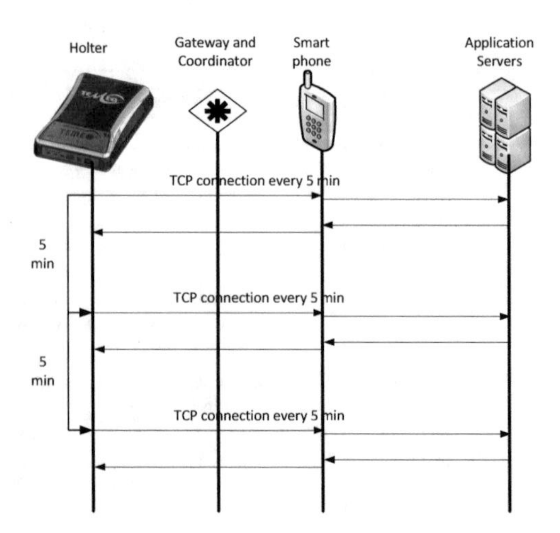

Fig. 4. Use-case scenario with ECG measurement using holter

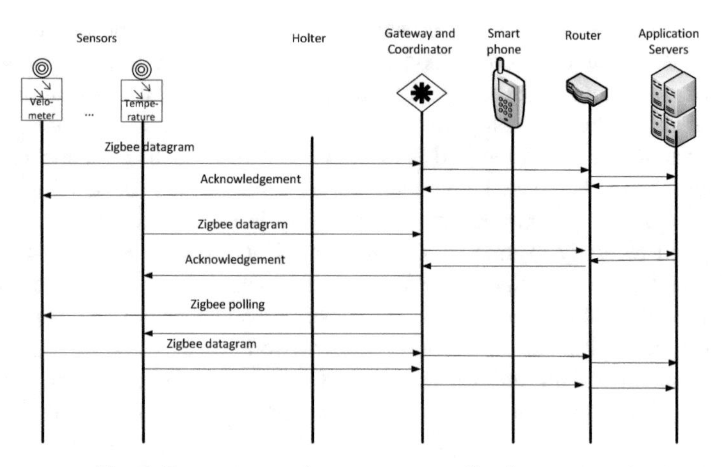

Fig. 5. Sensor report via gateway to application server

5 ZigBee Experiment Setup

The experiments with ZigBee network are setup in the lab and simulated using OMNET++. The use-case scenario tested is polling procedure from Fig. 5. Life experiment picture could be seen in [3]. It uses up to 10 sensors without coordinator. Quality of Service tests organized in the network are summarized on Table 1. The network consists of up to three radio channels. They could hear each other perfectly or in case of interference or high distance, they could retransmit the data from other sensors. Up to four retransmissions are setup during the experiment with read and write operations. The data transmitted is between 1 to 120 bytes using ZigBee datagrams at transmission intervals between 20 and 1000 ms for the third channel. Polling is regular at 200 ms in the sensor channel 1 and at 500 ms in sensor channel 2. The deterministic nature of the polling at regular intervals is worst-case scenario. We estimate throughput, loss, and delay. Table 2 presents main results from life experiment and simulation. Last five experiments correspond to the mesh topology and more than one bytes to be forwarded. This is the reason for high amount of errors [25]. The holter from BAN is considered to be traffic source that sends up to 120 bytes in regular time intervals.

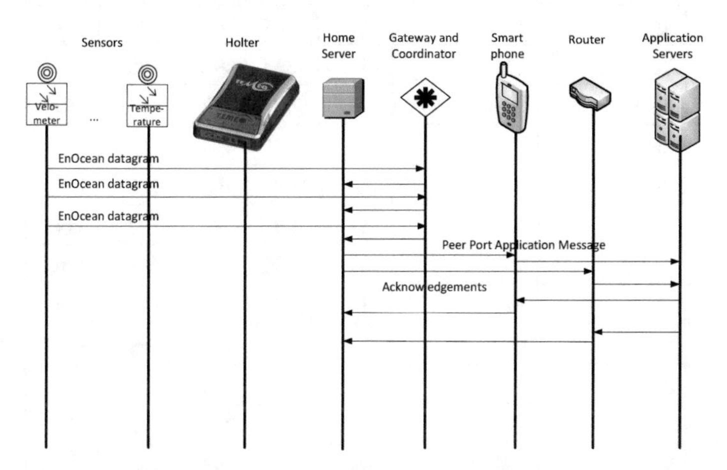

Fig. 6. Sensor report via gateway and home server to application server using peer port

Table 1. Experiment setup parameters

No	Read/ write	Sensor channel 1		Sensor channel 2		Sensor channel 3	
		Bytes to transmit	Interarrival time [ms]	Bytes to transmit	Interarrival time [ms]	Bytes to transmit	Intearrival time [ms]
1	Read	1	200	20	500	0	0
2	Read	60	200	20	500	0	0
3	Read	120	200	20	500	0	0
4	Read	120	200	20	500	120	1000
5	Read	120	200	20	500	1	10
6	Read	120	200	20	500	120	10
7	Write	1	200	20	500	0	0
8	Write	60	200	20	500	0	0
9	Write	120	200	20	500	0	0
10	Read	1	200	20	500	0	0
11	Read	60	200	20	500	0	0
12	Read	120	200	20	500	0	0
13	Read	120	200	20	500	120	10
14	Write	120	200	20	500	0	0

Table 2. Experiment results

No		Errors	Lab experiment Time to send [ms]			Simulation Time to send [ms]		
			min	max	mean	min	max	mean
1	Read	0	138	376	152	104	440	192
2	Read	3	344	1505	360	180	911	290
3	Read	17	471	3535	490	268	3473	450
4	Read	19	479	647	493	276	3590	530
5	Read	15	478	3887	499	469	4587	570
6	Read	28	422	569	493	510	4790	587
7	Write	0	157	239	172	137	3342	261
8	Write	3	365	3567	389	499	3704	640
9	Write	19	511	4575	577	844	4049	1018
10	Read	0	217	3453	458	168	3387	589
11	Read	945	356	5146	661	278	3789	732
12	Read	998	197	560	742	340	3973	850
13	Read	234	390	4719	562	455	4367	643
14	Write	401	548	5800	1694	634	4897	1890

6 Conclusion and Future Work

This paper presents an idea for combined sensor-to-cloud AAPELE platform that is reliable enough to send personal environment information. We set experiments in the lab and simulate the structure on OMNET++. All experimental results demonstrate the vitality of the proposed solution and its reliability. The high number of error in mesh structures is due to the high amount of data to be transmitted. The service could not work in real time, i.e. with reaction in less than 150 ms end-to-end. It is excellent for environment monitoring and management in near-real-time. In non-prioritized environments, the additional delay could be significant. The future work aims to map Quality of Service and Quality of Experience parameters over cross-layer approach for distributed quality management [24]. The lab and simulation experiments with many sensors and variable amount of data to be sent are also under analyses [7,8].

Acknowledgements. Our thanks to ICT COST Action IC1303: Algorithms, Architectures and Platforms for Enhanced Living Environments (AAPELE), project No ИФ-02-9/15.12.2012, Gateway Prototype Modelling and Development for Wired and Wireless Communication Networks for Industrial and Building Automation and project on irrigation controller development.

References

1. Salman, S., Wang, Z., Colebeck, E., Kiourti, A., Topsakal, E., Volakis, J.L.: Pulmonary Edema Monitoring Sensor With Integrated Body-Area Network for Remote Medical Sensing. Antennas and Propagation, IEEE Transactions on , vol.62, no.5, pp.2787,2794, May, doi: 10.1109/TAP.2014.2309132 (2014)
2. Jog, Y., Sharma, A., Mhatre, K., Abhishek, A.: Internet Of Things As A Solution Enabler In Health Sector. International Journal of Bio-Science and Bio-Technology Vol.7, No.2, pp.9-24 http://dx.doi.org/10.14257/ijbsbt.2015.7.2.02 (2015)
3. Goleva, R., Stainov, R., Savov, A., Draganov, P.: Reliable Platform for Enhanced Living Environment, First COST Action IC1303 AAPELE workshop Element 2014, in conjunction with MONAMI 2014 conference, Wurzburg, 24 Sept., 2014, 978-3-319-16291-12015, Springer International Publishing, http://dx.doi.org/10.1007/978-3-319-16292-8_23, pp. 315-328 (2015)
4. Ghayvat, H. Mukhopadhyay, S., Suryadevara, G. and N.: WSN- and IOT-Based Smart Homes and Their Extension to Smart Buildings, Sensors, 15, 10350-10379; doi:10.3390/s150510350, ISSN 1424-8220 (2015)
5. Smeets, R., Aerts, K., Mentens, N., Braeken, A., Segers, L., Touhafi, A.: Cryptographic key management architecture for dynamic 6LoWPAN networks. In: International Conference on Applied Informatics, https://lirias.kuleuven.be/handle/123456789/436437 (2014)
6. Kumar, P., Porambage, P., Ylianttila, M., Gurtov, A.: A Mobile Object-Based SECRET Key Distribution Scheme for Wireless Sensor Networks. In: IEEE Conference on Ubiquitous Intelligence and Computing and International Conference on Autonomic and Trusted Computing (UIC/ATC), pp. 656–661 (2013)
7. Kovatcheva, E., Nikolov, R., Madjarova, M., Chikalanov, A.: Internet of Things for Well-being – Pilot Case of a Smart Health Cardio Belt. XIII Mediterranean Conference on Medical and Biological Engineering and Computing 2013, 2014, 978-3-319-00845-5, IFMBE Proceedings, Roa Romero, Laura M. (eds), http://dx.doi.org/10.1007/978-3-319-00846-2_302, Springer International Publishing, 2014-01-01, pp. 1221-1224 (2014)
8. Mateev, H., Simova, I., Katova, T., Dimitrov, N.: Clinical Evaluation of a Mobile Heart Rhythm Telemonitoring System, ISRN Cardiology, 10.5402/2012/192670 (2012)
9. Agarwal, A., Agarwal, M., Vyas, M., Sharma, R.: A Study of ZigBee Technology, International Journal on Recent and Innovation Trends in Computing and Communication ISSN 2321 – 8169 Volume: 1 Issue: 4 287 – 292 (2013)
10. Gurjit, K. Ahuja, K.: QoS measurement of Zigbee home automation network using various modulation schemes, International Journal of Engineering Science and Technology (IJEST), ISSN : 0975-5462 Vol. 3 No. 2 Feb, pp. 1589-1597 (2011)
11. EnOcean_Equipment_Profiles_EEP_V2.5, EnOcean Serial Protocol, March 4 (2013)
12. Boonma, P., Suzuki, J.: Self-Configurable Publish/Subscribe Middleware for Wireless Sensor Networks, 978-1-4244-2309-5/09, IEEE (2009)
13. Buford, J., Yu, H., Lua, E.: P2P networking and applications. Morgan Kaufmann, USA, (2009)
14. Stainov, R., Goleva, R., Genova, V., Lazarov, S.: Peer Port Implementation for Real-time and Near Real-time Applications in Distributed Overlay Networks, 9th Annual International Conference on Computer Science and Education in Computer Science 2013 (CSECS 2013), 29 June, 2 July, Fulda-Wuertzburg, Germany, pp. 87-92 (2013)

15. Shiakallis, O., Mavromoustakis, C., Mastorakis, G., Bourdena, A., Pallis, E.: Traffic-Based S-MAC: A Novel Scheduling Mechanism for Optimized Throughput in Mobile Peer-to-Peer Systems, International Journal of Wireless Networks and Broadband Technologies (IJWNBT), volume 4, number 1, pp. 62-80, IGI Global (2015)
16. Tung, H., Tsang, K., Tung, H., Rakocevic, V., Chui, K., Leung, Y.: A WiFi-ZigBee Building Area Network Design of High Traffics AMI for Smart Grid, Smart Grid and Renewable Energy, 3, 324-333 http://dx.doi.org/10.4236/sgre.2012.34043, Published Online November (2012)
17. Huang, M. L., S. Lee, S.-C Park: A WLAN and Bluetooth Coexistence Mechanism for Health Monitoring System, 978-1-4244-2309-5/09/$25.00 ©2009 IEEE.
18. ZigBee Document 075360r15, ZigBee Health CareTM, Profile Specification, ZigBee Profile: 0x0108, Revision 15, Version 1.0, March 2010, Sponsored by: ZigBee Alliance.
19. Zigbee Home Automation Public Application Profile, ZigBee Profile: 0x0104, Revision 26, Version 1.1 (2010)
20. Felizardo, V., Gaspar, P., Garcia, N., Reis, V.: Acquisition of Multiple Physiological Parameters During Physical Exercise, International Journal of E-Health and Medical Communications, 2(4), 37-49, October-December (2011)
21. Garcia, N.M., Garcia, N.C., Sousa, P., Oliveira, D., Alexandre, C., Felizardo, V.: TICE.Healthy: A perspective on medical information integration. Biomedical and Health Informatics (BHI), 2014 IEEE-EMBS International Conference on , vol., no., pp. 464,467, 1-4 June, doi: 10.1109/BHI.2014.6864403 (2014)
22. Vitor G. R., Marilia C.: Enabling wireless cooperation in delay tolerant networks, Information Sciences, Volume 290, 1 January, Pages 120-133, ISSN 0020-0255, http://dx.doi.org/10.1016/j.ins.2014.08.035 (2015)
23. Ciobanu, R., Marin, R., Dobre, C., Cristea, V., Mavromoustakis, C.X.: ONSIDE: Socially-aware and interest-based dissemination in opportunistic networks. Conference Proceedings, Network Operations and Management Symposium (NOMS), pp. 1-6 (2014)
24. Nikolovski, V., Lameski, P., Joksimoski, B., Chorbev, I.: Cloud Based Assistive Technologies and Smart Living Environment System, Book Section, 2015, 978-3-319-16291-1, Mobile Networks and Management, Agüero, R., Zinner, T., Goleva, R., Timm-Giel, A., Tran-Gia, P. (eds), 10.1007/978-3-319-16292-8_26, Springer International Publishing, pp. 358-369 (2015)
25. Metter, C., Gebert, S., Lange, S., Zinner, T., Tran-Gia, P., Jarschel, M.: Investigating the Impact of Network Topology on the Processing Times of SDN Controllers, The Seventh International Workshop on Management of the Future Internet (IEEE/IFIP ManFI 2015), http://www.manfi.org/ (2015|)
26. Ruiz-Zafra, A., Benghazi, K., Noguera, M., Garrido, J.L.: Zappa: An Open Mobile Platform to Build Cloud-based m-Health Systems. IV International Symposium on Ambient Intelligence (ISAmI 2013), Advances in Intelligent and Soft Computing, Vol. 219, Springer-Verlag, , pp. 87-94 (2013)
27. Ji, Z., Ganchev, I., O'Droma, M., Zhao, L., Zhang, X.: A Cloud-Based Car Parking Middleware for IoT-Based Smart Cities: Design and Implementation. Sensors, Vol. 14 No. 12, pp. 22372-22393 (2014)

A Fuzzy Logic Approach for a Wearable Cardiovascular and Aortic Monitoring System

Cristina C. Oliveira[1,2], Ruben Dias[2] and José Machado da Silva[1,2]

[1] Faculdade de Engenharia, Universidade do Porto, Porto, Portugal
{cristina.oliveira,jms}@fe.up.pt
[2] INESC TEC, Porto, Portugal
rtdias@inescporto.pt

Abstract. A new methodology for fault detection on wearable medical devices is proposed. The basic strategy relies on correctly classifying the captured physiological signals, in order to identify whether the actual cause is a wearer health abnormality or a system functional flaw. Data fusion techniques, namely fuzzy logic, are employed to process the physiological signals, like the electrocardiogram (ECG) and blood pressure (BP), to increase the trust levels of the captured data after rejecting or correcting distorted vital signals from each sensor, and to provide additional information on the patient's condition by classifying the set of signals into normal or abnormal condition (e.g. arrhythmia, chest angina, and stroke). Once an abnormal situation is detected in one or several sensors the monitoring system runs a set of tests in a fast and energy efficient way to check if the wearer shows a degradation of his health condition or the system is reporting erroneous values.

Keywords: Electrocardiogram, Wearable, Fuzzy logic, Dependability

1 Introduction

Along with the progress of medical technologies, many countries are gradually becoming geriatric societies due to the rapid growth of the aging population. This increases the need for home health monitoring for securing independent lives of patients with chronic disorders or that have health care problems. The advances on sensors, wireless communications and information technologies have resulted in the rapid development of various wellness or disease monitoring systems, which enable extended independent living at home and improve the quality of life. Traditionally, clinical practice has been based on a post-diagnosis intervention basis (drugs, surgeries, prosthesis, etc.). Nowadays, and regardless of the patients' age, the health care community is trying to focus on prevention and wearable monitoring systems have been proposed to meet this task. Therefore, diseases tend to be prevented, rather than treated, after continuous vital signals monitoring, which provide information about the health status related with lifestyle and overall quality of life [1,2,3].

S. Loshkovska and S. Koceski (eds.), *ICT Innovations 2015*,
Advances in Intelligent Systems and Computing 399,
DOI: 10.1007/978-3-319-25733-4_27

Remote health monitoring can be used only if the monitoring device is based on a comfortable sensing interface, easy to use and customizable. Its interface must allow continuous remote control in a natural environment without interference or discomfort for the users. The textile approach to the implementation of sensing elements embedded in clothing items, allows for low-cost long-term monitoring of patients and to easily customize the sensor configuration according to the needs of each individual [4]. Applying this concept, it is possible to reduce health care costs maintaining the high quality of care, shift the focus health care expenditures from treatment to prevention, provide access to health care to a larger number of patients, reduce the length of hospital stays and address the issue of requirements for elderly population and/or chronically ill patients. It also allows accessibility to specialized professionals through telemetry, thus decentralizing the provision of health care.

Because these wearable monitoring systems are to be used for medical purposes (continuous monitoring, diagnosis, etc.), the reliability and safety of the system have to be perfectly controlled. Unfortunately, the complexity of these systems endlessly increases, making the existing techniques for dependability developed in aeronautics, space and automotive fields not totally appropriate for the medical case.

To overcome the lack of a dependability model for the development of complex pervasive medical monitoring devices, a fault tree analysis approach is used to identify the main risk of failure (see Fig. 1). A typical wearable device (hereafter the system) comprises a module to capture the biosignals, including the electrodes and the analogue front-end, a microcontroller, and a radiofrequency emitter to transmit the signal to a smartphone or a personal computer. In our approach the captured biosignals are received and analyzed within a smartphone. A rule based algorithm (fuzzy logic) decides whether the signals are normal or not. If not, it is diagnosed if the wearer shows an abnormal situation or instead the system is faulty. That is, the abnormality detected within the biosignals can be due to a wearer irregular state (pathological condition or intense physical activity) or due to a degradation of the system operation.

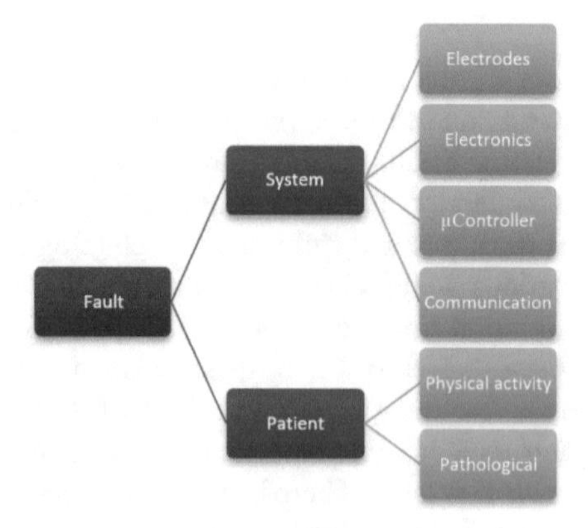

Fig. 1. Fault tree analysis of the wearable monitoring system

2 Combined Cardiac and Aortic Monitoring System

The combined cardiac and aortic monitoring system (SIVIC system) under development (Fig. 2) provides the synchronous measurement of the patient ECG (electrocardiogram) and of the pressure in the abdominal aneurysm sac, in order to have a more robust and reliable monitoring. Biologically compatible wireless pressure sensors, which show suitable linearity and sensitivity [5], are used to capture the intra-sac aneurysm pressure. An electronic readout unit (ERU) capable of energizing the pressure sensors and capture the pressure data is placed in the chest of the patient. This unit provides also the monitoring of a 12-lead ECG using textile dry electrodes [6]. The electronic unit and the electrodes are built in a customized clothing.

Data is transmitted to a smartphone for further processing, data display, and eventual communication with a healthcare center.

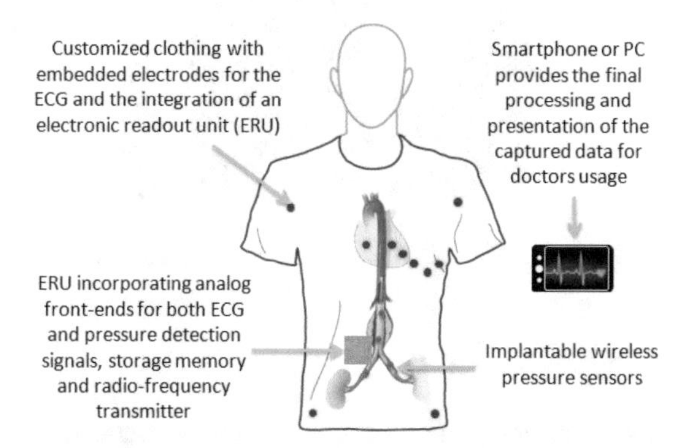

Fig. 2. Wearable ECG data capture and transmitter module

The 12-lead ECG data acquisition and transmission (DAT) module prototype that was developed is a circular board (30 mm Ø) with an ECG acquisition analogue front-end based on the low-power (0.75 mW/channel) Texas Instruments 24-bit ADS1298 chip and a PAN1740 Bluetooth Low Energy (BLE) module from Panasonic. The board includes also an I2C EEPROM and a DC/DC converter to supply a regulated 3.3 V. The PAN1740 is a small (9 x 9.5 x 1.8 mm) BLE single mode module based on the Dialog DA14580 SoC with an advertised power consumption of 4.9 mA when transmitting/receiving. This SoC includes a 32 bit ARM Cortex M0 microcontroller (μC) operating at a 16 MHz frequency, that is used to perform all the necessary processing operations, thus saving the cost of an external μC, the additional PCB area and power consumption. The EEPROM is used to save the application code during the developing phase. In the final version it can be removed and the code can be saved in the One-Time Programmable (OTP) memory present on the BLE module.

Wireless ECG monitoring systems with a high number of leads (e.g. 12-lead) are usually designed for clinical usage, being systems with a lower number of acquisition channels (e.g. 1 to 3 leads) commonly used in ambulatory cases [7,8]. Our system was designed having in mind its use in both clinical and ambulatory scenarios and thus the number of ECG data acquisition channels is reconfigurable. Inputs not used to capture

ECG signals can be used to acquire other biosignals. Figure 3 shows the T-shirt cardiac monitoring system being proposed.

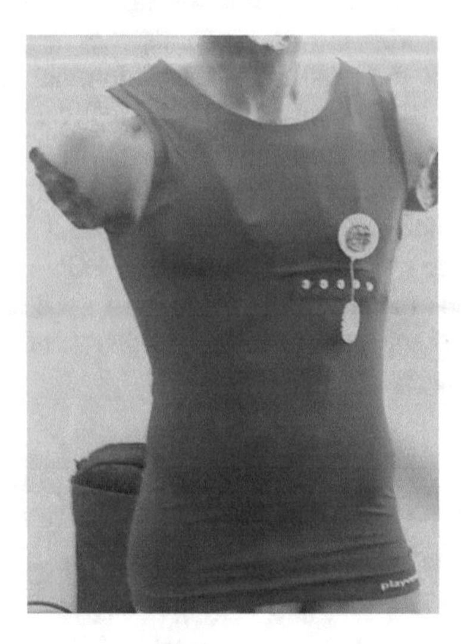

Fig. 3. The SIVIC T-shirt and data acquisition module

3 Data Fusion for Diagnosis

The ECG contains important hemodynamic information, such has the heart rate (HR). During an ECG cycle three main events take place: the P wave (contraction of the atria), the QRS complex (corresponding to the contraction of the left ventricle) and the T wave (relaxation of the ventricles). Their morphologies (amplitude and interval/segment length) will vary in accordance to the physiological condition.

The HR, given in beats per minute (bpm), is the interval between two consecutive R-waves in the QRS complex. Noise contamination such as baseline wander, power line interference, and muscle activities can corrupt the signal and reduce the clinical value of an ECG recording. Since wearable devices are more affected by noise, filtering of the ECG is a necessary pre-processing step to ensure a reduction of the noise components while preserving the QRS complex shape. The Pan-Tompkins algorithm is used for ECG filtering and the HR calculation [9].

The availability of different sensors in wearable systems allows for fusing the respective data to formulate better decisions from the captured data. Other biosignals, such as the blood pressure (BP), defined by the systolic (maximum) and diastolic (minimum) pressures, can provide important information about the patient condition, eventually affected by physical activity or diseases. Accelerometers enable tracking the wearer activity, i.e. if he is sitting, walking or running, which will influence the heart activity. The SIVIC system also includes an electrode-skin impedance measuring circuit, which allows detecting if the electrodes are connected to the patient or are loose/disconnected.

Signals that can be measured with the SIVIC system, the extracted features, and the patient/system condition inferred from the respective classification are summarised in Table 1.

Table 1. Data fusion model for the measured signals

Signals	Features	Classifier
ECG	HR I HR II HR III	Normal/Abnormal
Blood Pressure	Systolic Diastolic	Hypotensive/Normal/Hypertensive
AAA Sac Pressure	Mean Pressure	Endoleak
Accelerometer	Motion	Resting/Walking/Running
Electrode-Skin Impedance	Resistance	Connected/Disconnected

Data fusion techniques have been applied as a means for a combined analysis of several physiological signals that can potentially provide additional information on a patient's condition. Kenneth et.al performed the fusion of ECG, blood pressure, saturated oxygen content and respiratory data for achieving improved clinical diagnosis of patients in cardiac care units [10].

Table 2. Fusion rules for patient condition diagnosis

Signals	Condition	Rule	
ECG	Normal	HR between 60 and 100 bpm	
	Asystole	No QRS for at least 4 seconds	
	Extreme Bradycard	HR lower than 40 bpm for 5 consecutive beats	
	Extreme Tachycardia	HR higher than 140 bpm for 17 consecutive beats	
Blood Pressure (mmHg)		Systolic	Diastolic
	Normal	90-139	60-89
	Hypotension	<90	<60
	Hypertension	>140	>90
AAA Pressure	Normal	Low pressure (~40 mmHg)	
	Endoleak	Sistemic pressure	

In our case, as a first approach, a fuzzy logic system is used for the data fusion due to its probability assignment based on rules. Since the values of the features extracted from the biosignals can be assigned in regions well defined in the medical literature, the rules creation is relatively straightforward (Table 2).

3.1 Fuzzy Logic

The fuzzy logic system comprises 4 main components: fuzzy rules (knowledge base), fuzzy sets, fuzzy inference engine and defuzzification (Fig. 4) [11]. The inputs of the fuzzy logic system are the features previously extracted from the measured signals (Table 1). The outputs are the *Patient Status*, *System Status* and the *Global Status*, which can be normal or faulty - i.e., either the patient has a health condition or the monitoring system is malfunctioning. The outputs are determined based on the input values of the fuzzy sets and the assigned rules for each output. The rules to define the Patient Status are based on medical information, here collected from the literature, the rules for the *System Status* are defined from the system specifications, and the rules for the *Global Status* include both.

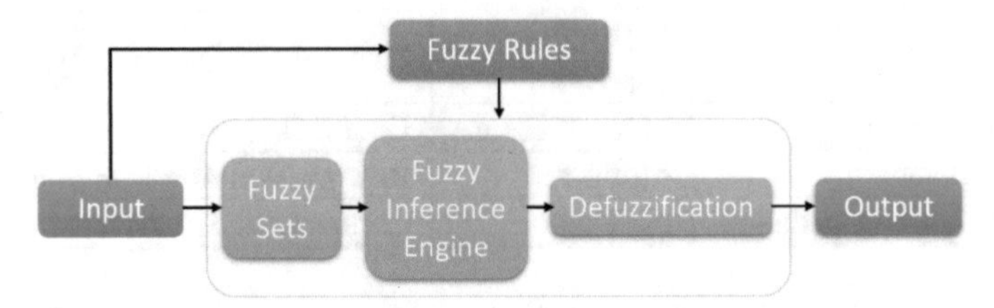

Fig. 4. Block diagram of fuzzy logic system

The fuzzy sets include the HR for each channel, the blood pressure (systolic and diastolic), and can also include the contact resistance and the acceleration if these data are available.

The trapezoidal curve (1) was chosen for the membership function. This is a function of a vector, x, and depends on four scalar parameters *a*, *b*, *c*, and *d*. The parameters *a* and *d* locate the "feet" of the trapezoid and the parameters *b* and *c* locate the "shoulders".

$$\mu_{trapezoidal}(x) = \begin{cases} 0, & x < a or x > d \\ \dfrac{x-a}{b-a}, & a \le x \le b \\ 1, & b \le x \le c \\ \dfrac{d-x}{d-c}, & c \le x \le d \end{cases} \tag{1}$$

Table 2 shows the normal values for the HR and BP, and some examples of pathologies.

4 Results

Data from the MIT Multiparameter database (MGH/MF) was used to test the fuzzy logic system using Matlab [12,13]. The features from ECG signals (leads I, II and V) and the arterial blood pressure (ART) were extracted and feed to the fuzzy logic system. The ECG provides the HR information and the ART waveform is used to know the systolic and diastolic pressures.

The fuzzy logic was evaluated for 3 situations:

1. The recorded signals have good quality, i.e. the signal-to-noise-ratio (SNR) is good enough to identify relevant features, but the patient's blood pressure is very high (record MGH085 from the MGH/MF database). The System Status is ok, but the Patient Status indicates a health problem. Result: Patient Status: 14; System Status: 86; Global Status: 86.
2. Atrial flutter, or arrhythmia, is an abnormality of the heart rhythm resulting in a rapid and sometimes irregular heartbeat. Atrial flutter is recognized on an ECG by presence of characteristic flutter waves at a regular rate of 240 to 440 beats per minute (Fig. 5). In this case the HR is calculated using lead V, and the ART waveform is also used for a more reliable HR estimation, since these signals are related. Result: Patient Status: 14; System Status: 86; Global Status: 86.

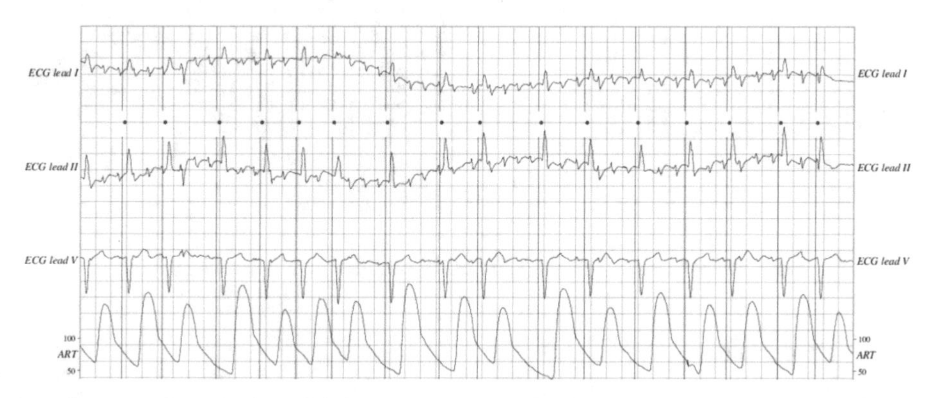

Fig. 5. MGH023 record: Atrial flutter (Grid intervals: time 0.2 s, ECG 0.5 mV, ART 25 mmHg)

3. Sinus tachycardia is a heart rhythm originating from the sinoatrial node with an elevated rate of impulses, defined as a rate greater than 100 bpm in an average adult. The calculated HR from each channel indicates the patient has tachycardia (MGH010 record) Result: Patient Status: 14; System Status: 86; Global Status: 81.

After validating the fuzzy logic system with a database that contains annotations from physicians, the SIVIC wearable system was used to acquire the ECG signal of lead I. The smartphone receives the acquired data via Bluetooth, filters the received signal and calculates the HR and SNR. These features (HR and SNR) are used by the fuzzy logic system to monitor the patient and the wearable system. When a degradation occurs in the patient or system, the smartphone detects the fault and requests for further tests to the monitoring system in order to determine the cause and, if possible, to correct the fault. Figure 6 displays ECG waveforms acquired with the SIVIC system. On the left side of the figure the ECG waveform presents a normal sinus rhythm. On the right side of Fig. 6 the ECG waveform is corrupted with noise and the monitoring system is unable to calculate a reliable HR, since the SNR is high. A possible cause for this situation is a loose electrode, which could be determined by measuring the electrode-skin impedance. Since this is a very common problem in wearable devices, the SIVIC system periodically records the impedance of the textile electrodes and stores this value for each user. When the problem in the signal was detected the smartphone sent a request to the SIVIC system to perform an impedance measurement, and received a value of 13.911 MΩ, which was much higher than the recorded impedance values for the wearer under observation (around 1 MΩ). In this situation the smartphone issues a warning for the user to readjust the electrodes embedded in the t-shirt.

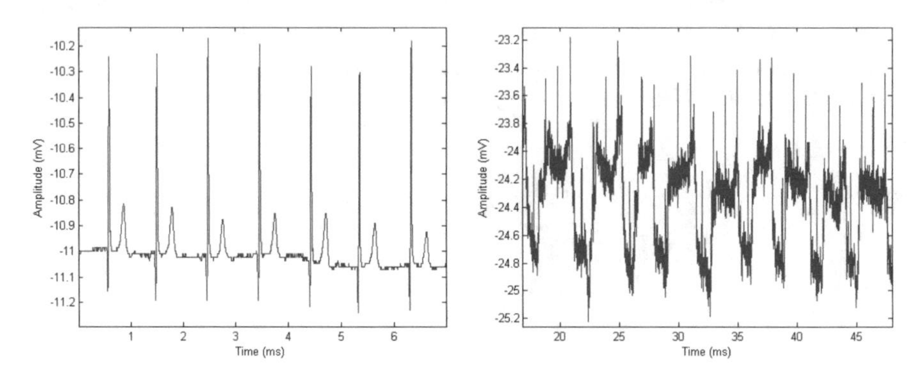

Fig. 6. Normal ECG (left side) and ECG corrupted with noise (right side)

5 Discussion

When the data fusion model detects that the System Status is degraded, further tests can be performed by the system to determine the cause. The smartphone sends an order for specific tests to be performed depending on the signals features. For instance if an ECG channel presents a behavior similar to the atrial flutter condition, but the remaining channels are normal, the cause of the flutter could be caused by the acquisition system, rather than the patient's heart. An oscillation in the ECG amplifier could cause such flutter in the signal. A simple test would be to connect both inputs of the amplifier and observe if the flutter persists. If not, it could be the case the signal is really displaying a health condition that is more visible in this particular ECG channel.

On the other hand, the data fuzzy model is flexible in the sense that further inputs can be added to the system providing extra information regarding the patient and the system. For instance environmental sensors, like accelerometers, can be added to monitor the patient activity. If motion is detected at the same time the ECG signal is degraded, the system can determine the degradation of the biosignal as temporary and not related with any fault from the electronics or the electrodes.

6 Conclusion

The advances on sensors, wireless communications, and information technologies have promoted the rapid development of various wearable patient monitoring systems. The availability of wearable vital signals monitoring systems allows for securing independent lives of patients with chronic disorders or who require a permanent vigilance, while improving their daily quality of life. The work presented herein shows how data fusion, notably fuzzy logic, can be explored to improve the dependability of a cardiovascular monitoring wearable system, after providing a means to, on the fly, diagnosing whether deviations detected in the acquired signals are due to a disease or condition of the patient, or actually to a fault in the system. It is also a tool which can help, in the electronics design stage, the process of identifying test operations needed to improve the system's diagnosability.

Acknowledgment. This work is financed by the ERDF - European Regional Development Fund through the COMPETE Programme (operational programme for competitiveness) and by National Funds through the FCT – Fundação para a Ciência e a Tecnologia (Portuguese Foundation for Science and Technology) within project SIVIC PTDC/EEI-ELC/1838/2012 (FCOMP-01-0124-FEDER-028937), and grant contract SFRH/BD/81476/2011(first author). José Machado da Silva is a member of the Management Committee of ENJECT (EU COST Action TD1405) the European Network for the Joint Evaluation of Connected Health Technologies.

References

1. Park, S., Jayaraman, S.: Enhancing the quality of life through wearable technology. Engineering in Medicine and Biology Magazine, IEEE. 22, 41–48 (2003).
2. Paradiso, R.: Wearable health care system for vital signs monitoring. In: 4th International IEEE EMBS Special Topic Conference on Information Technology Applications in Biomedicine, 2003. pp. 283–286 (2003).
3. McAdams, E., Krupaviciute, A., Gehin, C., Dittmar, A., Delhomme, G., Rubel, P., Fayn, J., McLaughlin, J.: Wearable Electronic Systems: Applications to Medical Diagnostics/Monitoring. In: Bonfiglio, A. and Rossi, D.D. (eds.) Wearable Monitoring Systems. pp. 179–203. Springer US (2011).
4. Paradiso, R., Loriga, G., Taccini, N.: A wearable health care system based on knitted integrated sensors. IEEE Transactions on Information Technology in Biomedicine. 9, 337–344 (2005).
5. Oliveira, C., Sepulveda, A., Almeida, N., Wardle, B., Machado da Silva, J., Rocha, L.: Implantable Flexible Pressure Measurement System Based on Inductive Coupling. IEEE Trans Biomed Eng. (2014).

6. Oliveira, C.C., Machado da Silva, J., Trindade, I.G., Martins, F.: Characterization of the electrode-skin impedance of textile electrodes. In: 2014 Conference on Design of Circuits and Integrated Circuits (DCIS). pp. 1–6 (2014).
7. Delano, M.K., Sodini, C.G.: A long-term wearable electrocardiogram measurement system. In: 2013 IEEE International Conference on Body Sensor Networks (BSN). pp. 1–6 (2013).
8. Amour, N., Hersi, A., Alajlan, N., Bazi, Y., AlHichri, H.: Implementation of a Mobile Health System for Monitoring ECG signals. Academy of Science and Engineering, USA (2015).
9. Pan, J., Tompkins, W.J.: A Real-Time QRS Detection Algorithm. IEEE Transactions on Biomedical Engineering. BME-32, 230–236 (1985).
10. Kenneth, E., Rajendra, A.U., Kannathal, N., Lim, C.M.: Data Fusion of Multimodal Cardiovascular Signals. In: Engineering in Medicine and Biology Society, 2005. IEEE-EMBS 2005. 27th Annual International Conference of the. pp. 4689 –4692 (2005).
11. Zadeh, L.A.: Fuzzy Logic. Computer. 21, 83–93 (1988).
12. Goldberger, A.L., Amaral, L.A.N., Glass, L., Hausdorff, J.M., Ivanov, P.C., Mark, R.G., Mietus, J.E., Moody, G.B., Peng, C.-K., Stanley, H.E.: PhysioBank, PhysioToolkit, and PhysioNet Components of a New Research Resource for Complex Physiologic Signals. Circulation. 101, e215–e220 (2000).
13. Welch, J.P., Ford, P.J., Teplick, R.S., Rubsamen, R.M.: The Massachusetts General Hospital-Marquette Foundation Hemodynamic and Electrocardiographic Database - Comprehensive collection of critical care waveforms. J Clinical Monitoring 7(1), 96-97 (1991).

Next Generation ICT Platform to Harmonize Medical, Care and Lifestyle Services

Mario Drobics, Karl Kreiner, Helmut Leopold

AIT Austrian Institute of Technology, Digital Safety & Security Department, Vienna, Austria
{mario.drobics,karl.kreiner,helmut.leopold}@ait.ac.at

Abstract. The key to active and healthy living/aging in the 21st century is to establish an individualized everyday-living environment that supports positive health behaviour and sustainable healthy lifestyle by means of applied ICT technology. Next generation ICT platforms have to support health-service as well as care and life-style service in a uniform as well as standardised way to enable integrated, scalable, and thus cost efficient solutions for the society.

The *KIT-Aktiv* service validation platform aims to promote healthy activities to a broad audience and enable its users to take control of their health. *KIT-Aktiv* provides a flexible ICT infrastructure to promote, support and monitor health related activities. By utilizing this infrastructure, different services and processes for a wide range of diverse user groups are supported. While school children can track their daily activities or organize a sport competition, older adults might monitor their prescribed exercise program to discuss it with their physician, later on. By integrating stakeholders from different domains like schools, sports, care-giving and medicine, the provided services do not only provide a value on their own, but can be incorporated in existing support or treatment processes. As the costs per user are very low (around 5€), the infrastructure can be scaled up to a large user group easily.

Keywords: Health monitoring, RFID, Tele-care services, Life-style management, Sensor networks, Internet of Things

1 Motivation

The process of aging is affected by different factors such as the living environment, lifestyle, and the presence or absence of chronic diseases [1]. These factors vary with age, but do not define age. Social and economic factors such as prior employment, education, residential area, as well as behaviours like smoking, physical exercise, activity in daily life, alcohol consumption, nutrition, and social contacts are important influences in the process of aging and are therefore described in the document "Keep fit for life" by the WHO including recommendations for elderly people [2,3].

An active lifestyle is particularly important at every age. It strengthens the body and can thereby improve quality of life as well as prevent accidents and injuries by strengthening the body and by creating a better awareness of the body [4,5,6]. Daily

S. Loshkovska and S. Koceski (eds.), *ICT Innovations 2015*,
Advances in Intelligent Systems and Computing 399,
DOI: 10.1007/978-3-319-25733-4_28

physical activity, however, is often decreasing, resulting in an increased risk for developing diseases later on.

To support positive health behaviour and sustainable healthy lifestyle in an everyday-living environment for a broad range of diverse users, it is necessary to provide different offerings so that each user can find a suit-able set of services for his/her needs. While for some users, regular physical activity will be sufficient, others may need to integrate more specific health data or a specific type of exercises. By providing an ICT-based infrastructure that is capable of managing different types of activities and health parameters combined with a low threshold access to this information, self-perception and motivation can be improved. Furthermore, interfaces for specific stakeholders like physicians, trainers, etc. ensure that this data can also be utilized in other health related processes and decisions.

In order to achieve cost efficient and scalable solutions, which enable early market acceptance, it is mandatory to build next generation tele-medicine service platforms on the one hand with a strong focus on standardization for sensors and end-user equipment as well as IT services and modules [7], while on the other hand it is essential to consider the requirements of different application domains within a harmonized modular IT architecture: medical services, care services, as well as lifestyle services. These different application domains are driven by heterogeneous requirements through complementary stakeholders and actors.

2 Approach

2.1 Application Domains

To derive the design of the underlying communication platform for tele-medicine (eHealth) and elderly care (AAL) with its essential disruptive effects on processes, markets and businesses models it was necessary to investigate and harmonize three areas:

1. the technology platform had to be built on generic building blocks for a broad range of applications in eHealth, AAL and even life-style to build and to enable an economy of scale for a next generation enhanced living environment (ELE)
2. processes, data structures, IT architectures, user-interfaces had to be designed and validated carefully in extensive proof of concept projects
3. the interaction with the end-users – which are at the end very complex eco-system – had to be analysed and elaborated in in depth discussions and proof of concept projects: medical specialists, practitioners, nurses, caregivers, relatives, etc.

Furthermore, in order to prevent the creation of isolated services, a high interoperability and flexibility of the underlying ICT systems is necessary. Thus, for our system design we investigated not only the concrete use-case of health prevention, but utilized experiences and evaluation results from a much broader range of different application scenarios:

- Diabetes Mellitus: In 2010 the Austrian Insurance Institution for Railways and Mining Industry started a program called *"Gesundheitsdialog-Diabetes"* (health-dialog diabetes). This program in-tended to support patients in using tele-medicine to get in a dialog with the health institutions and physicians (preferably their local practitioner) to enable a reliable and lasting individual support. For the first proof-of-concept an IT platform was set up based on a proprietary python implementation in 2011 [7]. In 2014, the system was used actively by 526 patients and 77 physicians. Overall, 269.083 transmissions had been recorded in 2014, resulting in 73.628 monitoring days. With an increasing amount of us-ers and the need to cover other diseases within the same system as well, it became necessary to move to a new system with greater flexibility and scalability.
- Chronic heart failure: To support the tele-medical treatment of heart failure the province of Tirol in Austria decided in 2013 to develop a proof-of-concept. Tele-monitoring systems appear to be effective in the vulnerable phase after discharge from hospital to prevent early readmissions. This POC implements a collaborative post-discharge HF disease management program (*HerzMobil Tirol* network) that incorporates physician-controlled tele-monitoring and nurse-led care in a multidisciplinary network approach. Thus, the underlying ICT system had to be flexible enough to deal with the complex care processes and fulfil the da-ta security guidelines of the public institutions [8].
- Fall prevention: Within the *iStoppFalls* project an AAL system to predict and prevent falls by monitoring mobility-related activities and other risk factors of falls in real-life was developed. Beyond continuous fall risk monitoring, this enabled tailoring individualized exercise programs coached by *iStoppFalls* [9]. The underlying ICT system had to handle the diverse amount of data coming from the sensors involved (accelerometer, video, questionnaires, games, etc.) and provide automatic feedback to the users and an easy to use interface for the health professionals.

2.2 Requirements for *KIT-Aktiv*

For *KIT-Aktiv* the general setting was a bit different to the previously mentioned use-cases, as the user has a more active role in this setting and interacts with a general infrastructure, used by anyone. Nonetheless, the system should also integrate in existing prevention and care processes, in order to support professional services as well as personal activities.

The general idea behind *KIT-Aktiv* [10] was therefore to provide a platform and infrastructure to support a wide range of health, care and fitness related services. This led to the following major requirements:

1. Provide easily accessible, low threshold interfaces
2. Provide specific interfaces depending on user, environment and role
3. Integrate a wide range of measurements from different devices
4. Provide secure data transmission and storage
5. Be scalable to a large number of users
6. Integrate in established service processes

The ICT platform with its architecture and processes had to fulfil requirements from different perspectives:

- Medical domain: telemedicine services for the management of chronical deceases like heart-failure, adiposity, etc. to support services for medical practitioners for example
- Care domain: care services to support application like increasing the mobility of elderly people by care organizations; and finally
- Life-style domain: services to support the fitness of kids or offer training support for the general public

Because of this close relation to different domains, it was important to incorporate the experience from other applications, as listed in the previous section. And most importantly, technology acceptance, especially in these application areas, very much depends on highest usability and lowest complexity for the users. Thus, an IT architecture which is following a modular architecture, open interfaces and virtualization of IT services and most importantly an ongoing validation in real-world application scenarios was a pre-requisite for an early technology acceptance [11].

3 Architecture

The architecture of *KIT-Aktiv* is composed of elements installed in the environment (public terminals, fitness-poles), the backend and the different front-end interfaces (see Fig. 1). External measurement devices where connected using different interfaces (RFID, Bluetooth). The central element used for identification is a secure RFID ID-chip, usually integrated in a wrist-band or key fob. This ID was used to identify the user on the different interfaces, sparing the need to remember and enter user-id and password.

The overall architecture is shown in Fig. 1, including the core elements (yellow) and external devices (green).

The **fitness-poles** where used to collect time and identity of the participants using an integrated RFID reader. To register at a pole, the users had to bring their RFID wrist-band close to the integrated reader. The reader was based on a micro-controller with an integrated RFID reader module and UMTS module. The data from the poles was then transmitted wirelessly to the backend, were the corresponding routes and distances are calculated. Additionally to the tracking data, regular status information was transmitted to the backend. Control commands could also be sent to the controller via SMS to read out the status information or reset the controller. The fitness-poles were equipped with an integrated solar-panel and a battery, so that they were completely energy autonomous.

The **public terminals** was built around an iPad and an integrated RFID reader. They provided access for the users to view their data and enter new measurements. This could be done by either entering them manually or by using a body scale which was connected using Bluetooth-LE. Identification on the terminal was done using the users RFID wrist-bands. During the login process, the terminals checked if the user has a personal accelerometer and read out this data automatically, if applicable.

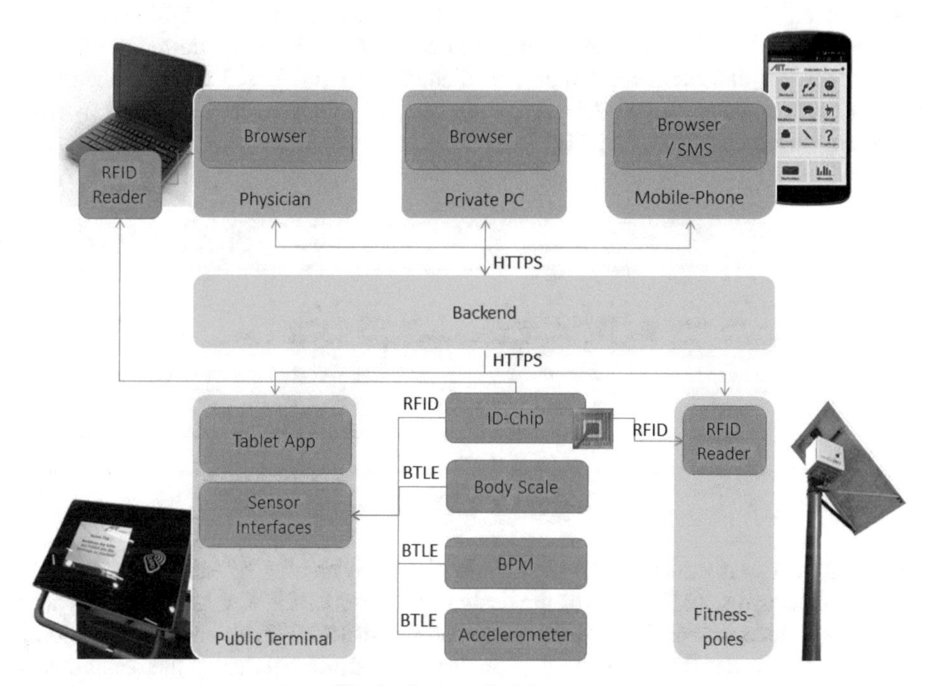

Fig. 1. System Architecture

The **backend** of *KIT-Aktiv* was based on the KIOLA platform, a modular framework for developing applications in the area of tele-health and ambient assisted living with a focus on collecting data from sensors. KIOLA was built using open-source technologies, such as the web-based framework Django and PostgreSQL as central data storage.

As Fig. 2 suggests, KIOLA is comprised of three layers: First, the core system provides basic functions for sensor data collection. In *KIT-Aktiv* two types of sensors have been used: Fitness poles placed around the community collecting vital parameters as well as distance walked by end-users. The sensor data model followed the principles of health-based observations, as they are suggested by the Health Level 7 (HL7) standard. Administrators could define an arbitrary number of observation profiles using a web-based interface. Observation profiles were then exposed through an automatically generated RESTful [12] application programmer's interface (API) to any sensors (e.g. the fitness poles used in *KIT-Aktiv*) capable of submitting data. Moreover, the core layer allowed tracking of any changes to data through an audit trail as well as a user-role model to define access to data. Next, a set of modules provided web-based user interfaces as well as APIs for (1) sensor management, used in *KIT-Aktiv* for configuration of the fitness poles (2) rule-based decision support (3) data visualization using interactive charts (4) organizer and calendar functions, used in *KIT-Aktiv* to advertise events (5) a search engine [13] for observations (6) data export (7) a notification engine for sending reminders using SMS, E-Mail and Google Cloud Messaging and (8) document management for PDF and word files. On top of these modules care plugins provided specialized user interfaces for different stake-holders (e.g. physicians and caretakers) addressing specific health do-mains, such as tele-health for chronic heart failure and diabetes or game-based fall prevention for elderly people. Aside from *KIT-Aktiv*, these care plugins have been successfully dep-loyed in various projects [7,8], [14].

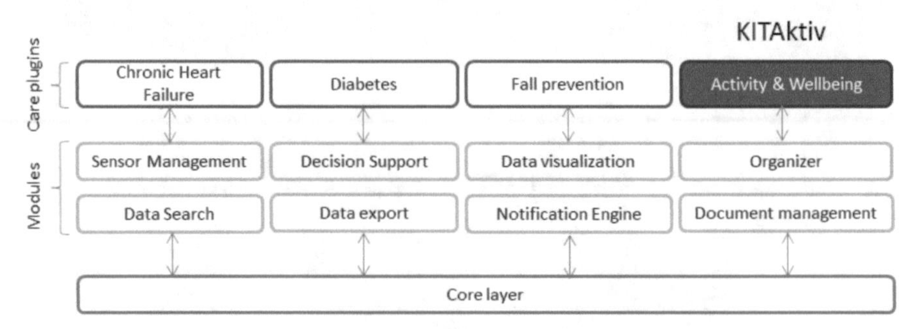

Fig. 2. KIOLA Framework

A key aspect to make KITAktiv accessible to a large number of end-users was mak-ing the platform scalable. One of the design principles of the KIOLA platform was a service-oriented approach following the principles of RESTful resources. As Fielding [15] points out, RESTful services are stateless by definition reducing the communica-tion overhead between servers when used in a clustered infrastructure. Thus, this ap-proach leverages horizontal scalability. Moreover, when it comes to data storage, the KIOLA platform used a hybrid approach between the Entity-Attribute-Value (EAV) data model and a classic relational approach. It has been designed to quickly move from one data model to another if performance proves to be insufficient.

Finally, with the rise of consumer health products, a huge variety of commercially available products (e.g. Fitbit, Withings) are available capable of tracking a person's health. However, once a user has decided for one product or another, it is hard to switch products, since this would require data export of existing data to another plat-form which is (1) sometimes not possible at all, (2) only available to premium cus-tomers or (3) requires technical skills that especially elderly users might not have. In contrast KITAktiv follows an open-access policy making it possible to integrate 3rd party vendors hardware where public APIs are available.

In order to provide end-users flexible access to the system and integrate other stake-holders as well, a **web-based front-end** was available. Depending on the role and rights of the user, different interfaces were provided. Currently, dedicated inter-faces for end-users, system-support, technician, and physicians/trainers are available. The system-support was intended to manage the user accounts and devices, but did not have access to any personal data. The technicians could monitor the status of the system and its components, while not having access to the user data. Physicians and trainers, finally, could obtain access to the measurements of a user, if this access was granted by the user. This could be done either only for a single session or for a certain period of time. In both cases, the user had to place its RFID writs-band on the provided RFID-reader to give his/her permission. To support also people without smartphones or web-access at home, personal reminders were sent via SMS, optionally.

4 Results

4.1 Living Lab

In order to harmonize the different requirements driven by different application scenarios with complementary stakeholders, a dedicated implementation as a living-lab had been set-up in the village of Grafenwörth in Lower-Austria starting in 2012. The local practitioner, the senior care centre, the primary- and secondary school and the municipal administration had been invited to participate in the design of the lab and to utilize it for their clients later on.

The project had been rolled-out in different phases. In the first phase, the general infrastructure, consisting of the public fitness routes and intelligent fitness poles had been set-up, consisting of 17 routes and covering 106 km. The routes had different difficulties, ranging from short walks (0.5km) to routes with 21km. Eight intelligent fitness poles were located at different start, end and intersection points of the routes, to enable a tracking of the activities. On these poles an overview of the different tracks was provided, while along the tracks signposts were provided.

In a second phase, the potential to utilize this infrastructure to support especially older adults had been investigated. In an end-user driven development process, public terminals and a new frontend had been developed.

In the third phase, the system had been rolled out to a couple of school classes from the local primary and secondary school. The school children used the system to track their daily activities and to perform exercises during their physical education lessons.

4.2 Validation

The *KIT-Aktiv* platform has currently been rolled out to two distinct user groups with specific use-cases: older adults and school children. Other users groups will be integrated within the next months.

In 2014, 21 older adults (between 51 and 85 years old) had been en-rolled to the system, to evaluate the general acceptance and usability. At the beginning of the trial, the users had a medical examination with their local practitioner to eliminate any health risks and to discuss an individual activity goal. They used the system for 4 month and where regularly supported by their physician. At the beginning and the end of the trial, qualitative interviews were carried out. Additionally, quantitative methods were applied to assess the socio-demographic data and usability, user experience, and acceptance. Overall, 150 hours of activity with a distance of 600 kilometres were recorded. It turned out, that shorter routes were preferred and that people liked to walk/run in groups to also have some social contact. In the final review, more social activities were re-quested, as well as improvements in the visibility of the way signs and a higher availability of the technical system. The latter was mainly caused by technical problems in the roll-out phase, which could be overcome later on [10].

In late 2014 and early 2015, the local primary and secondary school started to provide RFID wrist-bands to some of their classes. They used the system mainly to track their daily activities (e.g. the walk from the school to lunch). Additionally, specific sport activities where tracked to provide an exact measurement for competitions. The children and their teachers can access the data using the web front-end to compare results and obtain an overview on the total activity. Thus, the system helps to raise the awareness for physical activity. However, as these activities are currently still ongoing, no final conclusion can be drawn, yet.

4.3 Outlook

As a next step, it is planned to integrate a wider range of fitness activities and user groups. First of all, health support for people with chronic conditions will be added. Based on the existing KIT tele-monitoring solution [7,8], support for people suffering with diabetes mellitus or chronic heart failure will be provided. Furthermore, the local fitness centre will be integrated using an RFID reader to track fitness classes or strength training sessions. To increase participation and motivation, it is planned to pro-vide competitions and other game-based approaches.

5 Conclusion

In this paper we have shown, that *KIT-Aktiv* provides a flexible infra-structure for fitness- and health-data management. It has been rolled out in one village in Austria to two distinct user-groups. Due to the flexible architecture and system design, it was possible to cover both use-cases nicely within one platform. Evaluation has shown, that the system is well accepted by both current user groups, older adults and school children.

References

1. World Health Organization: Men Ageing and Health. Achieving health across the life span. 01WHO/NMH/ NPH 01.2., World Health Organization, Geneva (2001)
2. World Health Organization, Tufts University School of Nutrition and Policy: Keep fit for life: Meeting the nutritional needs of older persons. Boston, MA, 26–29 World Health Organization, Geneva (1998)
3. World Health Organization: Life course perspectives: coronary heart disease, stroke and diabetes. Key issues and implications for policy and research. Summary report of a meeting of experts. WHO/NMH/NPH/01.4, Noncommunicable Diseases Prevention and Health Pro-motion Department, Ageing and Life Course, World Health Organization, Geneva (2001)
4. Alesii, A., Mazzarella. F., Mastrilli, E., Fini. M.: The elderly and quality of life: Current the-ories and measurements. Ital Med Lav Erg, 28:3, Suppl Psicologia 2, 99-103 (2006).

5. Kaba-Schönstein, L.: Gesundheitsförderung I: Definition, Ziele, Prinzipien, Handlungsfelder und -strategien, In: Bundeszentrale für gesundheitliche Aufklärung (BZgA) (Hrsg.): Leitbegriffe der Gesundheitsförderung. Fachverlag Peter Sabo, Schwabenheim a. d. Selz (2006).
6. Matsuo M., Nagasawa, J., Yoshino. A., Hiramatsu, K.: Effects of Activity Participation of the Elderly on Quality of Life. Yonago Acta medica 2003; 46, pp. 17–24, Japan (2003).
7. Kastner, P., Lischnig, M., Tritscher, J., Eckmann, H., Schreier G.: DiabMemory - Proof of Concept für mHealth bei Patienten mit Diabetes Mellitus; In: Proc. eHealth 2011, pp. 275-280, OCG, Vienna (2011).
8. von der Heidt, A., Ammenwerth, E., Bauer, K., Fetz, B., Fluckinger, T., Gassner, A., Gran-der, W., Gritsch, W., Haffner, I., Henle-Talirz, G., Hoschek, S., Huter, S., Kastner, P., Kres-tan, S., Kufner, P., Modre-Osprian, R., Noebl, J., Radi, M., Raffeiner, C., Welte, St., Wise-man, A., Poelzl, G.: HerzMobil Tirol network: rationale for and design of a collaborative he-art failure disease management program in Austria; Wiener klinische Wochenschrift 126.21-22, pp. 734-741, Springer, Vienna (2014).
9. Gschwind, Y., Eichberg, S., Marston, H., Ejupi, A., De Helios, R., Kroll, M., Drobics, M., Annegarn, J., Wieching, R., Lord, S., Aal, K., Delbaere, K.: ICT-based system to predict and prevent falls (iStoppFalls): study protocol for an international multicenter randomized con-trolled trial; BMC Geriatr, 14, pp 1-13, BMC (2014).
10. Drobics, M., Hager, M.: *KIT-Aktiv* - Fit & Aktiv im Alter. In: Proc. 7. Deutschen AAL Kongress, VDE Verlag, Berlin (2014).
11. Drobics, M., Dohr, A., Leopold, H., Orlamünder, H., Standardized Communication in ICT for AAL and eHealth. In: Proc. 5. Deutscher AAL-Kongress, VDE Verlag, Berlin (2012).
12. Fielding, R.T.. Architectural styles and the design of network-based software architectures. Diss. University of California, Irvine (2000).
13. Kreiner, K., Gossy, C., Drobics, M.. Towards a light-weight query engine for accessing health sensor data in a fall prevention system, Studies in health technology and informatics 205, pp. 1055-1059 (2013).
14. Kreiner, K., De Rosario, H., Gossy, Ch., Ejupi, A., Drobics, M.: Play up! A smart knowledge-based system using games for preventing falls in elderly people, In: Proc. eHealth 2013, pp. 243-248, OCG, Vienna (2013).
15. Fielding, R. T., Taylor, R. N.: Principled design of the modern Web architecture. Transactions on Internet Technology (TOIT) , vol. 2/2, pp. 115-150, ACM (2002).

Online Offset Correction of Remote Eye Tracking Data: A Novel Approach for Accurate Gaze-Based Mouse Cursor Control

Chris Veigl, Veronika David, Martin Deinhofer, Benjamin Aigner

University of Applied Sciences (UAS) Technikum Wien, Vienna, Austria
{veigl,david,deinhofe,aignerb}@technikum-wien.at

Abstract. Camera-based eye- and gaze tracking systems have a wide range of application areas including cognitive science, biometrics, usability studies and Assistive Technology. Although the accuracy of remote eye trackers improved considerably in recent years, it is still impractical to use eye tracking systems for cursor control because of jitter and inaccurate positioning. This work presents a novel approach for online offset correction of the estimated gaze point, so that state-of-the art, affordable remote eye tracking devices can be used for cursor control in desktop applications without special GUI adaption. For gaze point correction, additional sensor systems or input devices can be combined with the eye tracking device. In evaluations with test users it could be shown that the hit rate of small targets presented at random screen locations could be increased from 54% to 89% which significantly improved the usability of gaze tracking based mouse cursor control.

Keywords: Assistive Technology, Accessibility, Motor Disabilities, Eye Tracking, Gaze Tracking, Cursor Control, Human Computer Interface

1 Introduction

Many people with disabilities worldwide are supported by Assistive Technologies (AT) [1]. Available AT devices and systems provide various assistive functionalities, thus improving the quality of life of people with special needs. Eye tracking is a well-established field in science with different methods for data acquisition (including EOG-, head-mounted or remote camera based systems) and applications in cognitive science, usability studies, marketing - to name just a few [2,3]. As camera-based eye tracking systems improved in accuracy and became more affordable in the recent years, these systems gained a huge impact in the Assistive Technology market. The great advantage of eye-based computer control is that even people with severely limited motor capabilities – resulting for example from Muscular Dystrophy, ALS, Cerebral Palsy or Quadriplegia – can control a computer in a very efficient way by using eye movements and eye gaze [4,5]. Compared to low-bandwidth input systems like single switch access, eye tracking based interaction improves the interaction efficiency significantly and thereby enables new use-cases as for example speech synthesis

S. Loshkovska and S. Koceski (eds.), *ICT Innovations 2015*,
Advances in Intelligent Systems and Computing 399,
DOI: 10.1007/978-3-319-25733-4_29

via selection of letters from an on-screen keyboard. Every remote eye tracking system uses a calibration procedure (for example a 9-point calibration) where the user follows a marker with the eyes and the system collects position data of the eye pupil center and/or glint point locations [6]. These data later serve as primary information for the estimation of the user's gaze direction and gaze point [7,8]. However, even well-calibrated eye- and gaze tracking systems suffer from inaccuracy (an offset between the actual gaze point and the estimated gaze point) due to several reasons – for example changing head position or rotation, changing lighting conditions or interfereence, inadequate calibration quality or changes in the user's gaze behavior [4]. A direct control of the mouse cursor from the estimated gaze point location is often not adequate for using standard GUI applications or for controlling the standard graphical desktop provided by an operating system. This is due to small offset errors of the estimated gaze point, whereby small interactive surfaces (like buttons or check boxes) cannot be reached. Additionally, jitter of the cursor is caused by different sources of noise entering the measurement chain. This jitter can even cause unpleasant feelings and nausea when using direct cursor control from estimated gaze data for a longer time. To avoid these problems, eye tracking devices are usually bundled with special software tools for Augmentative and Alternative Communication (AAC), on-screen keyboards, speech synthesis or computer control [4]. These applications often target a given use-case and provide large interaction surfaces on the computer screen which can easily be selected via eye gaze.

In this work, a novel approach for offset correction is presented, where the estimate eye gaze point provides coarse cursor control, and additional sensors are used for fine adjustment of the cursor position. For this approach, the user needs additional motor capabilities – although they may be very limited – for example slight head-, finger-, toe- or lip-movement. These additional data are processed by the flexible AsTeRICS open source framework (Assistive Technology Rapid Integration and Construction Set) [10]. Thus, different modalities for offset correction can be evaluated very effectively – when reasonable also directly at the user's site.

2 Methods

The AsTeRICS system (Assistive Technology Rapid Integration & Construction Set) [10] is a hardware and software framework, which targets to reduce the time, effort and costs of developing Assistive Technology solutions. It offers a flexible and affordable set of components that enables building assistive functionalities by connecting elements (so-called "plugins") in a graphical editor without programming. [11].

In course of this work, the AsTeRICS framework was extended with dedicated plugins for recent low-cost eye-tracking hardware including the EyeTribe tracker [8] and the Tobii EyeX development kit [7]. Both trackers are available for about 100 Euros, use a remote, infrared supported camera-based tracking method and require a USB 3.0 connection to the hosting PC. The data from the devices can be accessed via a provided SDK/API. EyeTribe provides a Java-API [8] which fits the AsTeRICS

OSGI ecosystem very well [10]. For the Tobii EyeX, a JNI layer was used to bridge between C++ and Java [7].

As an additional input method for users with severely limited motor capabilities, the so-called LipMouse device (Fig. 1) was developed. The LipMouse is a "zero-way" joystick comparable to existing commercial products for mouth control [13,14], but can be configured to very high sensitivity so that even slight lip movements are sufficient as an input modality. The LipMouse features 4 force sensitive resistors (FSRs) which measure forces (0.2 N – 20 N) applied to the mouthpiece in the directions up/down/left or right, a pressure sensor which can measure positive and negative pressure (-7 kPa – 7 kPa) applied to the mouthpiece via sip- and puff actions of the user and additional internal and external switches to trigger mode changes or special functions.

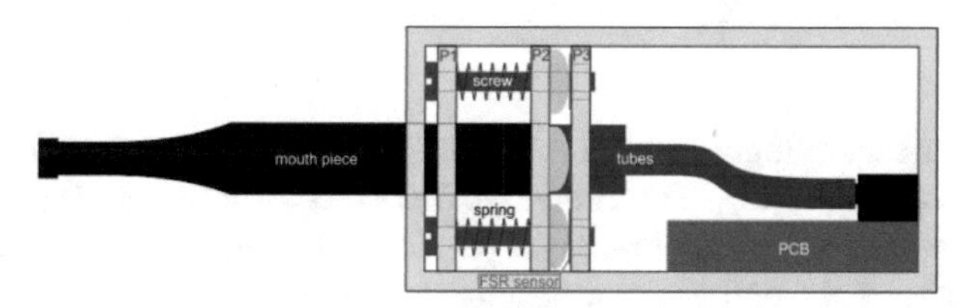

Fig. 1. The LipMouse alternative input device, schematics

The sensors are connected to an Atmel AVR microcontroller mounted on a Teensy++ evaluation board [15]. The microcontroller provides a built-in 10-bit ADC and USB capabilities. The firmware of the microcontroller processes all sensor measurements and creates a data packet which is compatible to the AsTeRICS CIM protocol [10] so that the sensor data can be transferred to the AsTeRICS framework where the sensor data fusion with the gaze tracking information takes place.

3 Implementation

In the AsTeRICS framework, a model is considered as the container that holds connected components (plugins) and produces a specific functionality. The components of each model are classified into 3 categories: sensors, processors and actuators. Sensors monitor the environment and transmit input information to the rest of the components of the model. Processors are responsible for receiving, processing and forwarding this information. Finally, actuators receive data and carry out accordingly the desired actions.

The basic idea of this work was to add one or more AsTeRICS processor plugin(s) which provide gaze point data from a gaze tracking device and at the same time accept additional data from desired sensor plugins which can be used for compensation of inaccuracies and offset correction. The processor plugin will then be able to perform different sensor fusion algorithms which can be evaluated with users.

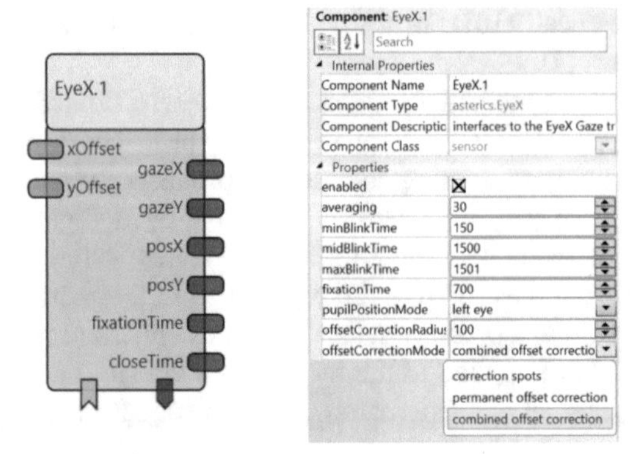

Fig. 2. Newly developed EyeX plugin with offset correction input ports and three offset correction modes

Two new eye tracker plugins for the AsTeRICS framework have been developed: the EyeX processor plugin (Fig. 2) and the EyeTribe processor plugin. Both plugins provide two offset correction input ports (*xOffset* and *yOffset*) where relative coordinate values for the current x and y cursor position can be fed into the plugin. Furthermore, a so-called event-listener port is provided by both eye tracker plugins, which allows setting the offset correction mode. Depending on the user's capabilities, desired activities could be used to trigger these events, as for example closing an eye-lid for a certain time or sipping/puffing into the mouthpiece of the LipMouse. For adjusting the x- and y-offset values, the user needs another input modality with two degrees of freedom, for example head-movements tracked by a web-cam or small forces applied to the LipMouse mouthpiece via lip-, finger-, or toe-movements in x or y direction. Three different algorithms for sensor data fusion and offset correction have been implemented for both eye tracker plugins: "Offset correction spots using linear approximation", "Permanent Offset Correction" and "Combined Offset Correction". These offset correction algorithms can be applied "online" – without interrupting a running gaze tracking session. This offers the advantage that users are able to perform the offset correction without assistance. In the following, the three offset correction approaches are explained in detail.

3.1 Offset Correction Spots with Linear Approximation

This approach allows adding specific correction spots at gaze locations where the standard calibration procedure of the eye tracking device obviously failed or decreased over time. As an alternative to repeating the whole calibration process, individual offset correction spots can be set at desired locations on screen (usually at a location where a small interaction surface or button cannot be reached when using only the eye tracker). Setting an offset correction spot is accomplished in three steps: 1) an incoming event at the eye tracker plugin's event listener port stops the gaze-based cursor control. 2) Offset correction values are accepted from the *xOffset* and *yOffset* input ports so that the user can adjust the current cursor position until it matches the gaze point. 3) Another incoming event indicates that the offset correction is done and the offset correction spot's location and the associated offset correction values can be saved. Subsequently, gaze-based cursor control is restarted. A linear approximation algorithm determines weighed offset values for the incoming gaze data which are calculated from the distance of the current gaze point to the previously added offset correction spots. The events for starting and stopping the creation of an offset correction spot can be triggered via desired input methods – for example via a momentary switch or via sip/puff activity at the LipMouse mouthpiece.

3.2 Permanent Offset Correction

In this mode, the plugin uses the *xOffset* and *yOffset* values which enter the plugin's input ports for permanent offset correction. The incoming values are added to the estimated gaze point x and y coordinates. This approach is useful when the user would like to add a fixed offset to all gaze point estimations. A practical example is an on-screen keyboard with key captions in the upper left corner of the key (not in the centre as, for example, the standard Windows-7 on-screen keyboard). In this case the user's attention is attracted to the key captions so that small inaccuracies of the gaze point easily lead to unwanted selection of neighbouring keys. When adding a small positive offset in x and y direction, this problem can be avoided.

3.3 Combined Offset Correction

This mode uses either the estimated gaze data from the eye tracking device or the additional *xOffset* and *yOffset* values which enter the plugin's input ports for adjusting the latest gaze point. This approach is useful for general desktop or application control purposes where many small interaction surfaces appear on various locations on the screen. Gross movements – where the cursor location differs largely from the last fixation point – are performed via eye tracking, where small adjustments are performed via the offset correction. Whenever non-zero offset values are received at the plugin's input ports, the plugin bypasses gaze point-based cursor control and the cursor position can be influenced by the *xOffset* and *yOffset* inputs. The plugin remains in offset correction mode until a significant difference between the current cursor location and the estimated gaze point is detected, which indicates that the user looks on a different screen location. This re-activates the gaze-point based cursor control for coarse movement. The combination of these two cursor control methods enables efficient fast movements via gaze on the one hand, and accurate cursor positioning without jitter on the other hand, thereby minimizing the disadvantages of both control variants.

4 Evaluation and Results

For the purpose of evaluation, the system was first tested in the AT laboratory of the University of Applied Sciences Technikum Wien by the authors and colleagues. The utilized setup can be seen in Fig. 3. In this evaluation phase, many adjustments and improvements to the software and hardware modules could be realized and prototypical AsTeRICS models were created.

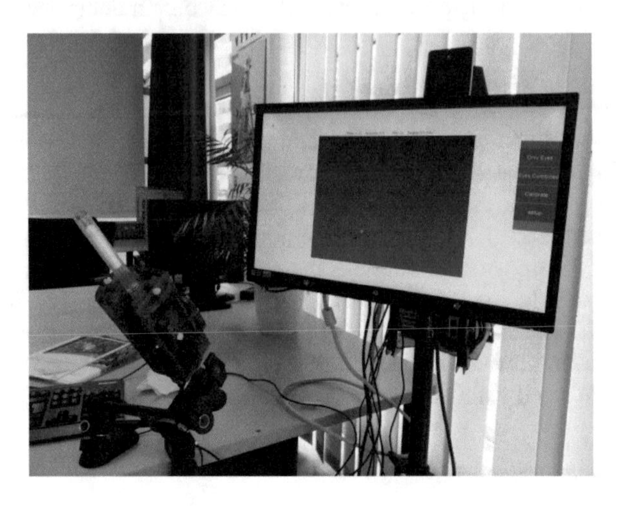

Fig. 3. Setup with Lipmouse sensor for fine tuning the gaze point estimation of a Tobii EyeX eye tracking device

Figure 4 shows a complete AsTeRICS model for gaze based mouse cursor control with online offset correction using the LipMouse device. In this model, the LipMouse sensor and the Tobii EyeX tracking devices are used in combined offset correction mode. The EyeX plugin provides the interface to the Tobii EyeX hardware and the implementation of the sensor fusion algorithms. In the AsTeRICS model, the EyeX plugin connects to the LipMouse plugin and to the Mouse plugin. The X and Y output values of the LipMouse plugin are routed to the EyeX plugin via two Adjustment-Curve plugins. These plugins apply an adjustable gain curve to the sensor values of the LipMouse. Thus, the sensitivity of the LipMouse in x and y direction can be adjusted to the preferences of the user. The pressure values (sip/puff activities of the user) are fed into a threshold plugin, so that pressure values exceeding a certain level create a click-event. This click-event is then routed into the Mouse plugin. Thus, the user can perform a left or right click by sipping or puffing into the mouthpiece.

Using this configuration, qualitative and quantitative measurements have been performed, including usability tests of the system for various standard computer applications and evaluation of the mouse cursor control efficiency (positioning speed and accuracy).

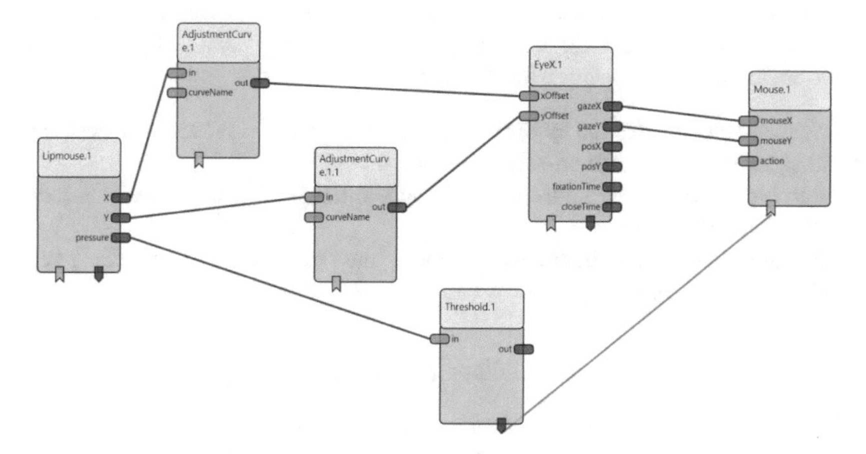

Fig. 4. AsTeRICS model for combined offset correction of gaze data via the LipMouse sensor

4.1 Qualitative Measurement: Versatility of Cursor Control

The scenario depicted in Fig. 5 shows a 14 year-old muscular dystrophy patient using the system for computer control and gaming. The user relies on a device for breathing support and thus cannot actuate the LipMouse with his mouth or lips. As the user can still create minimal movements with his fingers, he decided to perform the offset correction by utilizing the LipMouse mouthpiece as a thumb-actuated joystick.

Fig. 5. A 14-year old muscular dystrophy patient using a Tobii EyeX eye tracker and the LipMouse device in combined offset correction mode for desktop control and gaming

Using the combined method described in the previous chapter, the user learned to use his computer mouse cursor for versatile desktop control in about 10 minutes. The tasks he could accomplish successfully include:

- Starting and closing desired applications using the GUI of a Windows-7 desktop
- Writing an email using the on-screen keyboard
- Performing left and right mouse clicks using additional switches – creating, renaming and deleting folders
- Using small buttons and interaction surfaces like "minimize", "maximize", "close window", "resize window"

4.2 Quantitative Measurement: Cursor Control Efficiency

In this series of measurements, the freely available Aimbooster benchmark tool [12] was applied to determine mouse positioning accuracy and speed. Aimbooster is a Flash-application which can be accessed online and provides numerous options for the creation of different mouse control- and mouse-clicking benchmark tasks. The target size, presentation time, target movement and many other parameters can be defined so that a repeatable cursor control assessment is possible. The provided "precision"-preset was adapted for presentation of non-moving targets of 20 pixel diameter which are displayed at random positions on a surface of 1000 x 800 pixels. The trial time was set to 2 minutes and the users were instructed to hit as many targets as possible. The utilized system hardware setup consisted of an "Intel Core I-5" computer with 8 GB RAM connected to a 22" screen with a resolution of 1920 x 1080 pixel. The Tobii EyeX tracker was mounted at the bottom of the computer screen and the test persons were sitting in about 60 cm distance (head to screen surface).

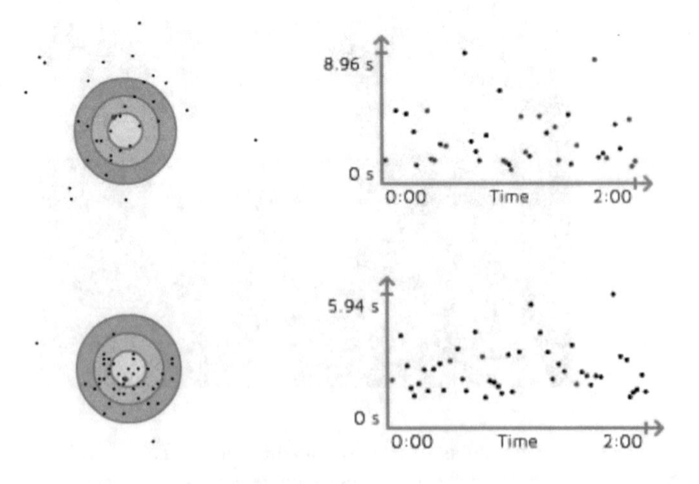

Fig. 6. The measurement results summary of a typical cursor positioning task using the Aimbooster tool (above: plain gaze based positioning, below: combined offset correction)

Figure 6 shows the results of a typical measurement session. The sighting discs presented on the left side show the click locations in and around the target during one session (2 minutes). The timelines presented on the right side show the duration for positioning and clicking each target.

Table 1. The cumulative accuracy and timing results of 3 users (5 sessions)

	Plain gaze based		Offset corrected	
	Hits	avg. time	Hits	avg. time
User A	20/39 (51%)	3.04 sec.	42/46 (91%)	2.58 sec.
User B	51/89 (57%)	1.32 sec.	43/50 (86%)	2.36 sec.
User C	38/70 (54%)	2.3 sec.	48/53 (90%)	2.70 sec.

Table 1 shows the cumulative results of 3 different users (male, aged 20-40, without motor disability) can be seen. Each user performed 5 trial sessions, which were averaged. The quantitative measurements show that the click-efficiency and overall interaction performance increased significantly by applying the LipMouse-based offset correction, compared to the plain eye tracking based cursor control. When the results of all test trails are taken into account, the average click accuracy (target hit rate) of a 20-pixel-diameter target could be increased from 54% to 89%, where the average delay of the click-time caused by the offset correction process was just 320 milliseconds (or 14%). The reason why the average click-time for User A was significantly higher in "plain gaze" mode was that User A tried to compensate an inaccurate cursor position by looking beside the target spot (this increased the click-time but not the hit rate). These results suggest a significant increase in usability of eye tracker based cursor control for computer use.

5 Conclusions and Future Work

The method presented herein combines gaze point data delivered by various state-of-the-art eye tracking devices with sensor data from alternative sensor systems suitable for users with severe motor disabilities. Via three different offset correction approaches implemented in the flexible AsTeRICS open source framework, the sensor data can be combined with the gaze point data so that inaccuracies and jitter of the gaze tracking based cursor control can be minimized or even avoided. The achieved results acquired in user evaluations with 4 different test users are promising: a test person with muscular dystrophy could use the system for controlling a graphical desktop environment without special software or magnification tools, and 3 test persons could improve the hit rate of randomly presented small targets significantly.

In future studies we plan the evaluation of the system with more users with motor impairments and the collection of quantitative to undermine the results of this study which suggest that online offset correction can considerably increase the computer interaction efficiency of those users.

Acknowledgement. This work is supported by the European Commission as part of the Prosperity4All (Large Contribution) EU project funded by the Seventh Framework Programme – under grant agreement no 610510.

References

1. Eurostat: Population and Social Conditions: Percentual Distribution of Types of Disability by Sex and Age Group, http://epp.eurostat.cec.eu.int
2. Duchowski. A.: Eye Tracking Methodology: Theory and Practice, 2nd ed., Springer, London (2007)
3. Narcizo. F.B., de Queiroz. J.E.R., Gomes. H.M.: Remote Eye Tracking Systems: Technologies and Applications. In: 26th Conference on Graphics, Patterns and Image Tutorials. pp. 15-22. IEEE (2013)
4. Al-Rahayfeh. A., Faezipour. M.: Eye Tracking and Head Movement Detection: A State-of-Art Survey. IEEE Journal of Translational Engineering in Health and Medicine 1, 2100212 (2013)
5. Man. D.W.K., Wong. M.L.: Evaluation of computer-access solutions for students with quadriplegic athetoid cerebral palsy. American Journal of Occupational Therapy 61, 355–364 (2007)
6. Bates. R., Istance. H., Oosthuizen. L., Majaranata. P.: D2.1 Survey of De-Facto Standards in Eye Tracking, Communication by Gaze Interaction (GOGAIN), (2005). IST-2003-511598, http://www.cogain.org./reports
7. Tobii EyeX development kit and API, http://www.tobii.com/de/eye-experience/eyex/
8. The EyeTribe API reference, http://dev.theeyetribe.com/api/#api
9. Robitaille. S.: The illustrated Guide to Assistive Technology and Devices, pp.148-151. Demos Medical Publishing, New York (2010)
10. AsTeRICS - Assistive Technology Rapid Integration & Construction Set, http://www.asterics.eu
11. García-Soler. A., Diaz-Orueta. U., Ossmann R., Nussbaum. G., Veigl. C., Weiss. C., Pecyna. K.: Addressing accessibility challenges of people with motor disabilities by means of AsTeRICS: A step by step definition of technical requirements. In: Miesenberger. K. et al. (eds) ICCHP 2012. LNCS, Part II, vol. 7383, pp. 164-171. Springer, Heidelberg (2012)
12. Aimbooster mouse control benchmark tool, http://www.aimbooster.com
13. LifeTool IntegraMouse Plus, http://www.lifetool.at/assistive-technology/lifetool-developments/integramouse-plus.html
14. Quadjoy Hands-Free Mouth Mouse, https://quadjoy.com
15. Teensy++ microcontroller evaluation board, https://www.pjrc.com/teensy

Author Index

Printed in the United States
by Baker & Taylor Publisher Services